Mathematical Models and Their Analysis

The Harper & Row Series in Applied Mathematics
David J. Benney *Consulting Editor*

FORTHCOMING
Mark J. Ablowitz and A. S. Fokas *Complex Variables*
Gregory R. Baker *Introduction to Scientific Computing*
David J. Benney *Ordinary Differential Equations*
Sherwin A. Maslowe *Partial Differential Equations*

PUBLISHED
Frederic Y. M. Wan *Mathematical Models and Their Analysis*

Mathematical Models and Their Analysis

Frederic Y. M. Wan

University of Washington

HARPER & ROW PUBLISHERS, New York
Cambridge, Philadelphia, St. Louis, San Francisco,
London, Singapore, Sydney, Tokyo

1817

Sponsoring Editor: Peter Coveney
Project Editor: Ellen MacElree
Text Design: Barbara Bert/North 7 Atelier Ltd.
Cover Design: Edward Smith Design, Inc.
Text Art: Fineline Illustrations, Inc.
Production Manager: Willie Lane
Compositor: TAPSCO, Inc.
Printer and Binder: R. R. Donnelley & Sons Company
Cover Printer: Phoenix Color

Mathematical Models and Their Analysis

Library of Congress Cataloging-in-Publication Data

Wan, Frederic Y. M.
    Mathematical models and their analysis/by Frederic Y. M. Wan.
        p. cm.
    Includes index.
    ISBN 0-06-046902-1
    1. Mathematical models. I. Title.
QA401.W33 1989
001.4'34—dc19   88-13897
                                                    CIP

89 90 91 92   9 8 7 6 5 4 3 2 1

To My Mother Olga
  who knew the value of higher education

Contents

Preface

Mathematical modeling has always been an important activity in science and engineering. The formulation of qualitative questions about an observed phenomenon as mathematical problems was the motivation and an integral part of the development of mathematics from the very beginning. Nevertheless, when I began to organize lecture material for an undergraduate course on mathematical modeling in the early seventies, I could not find an appropriate text for models involving nonlinear differential equations. The present volume is the outgrowth of the lecture notes for several modeling courses of this type over the last fifteen years. A few texts on similar material have appeared since the course notes were first written. However, this book still contains enough novel features to offer another choice of material and presentation for instructors of mathematical modeling. Moreover, the material on natural resources in the last five chapters has not appeared in other mathematical modeling texts.

Special Features

One special feature of this book is the classification of mathematical models according to the principal concern of the model. For some models, we may be interested in the temporal evolution (including the steady-state behavior) of the phenomenon under investigation and its stability, or in controlling the evolution and guiding it toward a target in some optimal way. For other models, we may be interested in the propagation of a "signal" or diffusion of "energy" in space. In still others, we may be interested in deciphering some order in a random or chaotic process. The formulation and analysis of a mathematical model vary with the underlying mathematical process theme(s) of the phenomenon being investigated (Greenspan, 1973). This book cultivates the students' appreciation for the importance of model classification by process themes.

With the computing power available in today's computers, many complex phenomena that could only be studied by analyses of relatively simple mathematical models in the past can now be studied by computer simulation. However, none of the phenomena investigated here involves computer simulation. In fact, a special feature of this book is an attempt to get across by way of examples the idea that much can be learned from mathematical analyses of simple mathematical models for a given phenomenon. In the best scenario, computer simulation is used mainly to confirm quantitatively what we expect qualitatively from our mathematical analysis. In any case, scientific computing is shown to be more effective and more efficient if it is preceded by some mathematical or heuristic analysis.

Applied mathematics in general and mathematical modeling in particular are concerned with understanding specific phenomena. It is important that the results of our analysis of a model be properly understood in the context of the phenomenon being studied. A special effort is made throughout this book to give a qualitative interpretation of the mathematical results for a model. These qualitative conclusions are displayed in a highlight format, not only to call attention to the major results of our investigation but also to inculcate the importance of their interpretation in the minds of the students.

The qualitative conclusions about a phenomenon based on model analysis sometimes do not agree with the observational data or experimental results. In such cases, the mathematical model used must be improved by incorporating some of the features of the actual phenomenon not previously considered essential. The improved model would then be analyzed and the results interpreted again to check for accuracy. The whole cycle of formulation, analysis, interpretation, and assessment should be repeated until the new improved model is validated. This important practice is illustrated in each of the phenomena investigated in this book.

The Modeling Cycle

The model of planetary motion in Chapter 3, for example, is an improved version of the classical Newtonian model of Chapter 2. The improved model is needed to explain the precession of the perihelion of planet Mercury which is not compatible with the conclusion of the Newtonian theory. The improvement consists of retaining the relativistic effect which leads to a more complex nonlinear mathematical problem. For its solution, the students are introduced to two different perturbation methods. The stability of the periodic orbits of the relativistic model is analyzed in Chapter 4.

The analysis of traffic flow in Chapters 6, 7, and 10 is another example of repeated model improvements. In the particle-type models of Chapter 6, the instantaneous velocity control is seen to be always unstable; an acceptable model is found by the introduction of a time lag. Other linear and nonlinear particle models are also investigated for possible improvements on the model's stability performance. For crowded highways, the Lighthill-Whitman kinematic wave model in Chapter 7 is more informative. However, the theory becomes unsatisfactory for problems with shock formation. The improved model of Chapter 10, which includes the effect of "looking ahead," involves a higher-order partial differential equation but eliminates the possibility of a multivalued solution (shock overhangs) in the Lighthill-Whitham model.

Each of Chapters 5, 11, 12, 14, and 15 starts with a simple model; successive improvements are made to deal with new and more sophisticated questions. It is hoped that students will see from our approach the benefits of starting from a relatively simple paradigm to more and more elaborate models as the need arises. Substantive progress toward understanding of the phenomenon of interest can rarely be made by starting with the most general model possible.

Organization of the Text

As it is organized, this book can be used for students with only a strong background in calculus and ordinary differential equations. New mathematical techniques are introduced and developed for a particular model when needed. Chapters 6, 7, 8, and 10 have been used for a one-semester course for second-year students. Chapters 2 to 7, 10, 11, and 13 constitute the material for an upper-level semester course in an honors program. Chapters 2 to 7 and 11 have formed a one-quarter course for third-year students. The entire book is currently read in a one-quarter first-year graduate course for applied mathematics students who are also required to register for the Applied Mathematics Clinic to get active learning experience on mathematical modeling. Finally, it can be used for an undergraduate course for students without any background on PDE by omitting Chapters 7 to 11. In order to limit course prerequisites to calculus and elementary differential equations, I have avoided mathematical models involving (continuous time) stochastic processes. With some preparatory material on the needed probability theory, interesting stochastic models related to the deterministic ones already discussed may also be included. With some regret, I have decided not to include some models on biological phenomena and to refer anyone interested in mathematical biology to J. D. Murray (1977), L. Segel (1984), and L. Edelstein-Keshet (1988).

Acknowledgments

I am indebted to the following reviewers for their careful reading of and valuable comments on various drafts of the manuscript: Jan Boal, Georgia State University;

Bernard Fusaro, Salisbury State College; John Gregory, Southern Illinois University; and Jay Walton, Texas A & M University. Many of their helpful suggestions have been incorporated into the manuscript.

It should be clear from its contents that this book has benefited considerably from Lin and Segel (1974) and Clark (1976). Some material developed jointly with Alar Toomre for a second-year ODE course some years ago can be found in Chapter 4 and in the Exercises of Chapter 2. Bill Criminale, Joel Feldman, Charles Lange, Don Ludwig, Robert Miura, Brian Seymour, and Hubertus Weinitschke have taught modeling courses using different versions of the course notes for this book. I am grateful for their feedback, particularly for the list of typos and other errors from Feldman on the notes he used. Lilly Harper typed the many drafts of the entire manuscript and cheerfully put up with my seemingly endless revisions. I would like her to know how much her work and extra effort have been appreciated. No words, however, can convey my appreciation for my wife, Julia, who has regularly taken time from her own busy professional schedule to be my most helpful advisor and constructive critic for the last 28 years, including the writing of this preface.

Frederic Y. M. Wan
Bainbridge Island

Mathematical Models and Their Analysis

1

Groping in the Dark

Introduction

The mathematical content of this chapter consists of an introduction to dimensional analysis and to dynamic programming. The mathematical prerequisites for the chapter are multivariate calculus and linear equations.

1.1 Mathematical Models and Mathematical Modeling

This book is about applied mathematics. Mathematical modeling of phenomena and situations is an integral part of applied mathematics. We will study a large number of mathematical models in this book, mainly those involving differential equations. We will see mathematical modeling at work. However, this is not a book which teaches the technique of mathematical modeling! Let me explain.

Applied mathematics is concerned with a better understanding of phenomena by the use of mathematical methods. The process toward this goal may be divided into several components. For any particular phenomenon, we begin by identifying a few important questions of interest and the main factors influencing the answers. We may have more questions or more contributing factors; but it is usually more productive to keep things simple initially and consider other issues after some success with the simpler problem. The identified questions are then quantified and related to the influencing factors by the laws of nature (such as Newton's laws of motion in the case of particle dynamics). At this point, we have a *mathematical model* of the phenomenon to be analyzed. The process of *formulating* the mathematical model is often called *mathematical modeling*. Next, we apply known mathematical methods to extract from the mathematical model information sought about the phenomenon. Occasionally, new mathematical methods

1

may have to be developed for the *analysis* of the model. Finally, the mathematical results obtained from our analysis will have to be interpreted and compared with observations and empirical data available for the modeled phenomenon. If the results from the model are not sufficiently realistic, we must return to the formulation phase of the process and revise the mathematical model. Even if the results are consistent or compatible with observations, we may still want to revise the model and make it more sophisticated by incorporating more questions or more influencing factors. This interpretive phase of the process is sometimes called *model evaluation* or *model validation*. A more detailed description and discussion of the cyclic stages in a quantitative study of a phenomenon can be found in Wan (1983, and references cited therein). Collectively, these cyclic stages are also called mathematical modeling.

For a long time, textbooks on applied mathematics had been mainly concerned with the mathematical methods most often needed for the *analysis* phase, reflecting the type of applied mathematics courses taught in our colleges and universities up to the early sixties. In the late sixties, courses on the modeling (formulation) phase of applied mathematics sprang up all over the country. The development seems to have been initiated by a desire for a better balance in the teaching of the theory and application of mathematics in our colleges and universities. It has continued to grow over a period of more than twenty years, sustained by a developing need for quantitative analyses in many scientific endeavors in the biological, social, resource, environmental, and management sciences. Physicists and engineers have traditionally done their own mathematical modeling and analysis. However, the practitioners in the biological and social sciences are not as equipped to enter the arena of quantitative investigation. They tend to work with mathematicians on the formulation and analysis of mathematical models for their more complex problems. The involvement of mathematicians in these areas began to change their more traditional role in the application of mathematics. They must now concern themselves with the modeling aspect as well, not just with the solution of well-formulated mathematical problems in physical science and engineering. This new role has for some time provided the interested faculty member a good justification for promoting courses in modeling at the undergraduate level. Along with the courses comes a demand for texts on mathematical modeling. A good number of them have appeared over the last twenty years.[1]

---

[1] Professor T. Berry of the University of Manitoba has compiled a list of more than seventy books on modeling; nearly all published within the last ten years.

1.2   Teaching Mathematical Modeling

Through our calculus courses we encounter some of the best mathematical modelers in the history of mathematics: Newton, the Bernoullis, and Euler, just to name a few. Serious students of the mathematical sciences will learn about the work of many other successful modelers, past and present. It is therefore a given that individuals can become proficient in mathematical modeling. Whether they can learn mathematical modeling in a conventional classroom environment is another matter. Teaching modeling to mathematics students has a number of formidable obstacles. We mention here three important ones:

1. There are no general recipes which can be taught in lectures for direct applications to different classes of modeling problems.

2. The overwhelming complexity of a real life situation does not lend itself to immediate success in a do-it-yourself environment.

3. The science required for successful modeling varies a great deal from one problem to another.

Unlike conventional mathematical methods, such as separation of variables for linear partial differential equations (PDE), a student cannot get "recipes" for different classes of modeling problems. The only recipe which comes to mind is *dimensional analysis* which will be discussed further in Sections 1.4 to 1.7. It is true that there are also some general working principles or rules of thumb for modeling which apply to most modeling problems. One group of general principles which I often share with my students consists of the following:

1. Start with the simplest relevant model and gradually incorporate more features as the phenomenon becomes better understood through the results for the simpler models.

2. If the model is nonlinear, it may pay to first study the corresponding linearized problem, but keep in mind that many nonlinear phenomena cannot be captured by a linear theory.

3. Be flexible; if you cannot make progress with one approach, be willing to try a different approach to the problem.

4. Be imaginative; you may have to model by analogy (e.g., the dendrite of a neuron is not a conducting cable; but it has been modeled successfully by the one-dimensional cable equation).

5. Old mathematics may be useful for models in new areas; it pays to consider how mathematical methods originally developed for other areas of application can be helpful to your new problem.

Elaboration of these principles and their illustrations by specific examples can be found in Wan (1983). Still, they are just too general to be helpful to inexperienced individuals in modeling specific problems of their own. The process of arriving at a simple but reasonably rich model for an urban transportation problem (Wan, 1983) cannot be translated into one which would give a simple but reasonably rich model for the optimal logging schedule for a forest of Douglas firs! Thus, any case study approach to teaching mathematical modeling generally would not, by itself, teach students the modeling process.

Many students in modeling courses have contended that they must be actively involved in a modeling problem if they are to learn mathematical modeling. They feel that it is simply not possible to learn it by attending lectures. There is, of course, a lot of truth to this point of view. There have been courses on mathematical modeling so organized to allow students to do individual or team modeling. These have experienced varying degrees of success. A fairly typical experience with these courses is described in Cross and Moscardini (1985). In courses which require modeling of real life situations, many students are overwhelmed by the complexity of phenomenon to be modeled. Without some guidance, they would not be able to do anything meaningful. On the other hand, any suggestion from the instructor would be taken as correct and binding and would be pursued to the limit. Finally, without some tangible results from the initial burst of activities, student interest would quickly wane. These observations seem rather discouraging but can be used constructively for designing a more realistic and reasonable course on mathematical modeling with active participation by the students.

One of the reasons why students of mathematics have often been overwhelmed when they have to model a phenomenon or situation on their own is a lack of the knowledge in science needed for the problem. For instance, if you want to know how long to roast a turkey, you must have some idea about heat conduction or diffusion. Unfortunately, the modeling problem confronting a student invariably involves a different area of science than those encountered in the instructor's illustrative examples. Mathematical modeling will always involve both mathematics and some area of science. No student can hope to become proficient in mathematical modeling without the ability (and willingness) to acquire the needed science for a particular modeling problem. This ability is founded on a good background in some area of science or theoretical engineering. Aside from specific scientific knowledge of the field, such a background also prepares a student in scientific procedures and gives the student the confidence and framework needed to learn another field of science. Any mathematical modeling course should therefore have an advanced-level prerequisite in science or engineering. In view of its historical relation to the development of applied mathematics, an upper-level undergraduate (or first-year graduate) course in analytical mechanics or continuum mechanics would be very helpful. Without some science prerequisite, the mathematical models which can be discussed in a lecture-type course are not likely to be rich enough in content to sustain the interest of serious students of modeling.

It is not clear whether or not mathematical modeling can be taught successfully to undergraduate students. However, it seems reasonably conclusive that it cannot be taught successfully, except to a very select few, by conventional lectures without active participation by the students in the modeling process. Nevertheless, a conventional lecture course on mathematical models has a place in an undergraduate applied mathematics curriculum as long as we keep in mind its limitations. Inexperienced students need to see a number of different mathematical models, their formulation, analysis, validation, and reformulation, etc., to build up a reservoir of techniques and approaches to mathematical modeling. With this reservoir, a student should be in a better position to come up with a reasonable approach to model a particular situation in a second course on mathematical models where the student is required to do the modeling. And there will always be some who can extrapolate the general modeling process from the material in this volume and become proficient in mathematical modeling without the benefit of a second course.

1.3   Process Themes in Applied Mathematics

Mathematical modeling is a part of quantitative investigation in science and engineering. The mathematical problem governing planetary motion in physics is a mathematical model. The mathematical theory for electromagnetics is another mathematical model. Whether we classify Newton and Maxwell as applied mathematicians or physicists, the models they developed are generally considered to be a part of physics. In planetary motion, we are interested in the temporal *evolution* of the position, velocity, and acceleration of the planets. Now, the temporal evolution of position, velocity, and acceleration is also the focus of our attention in a particle model of automobiles moving along a single-lane road or highway. More generally, the temporal evolution of some quantities (such as the fish population in the study of the dynamics and management of fisheries) is the focal point in the study of many other phenomena not governed by Newtonian physics. Temporal evolution is just one of several scientific processes operating in natural and social phenomena to be investigated with the help of mathematical modeling.

In electromagnetics, we are interested in the propagation of electromagnetic waves. *Wave propagation* is also the scientific process which occupies our attention in the Lighthill-Whitham theory of traffic flow appropriate to long and crowded highways. Both the Lighthill-Whitham theory and the particle model of traffic flow will be discussed later in this book (see Chapters 6 and 7). Here in this chapter, we confine ourselves to listing a number of scientific processes which have been identified in existing mathematical models:

1. Evolution (of dynamical systems)
2. Stability (of steady-state configuration)
3. Wave propagation
4. Diffusion
5. Optimization
6. Control
7. Chaos
8. Stochastics

Each of these processes operates in many models with unrelated scientific origins. At the same time, more than one process may operate in the same model. Often, we classify the type of mathematical models by the principal process involved; hence, we speak of wave propagation models, stochastic models, etc. It is important that the teaching of mathematical modeling includes the identification of these scientific process themes in different mathematical models. This can be done in a first lecture course on mathematical models. The knowledge will enable us to consider the application of techniques successful for one model to other models with the same scientific process theme. The illustration of how the different processes manifest themselves in different models is also a useful function for a first course on mathematical models.

This book is not intended to teach mathematical modeling. Instead, it identifies and illustrates different types of mathematical models. Through specific models, the book shows the formulation-analysis-evaluation (-reformulation, etc.) cycle in action. It demonstrates how the same mathematical technique may apply to models from different areas of science and engineering. A good book would do all these well; but it must do more. It must convey an excitement about applied mathematics by showing readers impressive successes of the distant past, substantive progress in recent years, interesting projects from current activities as well as potential rewards in future challenges. If applied mathematics is not just mathematical physics or engineering mechanics, models must be chosen from more than one area of mathematical sciences. We have tried to structure the book to meet these objectives and constraints.

1.4   The Period of a Simple Pendulum

In mathematical modeling, the laws of nature governing a phenomenon decide how various factors and parameters influence the phenomenon. A good mathematical model usually manages to quantify this influence in a relatively simple and elegant way. In the remaining chapters of this volume, many exemplary models from different disciplines will be presented and analyzed to illustrate how we may benefit in a substantial way from well-formulated mathematical models. Too often, however, the relevant laws governing the phenomenon of interest are either not known to the investigator or they are too complex for mathematical modeling. For these situations, the method of *dimensional analysis* is invaluable. The present introductory chapter seems to be an appropriate place to discuss this basic general technique for the analysis of complex phenomena. In this section, we use the example of the simple pendulum to introduce the basic features of dimensional analysis. The general technique is then described in Section 1.5. A nontrivial application of this technique is given in Section 1.6. Even when the natural laws governing a phenomenon are known and tractable, dimensional analysis is still an important tool for the analysis of the relevant mathematical model. This will become evident in subsequent chapters.

For ease of describing the basic idea of dimensional analysis, we consider here the familiar simple pendulum with mass m and a weightless arm of length l. Let θ_0 be the initial angle made by the pendulum with its vertical position, positive counterclockwise, and let the pendulum mass have no initial velocity. The hinge at the pivot of the pendulum is assumed to be frictionless. Suppose we want to determine the period P of this pendulum. With our knowledge of physics, a straightforward procedure to determine P would be to formulate the equation of motion for the pendulum and solve the resulting initial-value problem for the angular position of the pendulum mass as a function of time. The desired pendulum period can then be obtained from the temporal evolution of the pendulum's angular position. Hence, it is not necessary to apply the method of dimensional analysis to this problem. On

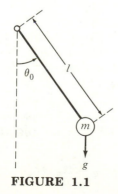

FIGURE 1.1

the other hand, the fact that we know the solution to the problem makes it an attractive illustrative example for dimensional analysis. Agreement with the known solution would indicate the effectiveness of dimensional analysis for the pendulum problem and give impetus to its more complete development.

So, let us suppose we do not know anything about Newtonian physics and only have an idea that the important factors for the determination of the period are m, l, θ_0, and gravitational acceleration g (with mg being the weight of the pendulum mass) so that

$$P = f(m, l, \theta_0, g) \tag{1.1}$$

If this is the extent of our knowledge of the physics of the problem, f would be an unknown function to be determined or estimated by suitable experiments.

As an alternative to time-consuming complex experiments involving four parameters, we note that all quantities in (1.1) have as their dimension some combination of mass (M), length (L), and time (T). With the dimension of a quantity Z denoted by $[Z]$, we have $[P] = T$, $[m] = M$, $[l] = L$, $[\theta_0] = 1$, and $[g] = LT^{-2}$. While the angular position of the pendulum θ_0 has no dimension $[\theta_0]$ may be taken to be $M^{a_1}L^{a_2}T^{a_3}$ with $a_1 = a_2 = a_3 = 0$, and θ_0 is said to be *dimensionless*. The other three factors m, l, and g are not dimensionless; nor is the unknown period. It is not difficult to see (and the general method to be described in Section 1.7 will show) that the combination $\Pi = P\sqrt{g/l}$ is dimensionless. Now, write (1.1) as

$$\Pi = \sqrt{\frac{g}{l}}\, f(m, l, \theta_0, g) \tag{1.2}$$

The key observation at this point is that *the right-hand side must be a dimensionless combination because the left side is dimensionless.*

For the right-hand side to be dimensionless, it cannot depend on m as there is no possible dimensionless combination of m, l, θ_0, and g. When we change the unit of measurement from grams to kilograms, the numerical value of the right-hand side for a particular combination of the four parameters would change if f depends on m. But the left side does not change with any change of scale in m and we expect the period of pendulum not to depend on the unit of measurement for m. In fact, the left side of (1.2) is dimensionless and hence does not change with any change of the three units of measurement (for mass, length, and time). By similar arguments, the right side of (1.2) can depend only on the dimensionless quantity θ_0 so that

$$\Pi = F(\theta_0) \quad \text{or} \quad P = \sqrt{\frac{l}{g}}\, F(\theta_0) \tag{1.3}$$

The function $F(\cdot)$ still has to be determined by experiments. However, the simplification resulting from the above dimensional analysis is already very substantial. Instead of trying to find the dependence of P on four different quantities g, m, l, and θ_0, we have now only to perform one set of experiments relating P to θ_0 for a fixed l (and any m). The dependence of P on l (and on the constant g) is known from the dimensional analysis.

The period of the simple pendulum is expected to be a well-behaved *even* function of θ_0. We may write

$$F(\theta_0) = F(0) + \frac{1}{2!} F''(0)\theta_0^2 + \frac{1}{4!} F''''(0)\theta_0^4 + \cdots \qquad (1.4)$$

where $(\ )' \equiv d(\ )/d\theta_0$. For very small values of θ_0, the period of the pendulum is seen to be effectively independent of θ_0. In this case, a single experiment suffices to determine the number $F(0)$ with the pendulum period given by $F(0)\sqrt{l/g}$. This is in agreement with the result from Newtonian mechanics, $P = 2\pi\sqrt{l/g}$, for small θ_0. For a moderately small θ_0, we may retain the quadratic term for better accuracy; two experiments would be needed for $F(0)$ and $F''(0)$ and so on. For a pendulum with an initial angular velocity ω_0, an additional dimensionless combination is possible and the dependence of P on the factors m, l, θ_0, g, and ω_0 becomes slightly more complicated (see Exercise 2 at the end of this chapter).

The above results show that dimensional arguments are often capable of providing useful information about a phenomenon of which very little is known. They may give explicit results or reduce the complexity of the analysis. We shall see in a later section that dimensional analysis plays other important roles when more is known about the phenomenon being investigated. For example, it could help to obtain a class of (similarity) solutions for difficult partial differential equations (see Section 10.4). Because of its importance in applied mathematics, the basic features of dimensional analysis will be outlined in the next section. A more impressive application of the technique (than the pendulum problem) will be described in Section 1.6.

1.5 Dimensional Analysis

In any modeling process, we usually identify a set of measurable scalar quantities (field variables and parameters) $\{u, W_1, W_2, \ldots, W_n\}$ for the description of an idealized and simplified version of the phenomenon of interest. The dimension of all quantities in the set is known and denoted by $\{[u], [W_1], \ldots, [W_n]\}$. For simplicity, suppose the model aims to determine the single scalar quantity u and suppose we know only that u is a function of all the W's,

$$u = f(W_1, W_2, \ldots, W_n) \tag{1.5}$$

The problem now is to find f.

Our experience with the simple pendulum suggests that we first identify a set of *fundamental dimensional units* (fdu's), $\{L_1, L_2, \ldots, L_m\}$, of the problem. The number and types of fdu's vary from problem to problem (but typically $m < n$). For example, we have $\{L_1 = M = \text{mass}, L_2 = L = \text{length}, L_3 = T = \text{time}\}$ for the pendulum problem, and $\{L_1 = D = \text{dollars}, L_2 = P = \text{labor population}, L_3 = T = \text{time}, L_4 = G = \text{goods}\}$ in some economic problems. We now *postulate*[2] that

All descriptive quantities in mathematical models have dimensions which are products of powers of the fdu's.

For example, the gravitational acceleration g has the dimension of length/time^2; we have therefore $[g] = L^1 T^{-2}$. In general we have for any quantity Z,

$$[Z] = L_1^{a_L} L_2^{a_2} \cdots L_m^{a_m} \tag{1.6}$$

for m real numbers $a_1, \ldots, a_m$. Z is said to be *dimensionless* if and only if $[Z] = 1$, that is, $a_1 = a_2 = \cdots = a_m = 0$. The numerical value of a dimensionless quantity is evidently unaffected by a change of units of measurement.

In dimensional analysis, the relationship between u and $\{W_k\}$ (as given by the expression 1.5) is assumed to remain unchanged for any system of measurement units. A change from one system of units to another involves a scaling (or rescaling) of each fdu which in turn induces a scaling of the descriptive quantities of the model. For example, in changing from cgs to mks units, $L_1 = M = \text{mass}$ is scaled by 10^{-3}, $L_2 = L = \text{length}$ is scaled by 10^{-2} and $L_3 = T = \text{time}$ is unchanged. In turn, this induces a scaling of a quantity such as energy E with $[E] = ML^2 T^{-2}$ by $10^{-3}(10^{-2})^2 = 10^{-7}$. The expression 1.5 is assumed to be *invariant* under an arbitrary scaling of any fdu, although the numerical value of individual quantities in the expression may be changed by the rescaling.

With no loss in generality, let $W_1, W_2, \ldots, W_m$ have independent dimensions and call them the *primary quantities*. The dimensions of the

[2] This postulate is sometimes deduced as a theorem from what may be considered more intuitive hypotheses (Bluman and Cole, 1974), but it is equally easy to verify the present postulate for a specific problem since the dimensions of all quantities involved are known.

secondary quantities $W_{m+1}, \ldots, W_n$ can then be expressed in terms of the dimensions of the primary quantities

$$[W_{m+j}] = [W_1]^{a_{m+j,1}} \cdots [W_m]^{a_{m+j,m}} \qquad (j = 1, 2, \ldots, n - m) \qquad (1.7)$$

Given that the relation (1.5) is to remain *unchanged* for any system of measurement units, the quantity u must also be a secondary quantity so that

$$[u] = [W_1]^{a_1} \cdots [W_m]^{a_m} \qquad (1.8)$$

Otherwise, u would have to depend on $W_1, \ldots, W_n$, and at least one other quantity involving a new fdu. The set of descriptive quantities for the model would then be incomplete.

We now form the dimensionless combinations

$$\Pi = \frac{u}{W_1^{a_1} \cdots W_m^{a_m}} \equiv \frac{u}{D} \qquad \Pi_j = \frac{W_{m+j}}{W_1^{a_{m+j,1}} \cdots W_m^{a_{m+j,m}}} \equiv \frac{W_{m+j}}{D_j} \qquad (1.9)$$

for $j = 1, 2, \ldots, n - m$, and then write (1.5) as

$$\Pi = \frac{1}{D} f(W_1, \ldots, W_n) = \frac{1}{D} f(W_1, \ldots, W_m, D_1\Pi_1, D_2\Pi_2, \ldots, D_{n-m}\Pi_{n-m})$$

$$\equiv F(W_1, \ldots, W_m, \Pi_1, \ldots, \Pi_{n-m}) \qquad (1.10)$$

where F is some new unknown function, $W_1, W_2, \ldots, W_m$ are of independent dimensions, and $\Pi, \Pi_1, \ldots, \Pi_{m-n}$ are dimensionless. With the relation 1.5 [and therefore (1.10)] *unchanged* for any system of units of measurements, we may change the measurement units so that the numerical values of all W's and Π's are unchanged except for an arbitrary change in W_1. It follows that $\partial F / \partial W_1 \equiv 0$ (Π does not depend on W_1) as the numerical value of Π would have to change otherwise. Similar arguments give $\partial F / \partial W_k \equiv 0$, $k = 1, 2, \ldots, m$, so that we have

$$\Pi = \phi(\Pi_1, \Pi_2, \ldots, \Pi_{n-m}) \qquad (1.11)$$

or in terms of the original quantities,

$$u = f(W_1, \ldots, W_n) = W_1^{a_1} \cdots W_m^{a_m} \phi\left(\frac{W_{m+1}}{D_1}, \frac{W_{m+2}}{D_2}, \ldots, \frac{W_n}{D_{n-m}}\right) \qquad (1.12)$$

This is the basic result in the theory of dimensional analysis and is known as the *Buckingham Π theorem*. The theorem asserts that

1A. A relation such as (1.5) in a mathematical model, which is unchanged for any system of measurement units, can be rewritten as a relation among dimensionless combinations of the original quantities in (1.5).

1B. The number of independent dimensionless combinations involved is equal to the difference between the number of original quantities and the number of fdu's for the model.

The above basic result in dimensional analysis is intuitively obvious and had been used by Fourier, Maxwell, Reynolds, and Rayleigh (just to name a few) long before the explicit formulation and formal proof of the Π theorem. More general versions of the theorem (see Lin and Segel, 1974 for example) as well as group theoretical treatments of the subject [see Bluman and Cole (1974) and references therein] have expanded the usefulness of dimensional analysis to other classes of problems, notably differential equations.

1.6   The Atomic Explosion of 1945

To emphasize the importance of dimensional analysis in applications beyond the textbook variety, we sketch in this section a dimensional argument for Sir G. I. Taylor's astounding deduction of the approximate energy released by the first atomic explosion in New Mexico. Taylor carried out his analysis in 1941. The motion picture records of the explosion by J. E. Mack, declassified in 1947, provided the experimental data needed for completing the deduction. However, the amount of energy released by the blast apparently was still classified in 1947!

An atomic explosion is idealized and simplified as the release of a large amount of energy E from a "point." The explosion sends off an expanding spherical fireball whose (outer) surface corresponds to a powerful shock wave. Let R be the spherical shock wave radius which increases with time. It is not unreasonable to expect this increase to vary with the amount of energy E released by the explosion, the ambient air density ρ_0, and the ambient air pressure p_0. We assume that these quantities are adequate for a proper description of the explosion and write

$$R = f(t, E, \rho_0, p_0) \qquad (1.13)$$

The fdu's in this case are again L_1 = mass, L_2 = length, and L_3 = time and the dimensions of the five quantities involved are

$$[R] = L_2 \qquad [t] = L_3 \qquad [E] = L_1 L_2^2 L_3^{-2}$$
$$[\rho_0] = L_1 L_2^{-3} \qquad [p_0] = L_1 L_2^{-1} L_3^{-2} \tag{1.14}$$

By the second part of the Π theorem, there is only one possible dimensionless combination among the four quantities t, E, ρ_0, and p_0 on the right side of (1.13) which may be taken as

$$\Pi_1 = p_0 \left[\frac{t^6}{E^2 \rho_0^3} \right]^{1/5} \tag{1.15}$$

while R may be made dimensionless by a multiplicative factor $[\rho_0/Et^2]^{1/5}$ to get

$$\Pi = R \left[\frac{\rho_0}{Et^2} \right]^{1/5} \tag{1.16}$$

By the first part of the Π theorem, we may write (1.13) as

$$\Pi = F(\Pi_1) \quad \text{or} \quad R = \left[\frac{Et^2}{\rho_0} \right]^{1/5} F(\Pi_1) \tag{1.17}$$

where F is an unknown function of Π_1.

In cgs units, we have $\rho_0 = 1.25 \times 10^{-3}$ g/cm^3 and $p_0 = 10^6$ g/cm-s^2. The energy released, E, is expected to be a very large number; an explosion of 50 pounds of TNT corresponds to an E of the order of 10^{15} ergs (g-cm^2/s^2). An atomic explosion is much more powerful than that with $E = 9 \times 10^{20}$ ergs for one kiloton explosion. From these figures, we see that Π_1 is a small number if t is not more than a second. For $\Pi_1 < 0.01$ (or $t < 0.07$ s for the 20-kiloton explosion), $F(\Pi_1)$ may be approximated by $F(0)$ so that (1.17) is approximately

$$R \simeq \left[\frac{E}{\rho_0} \right]^{1/5} F(0) t^{2/5} \tag{1.18}$$

which is the formula derived by Taylor (1950). Experiments using light explosives can be conducted to determine $F(0)$ which turns out to be approximately 1.

The approximate relation (1.18) may be written as

$$\tfrac{5}{2} \log R = \log t + \tfrac{1}{2} \log \frac{E}{\rho_0} \tag{1.19}$$

If $\tfrac{5}{2} \log R$ is plotted against $\log t$ for a fixed value of E/ρ_0, (1.19) gives a straight line with unit slope. A record of the radius of the spherical fireball for different times after the explosion may be similarly plotted to verify the unit slope linear relation between $\tfrac{5}{2} \log R$ and $\log t$. With the linear relation (1.19), we can then get from the same plot of the data an estimate of the corresponding value of E.

Mack's motion picture and other declassified photographic data provided Taylor with the information he needed to validate the approximate relation (1.19) and to estimate the energy released by that first atomic explosion. His estimate turns out to be remarkably accurate. It is also rather amazing that a straight-line log-log plot of the data was predicted theoretically by Taylor more than four years before the actual explosion! Dimensional analysis has become increasingly more useful in the study of explosions since that time [see, for example, R. M. Schmidt (1977), and other references by Schmidt and K. A. Holsapple cited in the Bibliography at the end of this book].

1.7   Construction of Dimensionless Combinations

In the two examples worked out above, the dimensionless combinations Π and Π_j were obtained by inspection. We now describe a systematic procedure for their determination in any problem.

For the case of the simple pendulum, any dimensionless combination among the variables must be of the product form $m^{x_1} l^{x_2} g^{x_3} \theta_0^{x_4} P^{x_5}$. This product has the dimension

$$M^{x_1} L^{x_2} (LT^{-2})^{x_3} (M^0 L^0 T^0)^{x_4} T^{x_5} = M^{x_1} L^{x_2+x_3} T^{x_5-2x_3}$$

For the combination to be dimensionless, we must have

$$x_1 = 0 \qquad x_2 + x_3 = 0 \qquad x_5 - 2x_3 = 0 \tag{1.20}$$

while x_4 is arbitrary (as it is already dimensionless). This system of three linear homogeneous equations for four unknowns has a one-parameter family of solutions which may be taken to be

$$x_1 = 0 \qquad x_2 = -x_3 \qquad x_5 = 2x_3 \tag{1.21}$$

One solution is therefore $\Pi = P\sqrt{g/l}$ corresponding to $x_4 = 0$ and $x_3 = \frac{1}{2}$. As long as we keep $x_4 = 0$, no new independent dimensionless combination is possible. An independent combination is simply $\Pi_1 = \theta_0$ corresponding to $x_3 = 0$ and $x_4 = 1$.

For the more general setting of Section 1.4, a dimensionless combination of $n - m$ secondary quantities $W_{m+1}, \ldots, W_n$ is the product

$$[W_{m+1}^{x_1} W_{m+2}^{x_2} \cdots W_n^{x_{n-m}}] = [W_1]^{\alpha_1}[W_2]^{\alpha_2} \cdots [W_m]^{\alpha_m} \qquad (1.22)$$

with

$$\alpha_k = x_1 a_{m+1,k} + x_2 a_{m+2,k} + \cdots + x_{n-m} a_{n,k} = 0 \qquad (1.23)$$

for $k = 1, 2, \ldots, m$. This linear homogeneous system of m equations for $n - m$ unknowns, $x_1, x_2, \ldots, x_{n-m}$, generally has an $(n - 2m)$-parameter family of solutions which can be obtained by the method of row-echelon reduction in elementary linear algebra. (Given that m is generally not a large number, Cramer's rule would also be adequate, though not as efficient.)

For the atomic explosion model, we have $m = 3$, $n - m = 5$, and $n - 2m = 2$. With (1.14), a dimensionless combination of the five secondary quantities t, E, ρ_0, p_0, and R is

$$[t^{x_1} E^{x_2} \rho_0^{x_3} p_0^{x_4} R^{x_5}] = [L_1]^{x_2 + x_3 + x_4}[L_2]^{2x_2 - 3x_3 - x_4 + x_5}[L_3]^{x_1 - 2x_2 - 2x_4} \qquad (1.24)$$

Hence, we have

$$\alpha_1 = x_2 + x_3 + x_4 = 0 \qquad \alpha_2 = 2x_2 - 3x_3 - x_4 + x_5 = 0$$
$$\alpha_3 = x_1 - 2x_2 - 2x_4 = 0 \qquad (1.25)$$

The two-parameter solution of (1.25) may be written as

$$x_1 = -2x_3 \qquad x_2 = -x_3 - x_4 \qquad x_5 = 5x_3 + 3x_4 \qquad (1.26)$$

The two dimensionless combinations may therefore be taken in the form given by (1.15) and (1.16). The second corresponds to $x_1 = \frac{1}{5}$ and $x_4 = 0$ while the first corresponds to $x_4 = 1$ and $x_3 = -\frac{3}{5}$. No other dimensionless combinations independent of these two are possible.

1.8   The Principle of Optimality

The development of the last few sections suggests that dimensional analysis is useful for the study of many phenomena. Unfortunately, this technique has nothing to say about many others. Take, for example, the following shortest route problem:

We have to get from site A to site J with three in-between stops. The options are B, C, or D for the first stop; E, F, or G for the second stop; and H or I for the third stop. The distance between any two consecutive stops is given in units of *leagues* in Figure 1.2. Find the shortest route.

Dimensional analysis apparently offers no help for this problem. There is, however, an obvious method for its solution. We simply calculate the total distance traveled for each of the 18 possible routes and pick out the one with the least total distance. This brute force approach would require a total of $4 \times 18 = 72$ additions and 18 comparisons. These calculations can be done nowadays in no time at all on a programmable computer (or pocket calculator) or, with a little patience, by hand.

For substantially larger versions of the problem, the same approach would require hours or days on the fastest computer available today. Medium-size versions of this problem are also solved thousands of times each week for different system parameters (such as the distance or other penalties between stops). In either case, the cost of the computing required may exceed the resources available for obtaining the solution of the problem. Therefore, it behooves us to give the problem (or any other problem ready-made for computer simulation) some thought before putting it on the computer. Some preliminary analysis will in fact lead to a more efficient method of solution which substantially reduces the required computing, and therefore makes the solution process more cost-effective. For the given problem, the shortest route can in fact be found with only 18 additions and 11 comparisons.

For this more efficient method of solution, suppose we have arrived at the third (i.e., the last in-between) stop at either site H or I. In either case, there is no alternative; the next stop has to be J (Figure 1.3). From site H,

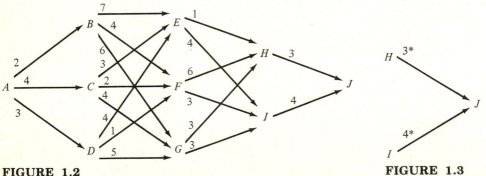

FIGURE 1.2

FIGURE 1.3

it would take another 3 leagues; from *I*, it would take 4 leagues. These conclusions require no addition or comparison.

Now, suppose we have arrived only at the second stop: site *E*, *F*, or *G*. If we should be at *E* and choose to go to *H*, then the shortest route to *J* would take $1 + 3^* = 4$ leagues. An asterisk in this section indicates the minimum distance from the next stop to *J*. On the other hand, if we should go (from *E*) to *I* instead, then the shortest route to *J* would be $4 + 4^* = 8$. A simple comparison indicates that we should not go to *I*; the best strategy from *E* would be to go to *H* (see Figure 1.4a). Similar considerations indicate that we should go to *I* from *F* and to *H* from *G* with a total distance to *J* equal to $3 + 4^* = 7$ and $3 + 3^* = 6$, respectively (see Figure 1.4b and c). Note that we arrived at the best strategy for all three sites of this second stop with a total of six additions and three comparisons.

For the first stop with site options *B*, *C*, and *D*, we repeat the process to obtain the best strategy for each site. The best route from *B* to *J* would be to go to either *E* or *F* for a total distance of 11 leagues (Figure 1.5a). The best route from *C* would be to go to *E*, for a total of 7 leagues to *J* (Figure 1.5b). From *D*, the best route to *J* would be to go to either *E* or *F* for a total of 8 leagues (Figure 1.5c). The operations needed for these solutions consist of nine additions and six comparisons, a third of each for each site.

We are now ready for the solution of the original problem. From the starting point *A*, we conclude after three additions and two comparisons that the best route to *J* would be to go to *C* or *D* for a total distance of 11 leagues (Figure 1.6). The total operations for all the stops add up to 18 $(= 6 + 9 + 3)$ additions and 11 $(= 3 + 6 + 2)$ comparisons. These are notably less than the 72 additions and 18 comparisons needed for the brute force method. The gain in efficiency goes up geometrically for more complex problems. Note that the shortest route generally does not head for the nearest next stop; such a route has a total distance of 13 leagues. Sometimes, it pays to sacrifice a little at the start for closer stops later. In Exercise 9, we see that the backward nearest neighbor approach (working backward from the end site) is also not optimal in general.

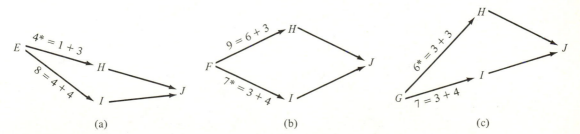

(a)   (b)   (c)

FIGURE 1.4

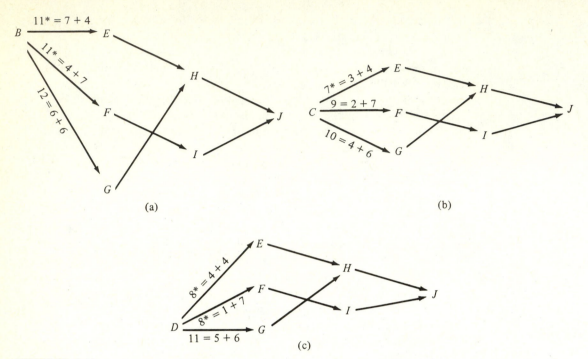

FIGURE 1.5

With high-speed computing facilities readily available to our scientists and engineers, there is a tendency for an individual to do "computer simulations" for all but the simplest problems. The above shortest route example shows that it often pays handsomely to give some thought to the simulation procedure. The efficiency gained corresponds to a reduction of computing cost which is sometimes critical to the feasibility of the solution process (given limited computing funds).

Also important is the fact that there may be a significant gain on a different front. In our efficient solution process for the shortest route problem, the following striking feature forces itself to our attention:

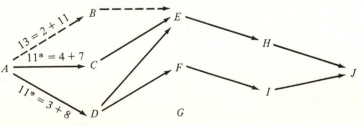

FIGURE 1.6

The optimal strategy from a given stop does not depend on how we arrive at that stop.

It is this so-called *Markov* property which allows us to work backward from the end stop, eliminating the nonoptimal route at each stop and thereby the associated calculations thereafter. This property, the analog of which plays a dominant role in probability theory, is common to a very large class of problems cutting across many disciplines. In this broader context, the property is often stated in the form of the following *Principle of Optimality* for systems evolving through a discrete number of stages (stops) and being in one of a number of possible states (sites) at each stage:

1C: An optimal strategy must have the property that regardless of how we enter a particular state (site) of the system, the remaining decisions must constitute an optimal (sub-) policy for leaving that state.

The solution process for the optimal strategy which makes use of this principle is known as *dynamic programming* [see Hillier and Lieberman (1974)]. Dynamic programming has been found useful in many other classes of optimization problems such as integer programming, knapsack problems, etc. (see the exercises at the end of this chapter). The extension of the dynamic programming technique to the solution of these problems makes the analysis of the shortest route problem valuable well beyond the gain in computing efficiency for that one single problem. For this reason alone we can never emphasize enough the following message:

Think before you compute!

Whenever possible, we should strive for a complete understanding of the qualitative features of the solution of a problem. Ideally, machine computation should be done only to confirm quantitatively what we already know qualitatively. This book is designed to promote this point of view.

EXERCISES

1. Use the procedure described in Section 1.7 to determine the independent dimensionless combinations for the simple pendulum problem of Section 1.4. (You may exclude the dimensionless quantity θ_0 if you wish.)

2. For a simple pendulum with an initial angular velocity $\omega \neq 0$, determine all the independent dimensionless combinations and a relation which gives the period P in terms of other factors similar to (1.3).

3. A spherical raindrop falls from a motionless cloud. Its terminal velocity v is expected to depend on its size (characterized by its radius r), density ρ, the gravitational acceleration g, and the viscosity of air μ, with $[\mu] = ML^{-1}T^{-1}$. (Viscosity measures resistance to motion caused by collisions between moving molecules.) Determine this dependence as explicitly as possible, ignoring other factors such as surface tension.

4. An object of mass m (e.g., a bomb) is dropped vertically from a height h above ground level with zero initial velocity. We are interested in its time in flight t before hitting the ground. Suppose we do not know any laws of physics but suspect that t is to depend on m, h, the gravitational acceleration g, and the shape of the object characterized by a dimensionless number s. Use dimensional analysis to determine t. (The effect of air drag is assumed to be incorporated in s.)

5. In tests for fuel economy, cars are driven at constant (optimum) speed on a level highway. With no acceleration, the force of propulsion must be in equilibrium with the force from air resistance. The variables affecting the propulsion force F_p are C_r (amount of fuel burned per unit time), K (energy contained in each gallon of gasoline), and v (speed), given by $[F_p] = MLT^{-2}$, $[C_r] = L^3T^{-1}$, $[K] = ML^{-1}T^{-2}$, and $[v] = LT^{-1}$. Determine F_p as a function of C_r, K and v.

6. In excavation and mining operations, the size of a crater resulting from a given explosive is of interest. Use dimensional analysis to relate the crater's hemispherical volume V produced by a spherical explosive located at some depth h in a given soil medium. Assume the crater volume to depend only on the radius r, mass m and energy E of the explosive, the gravitational acceleration g, and the mass density of the soil ρ. Determine the volume of the crater as a function of these parameters.

7. There is a rule of thumb for roasting a turkey in an oven: Set the oven at 400°F and allow 20-min cooking time for each pound of turkey weight. Determine whether or not this is a good rule.

8. Consider a simple pendulum with a frictional damping force $F = kv$ where v is the speed of the pendulum and k is a proportional constant of dimension MT^{-1}. Deduce how the period of this pendulum depends on the input factors: m, l, g, θ_0, and k.

9. Find the shortest route from A to J in Figure 1.7.

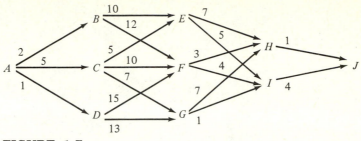

FIGURE 1.7

10. A company has three factories, all being considered for possible expansion. The total capital (in millions of dollars) available for expansion is 5. Each factory has a set of alternative plans for expansion as given below. Each plan involves a capital allocation C for the factory and an estimated revenue R from its expansion. Use a backward recursion to find an allocation of the available capital to the three factories which gives the highest total estimated revenue for this capital budgeting problem. (*Note:* There are three plans for factory 1, four plans for factory 2, and two plans for factory 3.)

Plan	Factory 1		Factory 2		Factory 3	
	C_1	R_1	C_2	R_2	C_3	R_3
1	0	0	0	0	0	0
2	1	5	2	8	1	3
3	2	6	3	9	—	—
4	—	—	4	12	—	—

11. Three different kinds of packaged cargos have different unit weights and unit values as given below. Load a vessel with a combination of cargo packages which gives the maximum total value without exceeding the total weight limit of 50 lb. Fractions of cargo packages are not allowed.

Cargo	Unit weight	Unit value
1	10	17
2	41	72
3	20	35

12. Use the method of dynamic programming to find a combination of nonnegative integers $m_1, \ldots, m_4$ which gives the maximum value of $M \equiv (m_1 + 2)^2 + (m_2 m_3) + (m_4 - 5)^2$ with $m_1 + m_2 + m_3 + m_4 \leq 5$.

Evolution of Dynamical Systems

The Earth and other planets in the solar system orbit around the sun. For many human endeavors such as space flight navigation, it is important to know the motion of these planets in some detail. The solar system is only one of many *dynamical systems* whose time-varying behavior has been or should be investigated. For our purpose, it suffices to define a dynamical system as any phenomenon the temporal changes of which are governed by a system of ordinary differential equations (ODE). On a smaller scale, analyses of the motion of automobiles along highways were initiated in the mid-fifties to obtain information useful for the design of safer and less congested highways. At still smaller scales, the growth and movement of fish populations are being investigated in order to develop an optimal fishery management policy; the temporal fluctuation of white blood cell counts in various types of leukemia patients is being investigated to help our search for appropriate treatments of these diseases, etc.

One aspect of the investigation of dynamical systems such as those mentioned above is concerned with the discovery of the natural laws which govern the behavior of these systems and with the use of these laws to predict the behavior of the systems at some future time. This aspect of the study of dynamical systems often requires mathematical deductions of different levels of sophistication and therefore engages the efforts of applied mathematicians. It is important that students of applied mathematics be exposed to this type of mathematical activity and be prepared to undertake challenging projects in this direction.

In the next two chapters, we illustrate, by way of planetary motion, the study of the *evolution* (temporal change) of dynamical systems. We begin with a brief discussion of Kepler's three laws for planetary motion in the usual framework of Newtonian physics. This discussion gives us an opportunity to review some useful techniques in ordinary differential equations and vector calculus. It also shows that *periodic motion* plays an important role in dynamical systems. The results from this highly idealized and simplified model of reality provide an amazingly accurate description of planetary motion. Still, the elegant Newtonian theory appears to have a tiny flaw. It does not predict the admittedly very small but definitely observable advance of the perihelion of the planet Mercury. We will show how a relativistic correction of the Newtonian theory eliminates the discrepancy between predictions based on classical Newtonian physics and actual observations. In the process, we will be exposed to two powerful perturbation methods developed in the twentieth century for obtaining approximate solutions for difficult problems in differential equations.

The classical theory of planetary motion is among the most impressive successes in the study of the evolution of different dynamical systems throughout the history of applied mathematics. Its delineation of a complex phenomenon is simple and elegant. Its predictive power for current and future applications is rich and deep. (We are fortunate enough to be in a position to see the theory at work in twentieth century space flights.) The need for the relativistic correction in one specific application provides us with a nontrivial demonstration of the applied mathematics cycle of formulation-analysis-evaluation (-reformulation, etc.) in action.

2

Here Comes the Sun

The Three Laws of Kepler

Vector calculus and nonlinear ordinary differential equations are involved in this chapter. However, the actual solution of the relevant ODE is deduced without requiring a prior exposure to any specific solution technique. The material in fact serves as a review of ODE for later chapters.

2.1 The Two-Body Problem

Consider the motion of a planet in the solar system which has a mass of m kilograms. Suppose at some reference time, $t = 0$, the planet is at a distance r_0 from the sun (which has a mass of M kilograms) and is moving with a certain (initial) velocity in a direction *not* radially along the line segment connecting the planet and the sun. To describe the motion of the planet, we treat it as a single mass particle and introduce a polar coordinate system in the plane defined by the radial direction and the direction of the planet's motion at $t = 0$ with the sun at the origin (see Figure 2.1). The position of the planet relative to the sun at a later time t is given by the position vector $\mathbf{r}(t)$ directed from the origin to the planet. Correspondingly, the velocity and acceleration vectors of the planet at time t are $\mathbf{v} \equiv \mathbf{r}^{\boldsymbol{\cdot}}$ and $\mathbf{a} \equiv \mathbf{r}^{\boldsymbol{\cdot\cdot}}$, respectively. Throughout this book, a dot as a superscript indicates differentiation with respect to t so that $(\ )^{\boldsymbol{\cdot}} \equiv d(\ )/dt$.

In this chapter, we focus our attention on the main factor responsible for the planet's motion, namely, the gravitational attraction of the sun, and ignore secondary effects such as the influence of other planets in the solar system which are considerably smaller in mass than the sun. In Newtonian physics, it is postulated that the acceleration of a planet is proportional to the force $\mathbf{F}$ acting on the planet. More precisely, *Newton's second law of motion* for a constant mass particle stipulates $m\mathbf{r}^{\boldsymbol{\cdot\cdot}} = \mathbf{F}$. Newton further postulated that the force acting on a planet is directly proportional to the

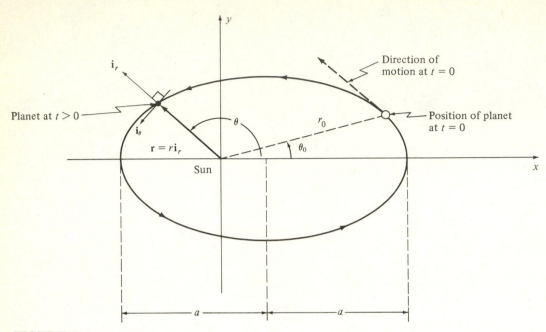

FIGURE 2.1

product of the masses, M and m, and inversely proportional to $r^2 \equiv |\mathbf{r}|^2$, where r is the distance between the sun and the planet. This force pulls the planet toward the sun so that

$$m\mathbf{r}^{\cdot\cdot} = -\frac{GMm}{r^2}\,\mathbf{i}_r \tag{2.1}$$

where $\mathbf{i}_r$ is the unit vector in the radial direction. The quantity G is a constant of proportionality whose numerical value depends on the choice of units for mass, length, and time, for example, $G = 6.67 \times 10^{-8}$ cm^3/g-s^2 in cgs units. The postulate on $\mathbf{F}$ is known as *Newton's law of universal gravitation* or the *inverse-square law*.

The vector equation (2.1) is equivalent to two scalar equations for the distance function $r(t)$ and the polar angle variable $\theta(t)$. These two quantities determine the position of the planet at any time t in the plane of the planet's motion. [As a consequence of (2.1) and the initial condition $z(0) = z^{\cdot}(0) = 0$, there is no out-of-plane motion.] With $\mathbf{r} = r\mathbf{i}_r = r(\cos\theta\,\mathbf{i}_x + \sin\theta\,\mathbf{i}_y)$, it is an exercise in vector calculus to deduce from (2.1) the following two scalar second-order differential equations:

$$2r^{\cdot}\theta^{\cdot} + r\theta^{\cdot\cdot} = 0 \tag{2.2}$$

and

$$r^{\bullet\bullet} - r(\theta^{\bullet})^2 = \frac{-MG}{r^2} \tag{2.3}$$

(see Exercise 1). Given the initial position and velocity vectors of the planet at some reference time $t = 0$,

$$\mathbf{r}(0) = \mathbf{r}_0 = r_0\mathbf{i}_{r0} \quad \text{and} \quad \mathbf{r}^{\bullet}(0) = \mathbf{v}_0 = v_{r0}\mathbf{i}_{r0} + v_{\theta0}\mathbf{i}_{\theta0} \tag{2.4}$$

with $\mathbf{i}_{r0} = \cos\theta_0\mathbf{i}_x + \sin\theta_0\mathbf{i}_y$ and $\mathbf{i}_{\theta0} = -\sin\theta_0\mathbf{i}_x + \cos\theta_0\mathbf{i}_y$, the two coupled nonlinear ODE (2.2) and (2.3) determine $r(t)$ and $\theta(t)$ for all later times ($t > 0$). Instead of obtaining explicit expressions for r and θ as functions of t, we will deduce from the two differential equations of motion, (2.2) and (2.3), the three empirical laws describing the motion of the planets proposed by Kepler in the early seventeenth century after laboring for 22 years over available observational data. For Kepler's laws, we must keep in mind that along the planet's orbit, the radial position of the planet is completely determined by its angular position. This is so because we may solve the relation $\theta = H(t)$ for t in terms of θ so that $t = T(\theta)$, and use it to eliminate t from $r = R(t)$ to get $r = R(T(\theta)) = r(\theta)$.

2.2   Kepler's Second Law

As r is not identically zero for all t, we can multiply (2.2) by r to get

$$2rr^{\bullet}\theta^{\bullet} + r^2\theta^{\bullet\bullet} = (r^2)^{\bullet}\theta^{\bullet} + r^2\theta^{\bullet\bullet} = (r^2\theta^{\bullet})^{\bullet} = 0$$

which can be integrated once to get

$$r^2\theta^{\bullet} = \frac{p_0}{m} \tag{2.5}$$

where p_0 is a constant of integration. Note that p_0 is determined by the position and velocity of the planet at $t = 0$ to be $p_0/m = r_0v_{\theta0}$. The quantity $mr^2\theta^{\bullet}$ is known as the *angular momentum* of the planet. Equation 2.5 tells us that the angular momentum of the planet is the same for all time, or

2A: Angular momentum is conserved.

Suppose that at time $t = t_1$, the planet is at point P_1 corresponding to $r = r_1$ and $\theta = \theta_1$, and it gets to a point P_2 corresponding to $r = r_2$ and $\theta =$

θ_2 at a later time $t_2 > t_1$. Recall that we may think of r as a function of θ for points on the orbit. Then the position vector $\mathbf{r}$ sweeps out a sector during the time interval $\Delta t \equiv t_2 - t_1$ with an area

$$\Delta A \equiv \left[\text{Area} \right]_{t_1}^{t_2} = \int_{\theta_1}^{\theta_2} \int_0^{r(\theta)} r \, dr \, d\theta = \frac{1}{2} \int_{\theta_1}^{\theta_2} r^2 \, d\theta = \frac{1}{2} \int_{t_1}^{t_2} r^2 \theta^\cdot \, dt$$

where $r(\theta)$ is the plane curve traced out by the planet's orbit. By (2.5), the integrand is independent of t and we get

$$\Delta A = \frac{p_0}{2m} \Delta t \tag{2.6}$$

Equation 2.6 is a mathematical statement of *Kepler's second law:*

> **2B:** The position vector of the planet's orbit sweeps out sectors of equal area in equal time intervals.

The formula (2.6) actually tells us a little more about this area than Kepler's original statement of the law; it gives a specific relation between this area and the (initial) angular momentum per unit mass of the particular planet.

2.3   Kepler's First and Third Laws

We now use (2.5) to eliminate $\theta^\cdot$ from (2.3) to get

$$r^{\cdot\cdot} - \frac{p_0^2}{m^2} \frac{1}{r^3} = - \frac{GM}{r^2} \tag{2.7}$$

Equation 2.7 is a second-order nonlinear ODE in which the independent variable t does not appear explicitly. Such an equation is said to be *autonomous*. The usual method of solution for this class of ODE is discussed in Exercise 3. Once we have $r(t)$, we can get $\theta^\cdot(t)$ from (2.5) and then $\theta(t)$ by a simple integration.

For the shape of the planet's orbit however, it suffices to have r as a function of θ. We can get $r(\theta)$ directly from (2.5) and (2.7) without first solving (2.7) for $r(t)$. This is accomplished by observing that

$$r^\cdot = \frac{dr}{d\theta} \theta^\cdot = \frac{p_0}{mr^2} \frac{dr}{d\theta} = - \frac{p_0}{m} \frac{d}{d\theta} \left(\frac{1}{r} \right)$$

where we have used (2.5) to eliminate $\theta^{\cdot}$, and correspondingly

$$r^{\cdot\cdot} = -\frac{p_0}{m}\frac{d^2}{d\theta^2}\left(\frac{1}{r}\right)\theta^{\cdot} = -\frac{p_0^2}{m^2 r^2}\frac{d^2}{d\theta^2}\left(\frac{1}{r}\right)$$

so that we can write (2.7) as

$$\frac{d^2}{d\theta^2}\left(\frac{1}{r}\right) + \frac{1}{r} = \frac{GMm^2}{p_0^2} \tag{2.8}$$

The solution of (2.8) is immediate as it is a linear second-order ODE with constant coefficients for $1/r$:

$$\frac{1}{r} = A\cos\theta + B\sin\theta + \frac{GMm^2}{p_0^2}$$

The constants of integration A and B are then determined by the initial conditions $r = r_0$ and $dr/d\theta = r_0 v_{r0}/v_{\theta 0}$ at $\theta = \theta_0$ (see Equation 2.4). For the case $v_{r0} = 0$, the resulting solution may be written as

$$r(\theta) = \frac{a(1 - e^2)}{1 + e\cos(\theta - \theta_0)} \tag{2.9}$$

where

$$\frac{p_0^2}{GMm^2} = a(1 - e^2) \quad \text{and} \quad \frac{p_0^2}{GMm^2 r_0} = 1 + e$$

For $0 < e < 1$, the formula (2.9) (or the corresponding expression for the case $v_{r0} \neq 0$ in Exercise 8) is the mathematical counterpart of *Kepler's first law:*

2C: The orbit of the planet around the sun is an ellipse.

The quantity e in the expression (2.9) is the *eccentricity* of the ellipse. The smaller e is the closer the ellipse is to a circle. The quantity a in (2.9) is the length of the *semimajor axis* of the ellipse (see Figure 2.1).

Let T be the time it takes for the planet to complete one revolution of its orbit. We know from (2.6)

$$\frac{p_0 T}{2m} = \text{area of the ellipse} = \pi a^2 \sqrt{1 - e^2}$$

But $\sqrt{1 - e^2} = (p_0/m)/\sqrt{GMa}$; therefore, we have

$$T^2 = \frac{4\pi^2 a^3}{GM} \tag{2.10}$$

which is the mathematical counterpart of *Kepler's third law:*

> **2D:** The square of the orbital period is proportional to the third power of the length of the semimajor axis of the elliptical orbit.

Similar to the case of the second law, Equations 2.9 and 2.10 contain much more information than Kepler's original statement of the first and third law. Moreover, Newton's results are quantitative and precise; they can be (and have been) used to make predictions on measurable quantities such as position and velocity which characterize the motion of the planet.

2.4 An Applied Mathematics Perspective

Kepler, the empiricist, spent nearly his entire professional life analyzing the data amassed by Tycho Brahe on the motion of the planets, and managed to infer from these data three rules which govern the behavior of the planets as they orbit around the sun. His accomplishment is epochal in the development of astronomy, a discipline which tries to understand the universe in which we live. By comparison, Newton's results on the same problem as given by (2.6), (2.9), (2.10) and others not presented here (e.g., expressions for r and θ in terms of time, etc.) are derived from his laws of motion and universal gravitation and are much more quantitative, more precise, and more substantial. There is a strong correlation between mathematical deduction and the high precision and depth in the conclusions drawn about a given phenomenon. The fact that Newton is a major figure in mathematics (as well as in physics) was no accident; he had to develop the mathematical tools needed to achieve the breadth and depth of his work in physics.

Newton's laws of motion and universal gravitation are contributions to physics. The deduction of the results for planetary motion from these physical laws is applied mathematics in action. The results for planetary motion themselves are again contributions to physics. The important point to be noted here is this: Applied mathematics is usually an integral part of many quantitative investigations. To strive for quantitative precision, modern curricula in science and engineering require students to be exposed to more and more mathematics. Even such relatively extensive exposure is often not enough to cope with many new problems which require new mathematical methods or the clever use of available mathematical techniques. It often requires a kind of experience in applied mathematics different from the materials in methods courses to meet these challenges. This book is designed to demonstrate this point with interesting examples from the past successes and present activities and, in the process, to introduce students to different mathematical ideas not encountered in conventional methods courses. Students may then pursue a more thorough study of some of these mathematical ideas in more advanced undergraduate or graduate courses in applied mathematics if they find these ideas interesting or useful.

Quantitative precision aside, there is another fundamental difference between the results of Kepler and Newton. Kepler's undeniably outstanding contribution consists of factual statements beautifully synthesized from observational data. Newton's remarkable results are meant to help us to understand these facts: Why is the orbit elliptical? Why the equal area rule? and, Why the "three-half-power" law for the orbital period? Newton provided the answers: The dynamics of all mass particles, be it a planet or an electron, are governed by his three laws of motion! Moreover, Newton's laws are also fundamental to mechanics of rigid bodies and deformable media, not just to mass particles or planetary motion.

Consequences of Newton's fundamental laws of physics are too numerous to be reported here. The formulation of the problem of planetary motion as an initial-value problem (IVP) in ODE is only one of them. It is an important one as it helped to stimulate the development of a different area of mathematics, namely mathematics for differential equations, from the prevailing mathematical activities of that period. We owe most of today's standard methods of solution for ODE to the interest in planetary motion and related problems in dynamical systems generated by Newton's contributions.

The influence of a mathematical model on the development of mathematical methods for its analysis was mentioned here for reasons beyond historical interest. It is intended to call the attention of students in applied mathematics to the fact that model formulation and the analysis of the mathematical model are two distinct stages in a quantitative study of any observed phenomenon; they require different kinds of mental processes and treatment. Rarely, if ever, is the formulation stage of a problem completely deductive. Newton's willingness to treat the planet and the sun as point masses and to neglect the effect of other planets was a bold experiment based on good scientific judgment. We do not have the time or the scientific background to dwell on the experimentation involved in the formulation stage of the problems encountered in this course. Nevertheless, we will try in this book to emphasize this distinctly different stage of an applied mathematical (or scientific) investigation and occasionally encourage students to set up a mathematical problem based on some neatly arranged observations. Exercises

6, 8–11 are designed for this purpose. Of course, the analysis or solution phase of these problems is also important and instructive.

A quantitative analysis of an observable phenomenon does not end with the solution of the mathematical problem. We must examine the solutions for its implications and check them with known facts to see whether the mathematical model is an adequate substitute for the real problem. Ultimately, we want to know whether or not the solution of the mathematical problem accurately describes the evolution of the actual phenomenon. Should Newton's idealization and simplification be too drastic for the resulting mathematical model to yield realistic results, a reformulation of the model would be necessary. We will address this aspect of our quantitative study of planetary motion in the next chapter.

EXERCISES

1. Derive the two scalar differential equations of motion for $r(t)$ and $\theta(t)$ from the vector equation

$$m\mathbf{r}\ddot{} = -\frac{GMm\mathbf{i}_r}{r^2} \qquad \mathbf{i}_r = \cos\theta\,\mathbf{i}_y + \sin\theta\,\mathbf{i}_y$$

2. Second-order ODE of the type $y'' = f(x, y')$ where primes indicate differentiation with respect to x are usually solved by setting $v \equiv y'$ [see Boyce and DiPrima (1976), p. 89]. Solve
   (a) $x^2 y'' + 2xy' - 1 = 0 \ (x > 0)$ (b) $xy'' + y' = 1 \ \ (x > 0)$
   (c) $y'' + x(y')^2 = 0$ (d) $2x^2 y'' + (y')^3 = 2xy' \ \ (x > 0)$

3. For second-order ODE of the type $y'' = f(y, y')$, we can eliminate the independent variable x by setting $v \equiv y'$ and then use y as the new independent variable [see Boyce and DiPrima (1976), p. 89]. Solve
   (a) $yy'' + (y')^2 = 0$ (b) $y'' + y = 0$
   (c) $y'' + y(y')^3 = 0$ (d) $2y^2 y'' + 2y(y')^2 = 1$
   You may leave your final answers in the form of an indefinite integral if necessary; but your solutions must contain two arbitrary constants of integration.

4. Solve the following IVP (initial-value problems):
   (a) $y'y'' = 2$, $y(0) = 1$, $y'(0) = 2$
   (b) $y'' - 3\epsilon^2 y^2 = 0$, $y(0) = 2$, $y'(0) = 4\epsilon$ ($\epsilon > 0$ is a known constant)
   (c) $(1 + x^2)y'' + 2xy' + 3x^{-2} = 0$, $y(1) = y'(1) = 0$
   (d) $y'y'' - x = 0$, $y(1) = 2$, $y'(1) = 1$

5. *Falling into the Sun.* If the earth stopped orbiting around the sun, how long would it take us to fry? [*Hint:* $r(t = 0) = r_0$ and $\theta^{\cdot}(t = 0) = 0$; then find T so that $r(t = T) = 0$.]

6. Going at top speed, grand prix driver X leads archrival Y by a *steady* 3 miles. Only 2 miles from the finish, X runs out of gas. Thereafter, X decelerates with time at a rate proportional to the square of her (instantaneous) speed and in the next mile X's speed exactly halves. Who wins? (*Note:* The course for this race is straight.)

7. Suppose the (initial) position and velocity vector of a planet at $t = 0$ (see Equation 2.4) are $\mathbf{r}(0) = r_0 \mathbf{i}_{r0}$ and $\mathbf{r}^{\cdot}(0) = v_{\theta 0} \mathbf{i}_{\theta 0}$.
   (a) Find $r(0), \theta(0), r^{\cdot}(0)$, and $\theta^{\cdot}(0)$.
   (b) Find p_0, a, and e (see Section 2.3).

8. Suppose (see Equation 2.4) $\mathbf{r}(0) = r_0 \mathbf{i}_{r0}$ and $\mathbf{r}^{\cdot}(0) = v_{r0} \mathbf{i}_{r0} + v_{\theta 0} \mathbf{i}_{\theta 0}$.
   (a) Find $r(0), \theta(0), r^{\cdot}(0)$, and $\theta^{\cdot}(0)$.
   (b) Find p_0, a, e, and ϕ when the solution of the IVP for (2.8) is written in the form:

$$r(\theta) = \frac{a(1 - e^2)}{1 + e \cos(\theta - \theta_0 + \phi)}$$

9. Find the general solution of (2.7).

10. *Agnew's Snowplow Problem.* One day it started snowing at a *heavy* and *steady* rate. A snow plow started out at noon, going 2 miles the first hour and 1 mile the second hour. What time did it start snowing?

11. *The Great Snowplow Chase.* A second identical snowplow starts sometime after noon in the wake of the first. If it ever catches up with the first plow, when, where, and how?

12. *The Triple Collision.* Three identical snowplows started at noon, 1 P.M., and 2 P.M., respectively. All three collided sometime after 2 P.M. When did it start snowing?

13. *Aging Springs.* Elastic springs weaken after repeated stretchings; therefore, the spring stiffness constant k is actually a slowly decreasing function of time. Find the displacement $y(t)$ from the "at rest" position of a spring-mass system in a gravity-free space with $k(t) = k_0/(1 + \epsilon t)^2$, where k_0 and ϵ are both positive constants, and with the initial condition $y(0) = 1$ and $y^{\cdot}(0) = 0$.

14. *Aging Dashpots.* Dashpots also get worn out with repeated usage so that the corresponding damping coefficient c is also a slowly decreasing

function of time. Find the displacement $y(t)$ from the "at rest" position of a spring-mass-dashpot system in a gravity-free space with $c = c_0/(1 + \epsilon t)$ and $k = k_0/(1 + \epsilon t)^2$, where c_0, k_0, and ϵ are all positive constants, and with the initial conditions $y(0) = 1$ and $y^{\cdot}(0) = 0$. Given k_0, ϵ, and the mass m, find the value of c_0 for which the motion of the mass is of constant amplitude. Is the motion periodic?

3

Slower than Light

The Precession of the Perihelion of the Planet Mercury

The basic idea of perturbation theory and the more sophisticated technique of Poincaré's method for elimination of secular terms are developed and applied in this chapter. The general method for a second-order autonomous ODE is needed to express its solution in terms of an elliptic integral.

3.1 The Relativistic Effect

Any scientific theory developed to explain an observed phenomenon must be tested against observations. Here the record of the simple and elegant Newtonian theory of planetary motion is remarkably good. The worst discrepancy known to date involves the planet Mercury. The point of the elliptical orbit of a planet closest to the sun is called the *perihelion,* the furthest point is called the *aphelion.* Astronomical data show that the perihelion of Mercury advances about 574 seconds of arc per century while the Newtonian theory predicts no such advance. Now, there are 324,000 seconds of arc in a right angle. Considering the scale and complexity of the problem, the theoretical predictions are still amazingly close to the observed facts in this case. Moreover, errors in these predictions are not unexpected since the influence of other planets in the solar system has been ignored in the conventional theory of Section 2.1. The solution of the corresponding many-body problem which takes into consideration the influence of other planets narrows the discrepancy in Mercury's orbit to only 43 seconds of arc per century. But the computations and observations involved are sufficiently accurate so that the source of the remaining discrepancy continued to be a mystery until the development of the *theory of general relativity.*

It had been known for some time that the sun also moves. What was not known is the rather complex effect this movement has on the observed motion of the planets, or more generally, the complex effect of the relative motion of two moving reference coordinate systems on the observations made in these two coordinate systems. The net relativistic effect is usually negligible unless we are dealing with motion comparable to the speed of light and/or with the accumulation of small effects over a long period of time. The advance of Mercury's perihelion is in fact an accumulation of small effects over time and must be analyzed by a relativistic theory. In this chapter, we discuss some computational aspects of the relativistic correction for the orbit of Mercury predicted by classical mechanics and demonstrate that this correction does account for the 43 seconds of arc per century discrepancy mentioned above. Our discussion will again be limited to the two-body problem in order to focus on the relativistic effect; to allow for the presence of other planets would only muddle up the issue. The two-body problem also keeps the mathematical details to a minimum.

To incorporate the effect of relative motion in orbital mechanics, we will work with a dimensionless variable $a(1 - e^2)/r$ where, as before, r is the radial distance from the sun to the planet at time t. For the Newtonian theory, we denote this quantity by w_0 and write the corresponding equation of motion (2.8) in dimensionless form as

$$\frac{d^2 w_0}{d\theta^2} + w_0 = 1 \tag{3.1}$$

with $a(1 - e^2) = p_0^2/GMm^2$. Evidently, we have from (2.9)

$$w_0 = 1 + e \cos(\theta - \theta_0) \tag{3.2}$$

For a theory which includes the relativistic effect, we denote the same quantity $a(1 - e^2)/r$ by w. Without a working knowledge of tensor calculus, it will not be possible for us to derive the equations of motion for w and the corresponding angular variable as we did for the nonrelativistic case. We merely state without proof [see Sokolnikoff (1964)] that (1) the angular momentum is still conserved, and (2) the quantity w is the solution of the nonlinear ODE:

$$\frac{d^2 w}{d\theta^2} + w = 1 + \epsilon w^2 \qquad \epsilon = 3\left(\frac{GMm}{p_0 c}\right)^2 \tag{3.3}$$

where c is the speed of light. From (2.5), we have $r\theta^{\cdot} = p_0/mr$ with r being of the order of magnitude of $a(1 - e^2) = p_0^2/GMm^2$ as given by (2.9). The orbital speed of the planet is therefore of the order of $p_0/ma(1 - e^2) = GMm/p_0$. It follows that $\sqrt{\epsilon}$ is a measure of the ratio of the planet's speed to the speed of light. As we know from observation, that ratio is very small; for example, we have $\epsilon = O(10^{-9})$ for the planet Mercury. {We will use regularly in this book the notation $y = O(x)$ to mean $|y| \leq C_0|x|$ for some constant C_0 not large compared to unity [Kevorkian and Cole, (1981)].}

The ϵw^2 term is now seen to represent the relativistic effect. This effect may be removed by taking $\epsilon = 0$ (corresponding to setting the speed of light equal to infinity) to recover the equation of motion (3.1) for w_0. With (3.3), we now see that the relativistic theory introduces only an additional term to the equation for $1/r$ of the Newtonian theory, a term small in magnitude compared to the other terms in the same equation. Therefore, we expect the classical Newtonian theory to be generally adequate for a numerically accurate description of planetary motion. However, in our effort to understand the source of the extremely small anomaly in that theory manifested in the precession of the perihelion of Mercury, we must obtain at least the approximate effect of the ϵw^2 term in the solution $w(\theta)$ of (3.3), however small this effect may be.

To get a first approximation of this small effect, we replace w^2 on the right side of (3.3) by w_0^2 which is expected to be nearly w in magnitude so that the error accrued to this replacement should be considerably smaller in magnitude than the error from omitting ϵw^2 altogether. The resulting linear ODE is

$$\frac{d^2 w}{d\theta^2} + w \simeq 1 + \epsilon w_0^2$$

$$= 1 + \epsilon \left[\left(1 + \frac{e^2}{2} \right) + 2e \cos(\theta - \theta_0) + \frac{e^2}{2} \cos 2(\theta - \theta_0) \right] \quad (3.4)$$

where we have used the double-angle formula to eliminate the $\cos^2(\theta - \theta_0)$ term. The solution of (3.4) is

$$w \simeq 1 + A \cos\theta + B \sin\theta + e(\epsilon\theta) \sin(\theta - \theta_0)$$

$$+ \epsilon \left[\left(1 + \frac{e^2}{2} \right) - \frac{e^2}{6} \cos 2(\theta - \theta_0) \right] \quad (3.5)$$

where A and B are constants of integration to be determined by the initial conditions.

Among the terms due to the relativistic effect in this solution, that is, terms multiplied by ϵ, those in brackets remain small, of order ϵ for all θ. In the remaining term, however, ϵ appears in the form of a factor $\epsilon\theta$ which increases without bound as θ (or equivalently t) increases. Consequently, it represents the dominant effect of relativity. For simplicity, we omit the bracketed terms in (3.5) as they are of little qualitative and quantitative importance, and rewrite the resulting expression for w in the form (as done in Section 2.3):

$$w \simeq 1 + e[\cos(\theta - \theta_0) + \epsilon(\theta - \theta_0)\sin(\theta - \theta_0)]$$

$$= 1 + e\sqrt{1 + (\epsilon\bar{\theta})^2}\cos(\bar{\theta} - \phi) \qquad \text{with } \bar{\theta} = \theta - \theta_0 \qquad (3.6)$$

where $\phi = \tan^{-1}(\epsilon\bar{\theta}) \simeq \epsilon\bar{\theta}$ for $|\epsilon\bar{\theta}| \ll 1$.

We see from (3.6) that the phase angle ϕ varies with θ (and therefore with time); thus, the "perihelion" of the orbit advances by an amount approximately equal to $2\pi\epsilon = 4.9 \times 10^{-7}$ radian after each revolution in the direction of the planet's motion. Technically, we say the perihelion *precesses*. Mercury has an 88-day revolution and goes through 415 revolutions per century. Therefore, we have

> **3A:** By the relativistic theory, the "perihelion" of Mercury's orbit precesses by 43 seconds of arc for each century.

The agreement between this result and known observations is astonishing!

We have put quotation marks around the word perihelion in the above paragraph because the solution (3.6) says that we no longer have an elliptical orbit and it really does not make sense to speak of a perihelion. But for the time scale of a century, $\epsilon\bar{\theta}$ is of the order of 10^{-4}, written as $\epsilon\bar{\theta} = O(10^{-4})$, so that $\sqrt{1 + (\epsilon\bar{\theta})^2} \simeq 1$. Therefore, the orbit is nearly elliptical and we may continue to think of the precession of the perihelion of an elliptical orbit.

3.2   The Perturbation Method

The ad hoc procedure for finding an approximate solution for (3.3) by taking advantage of $\epsilon \ll 1$ can be made systematic and more general to allow for higher-order corrections. We will describe the so-called (regular) *perturbation method* of solution which is extremely useful for differential equations beyond the study of planetary motion. There is usually a small parameter lurking around in most problems!

Besides being a function of the independent variable θ, the solution of (3.3) also depends on the parameter ϵ which appears on the right side of the ODE. We write $w = w(\theta; \epsilon)$ to denote this dependence on ϵ. It can be shown [see, for example, Lin and Segel (1974)] that w is an analytic function of ϵ

near the origin. In other words, w has a Taylor series expansion in ϵ about $\epsilon = 0$,

$$w(\theta; \epsilon) = w_0(\theta) + w_1(\theta)\epsilon + w_2(\theta)\epsilon^2 + \cdots \qquad (3.7)$$

where $w_0(\theta) = w(\theta; 0)$, $w_1(\theta) = w_\epsilon(\theta; 0)$, $w_2(\theta) = w_{\epsilon\epsilon}(\theta; 0)/2!$, etc., with $(\ )_\epsilon = \partial(\ )/\partial\epsilon$. We will have found the solution of (3.3) once the unknown coefficients w_k, $k = 0, 1, 2, \ldots$, in the *perturbation solution* (3.7) are determined.

To find the coefficients of the perturbation series, we substitute (3.7) into (3.3) and collect coefficients of the same power of ϵ to get

$$(w_0'' + w_0) + \epsilon(w_1'' + w_1) + \epsilon^2(w_2'' + w_2) + \cdots$$
$$= 1 + \epsilon(w_0^2) + \epsilon^2(2w_0w_1) + \cdots \qquad (3.8)$$

where $(\ )' = d(\ )/d\theta$. Equation 3.8 must be satisfied for all ϵ and in particular $\epsilon = 0$. We get by setting $\epsilon = 0$,

$$w_0'' + w_0 = 1 \qquad (3.9)$$

which, along with the initial conditions at $t = 0$, reproduces the nonrelativistic solution $w_0(\theta) = 1 + e\cos(\theta - \theta_0)$. Next, we differentiate both sides of (3.8) with respect to ϵ and then set $\epsilon = 0$ to get

$$w_1'' + w_1 = w_0^2 \qquad (3.10)$$

Repeated differentiations and evaluations at $\epsilon = 0$ give

$$w_{k+1}'' + w_{k+1} = \sum_{n=0}^{k} w_{k-n}w_n \qquad k = 0, 1, 2, \ldots \qquad (3.11)$$

With (3.11), we see that the coefficients w_k are determined successively by a sequence of *linear* second-order ODE with constant coefficients whose right-hand member is composed of previously obtained quantities. For example, having found w_0 from (3.9), the right-hand side of (3.10) becomes a known function and the ODE (3.10) itself can be solved easily for w_1. Having found w_1, we use it (as well as w_0) in (3.11) with $k = 1$ to get

$$w_2'' + w_2 = 2w_0w_1 \qquad (3.12)$$

which can be solved easily for w_2, etc. The important observation here is that:

> **3B:** The use of the perturbation series (3.7) reduces the original problem involving a *nonlinear* ODE to a sequence of problems each involving a *linear* ODE.

Linear problems are usually much more tractable as in the present case. Moreover, for $\epsilon \ll 1$, we have only to determine a few terms in the perturbation series (3.7) to get an accurate approximate solution of the problem.

We emphasize once more that the (regular) perturbation method is extremely useful for a wide range of problems with a small parameter. Even for cases where the solution of the relevant ODE is not known to be analytic in the small parameter, the method can still be exploited in the same way to obtain an "asymptotic" solution (see Kevorkian and Cole, 1981). Because of the importance of the method, it will be illustrated in more detail (including the treatment of the initial conditions) through a simple example in the next section. In the process, we will also gain a new perspective of the perturbation method which will be useful when we return to the subject of planetary motion.

3.3 A Simple Initial-Value Problem

For the initial-value problem (IVP)

$$y'' + (1 + \epsilon)y = 0 \qquad y(0) = a \qquad y^{\cdot}(0) = 0 \qquad (3.13)$$

we seek a (regular) perturbation solution

$$y(t; \epsilon) = y_0(t) + \epsilon y_1(t) + \epsilon^2 y_2(t) + \cdots \qquad (3.14)$$

Upon substituting (3.14) into (3.13) and collecting coefficients of the same power of ϵ, we get

$$(y_0^{\cdot\cdot} + y_0) + \epsilon(y_1^{\cdot\cdot} + y_1 + y_0) + \cdots + \epsilon^k(y_k^{\cdot\cdot} + y_k + y_{k-1}) + \cdots = 0 \quad (3.15)$$

$$y_0(0) + \epsilon y_1(0) + \epsilon^2 y_2(0) + \cdots = a \quad (3.16)$$

$$y_0^{\cdot}(0) + \epsilon y_1^{\cdot}(0) + \epsilon^2 y_2^{\cdot}(0) + \cdots = 0 \quad (3.17)$$

From these, we get a sequence of IVP:

$$y_0^{\cdot\cdot} + y_0 = 0 \qquad y_0(0) = a \qquad y_0^{\cdot}(0) = 0 \qquad (3.18)$$

and

$$\ddot{y}_k + y_k + y_{k-1} = 0 \qquad y_k(0) = 0 \qquad \dot{y}_k(0) = 0 \qquad (3.19)$$

for $k = 1, 2, \ldots$ by successive differentiations with respect to ϵ and evaluations at $\epsilon = 0$ as described in the last section.

The solution of the IVP (3.18) for y_0 is $a \cos t$; the solution for the IVP for y_1 is $-\frac{1}{2}at \sin t$, etc., so that

$$y = a[\cos t - \tfrac{1}{2}\epsilon t \sin t + O(\epsilon^2)] \qquad (3.20)$$

The perturbation solution (3.20) should be compared with the exact solution of the same problem:

$$y(t) = a \cos (\sqrt{1 + \epsilon}\, t) \qquad (3.21)$$

A conspicuous qualitative difference between (3.21) and (3.20) is that the exact solution is periodic in time while the perturbation solution is not, at least up to the terms determined. Moreover, the factor ϵt in the second term in (3.20) causes the perturbation solution (3.20) to grow without bound as t tends to infinity while the exact solution is bounded for all t. What was thought to be a small "correction term" therefore does not remain small for all t. A perturbation solution is said to be *uniformly valid* if the term involving ϵ^{k+1} remains small of order ϵ compared to the term involving ϵ^k for all values of the independent variable of interest (t in our case) and for all $k = 0, 1, 2, \ldots$. Evidently,

3C: The perturbation solution (3.20) is *not* uniformly valid.

Perturbation solutions of IVP are frequently not uniformly valid and are therefore useful only for sufficiently small values of the independent variable, such as for $|\epsilon t| \ll 1$ in (3.20). For this reason, the regular perturbation method is most useful for boundary-value problems (BVP) in differential equations since the (dimensionless) independent variable remains 0(1) for such problems. For the present problem, we are interested in the solution for all t and therefore have the obvious task of modifying our perturbation method to get a periodic approximate solution adequate for all t. We will do this in the next section.

We can insist that any approximate solution of (3.13) be periodic because we know the exact solution is periodic. What should we do if we do not have the exact solution as is often the case when we have to look for an approximate solution? For second-order systems, the phase diagram technique is often informative. For higher-order systems, local critical point analyses may be helpful. We will pursue this point further in Chapter 4. For the moment, we merely stress:

3D: The periodicity of the perturbation solution does not imply the periodicity of the exact solution and conversely.

For example, the perturbation solution of the IVP

$$y^{\cdot\cdot} + \epsilon y = 0 \qquad y(0) = 1 \qquad y^{\cdot}(0) = 0$$

is

$$y = 1 - \frac{1}{2!}\,\epsilon t^2 + \frac{1}{4!}\,\epsilon^2 t^4 + \cdots$$

Though the perturbation is not uniformly valid and the individual terms in the series are not periodic, the entire perturbation series sums up to the periodic exact solution $y = \cos\sqrt{\epsilon}\,t$. In contrast, the leading term of the perturbation solution of the IVP

$$y^{\cdot\cdot} + \frac{1}{(1 + \epsilon t)^2}\,y = 0 \qquad y(0) = 1 \qquad y^{\cdot}(0) = 0$$

is $y_0(t) = \cos t$, but the exact solution is not periodic. (Can you find the exact solution?)

It should be mentioned that the perturbation method also applies when the initial conditions involve the small parameter as well. For example, if $y(0) = a + b\epsilon$ in (3.13), then we would have $y_1(0) = b$ in (3.19) and therefore

$$y = a\left[\cos t - \epsilon\left(\frac{1}{2}\,t\sin t - \frac{b}{a}\cos t\right) + O(\epsilon^2)\right]$$

instead of (3.20).

3.4 Poincaré's Method

To find a uniformly valid perturbation solution for the simple IVP (3.13), let us take a closer look at its exact solution $y = a\,\cos(\sqrt{1 + \epsilon}\,t)$. We note in particular that the frequency, $\omega = \sqrt{1 + \epsilon}$, of the oscillatory solution depends on the parameter ϵ. On the other hand, the correction for the zeroth-order approximate solution (i.e., for the $\epsilon = 0$ case) by the regular perturbation method does not modify the frequency explicitly. This suggests that we may improve our perturbation scheme if we consider the solution to be a function of $\tau \equiv \omega(\epsilon)t$ instead of t itself where $\omega(\epsilon)$ is analytic at $\epsilon = 0$ with $\omega(0) = 1$ so that

$$\omega(\epsilon) = 1 + \omega_1\epsilon + \omega_2\epsilon^2 + \cdots \tag{3.22}$$

We have required $\omega_0 = 1$ to recover the correct solution for the $\epsilon = 0$ case. Correspondingly, an appropriate perturbation series for y is

$$y = Y_0(\tau) + \epsilon Y_1(\tau) + \epsilon^2 Y_2(\tau) + \cdots \tag{3.23}$$

with

$$y^{\boldsymbol{\cdot}} = \omega\dot{y} = (1 + \epsilon\omega_1 + \epsilon^2\omega_2 + \cdots)(\dot{Y}_0 + \epsilon\dot{Y}_1 + \cdots) \tag{3.24}$$

$$y^{\boldsymbol{\cdot\cdot}} = \omega^2\ddot{y} = (1 + \epsilon\omega_1 + \epsilon^2\omega_2 + \cdots)^2(\ddot{Y}_0 + \epsilon\ddot{Y}_1 + \cdots) \tag{3.25}$$

where a dot on top indicates differentiation with respect to τ.

Upon substituting (3.22)–(3.25) into the IVP (3.13), we get

$$(\ddot{Y}_0 + Y_0) + \epsilon(\ddot{Y}_1 + Y_1 + 2\omega_1\ddot{Y}_0 + Y_0) + \cdots = 0$$

$$Y_0(0) + \epsilon Y_1(0) + \epsilon^2 Y_2(0) + \cdots = a$$

$$\dot{Y}_0(0) + \epsilon[\dot{Y}_1(0) + \omega_1\dot{Y}_0(0)]$$

$$+ \epsilon^2[\dot{Y}_2(0) + \omega_1\dot{Y}_1(0) + \omega_2\dot{Y}_0(0)] + \cdots = 0$$

By successively differentiating both sides with respect to ϵ and evaluating the results at $\epsilon = 0$ (as was done for the regular perturbation method), we see that these three equations are equivalent to a sequence of IVP for Y_k, $k = 0, 1, 2, \ldots$. In particular, we have

$$\ddot{Y}_0 + Y_0 = 0 \qquad Y_0(0) = a \qquad \dot{Y}_0(0) = 0 \tag{3.26}$$

$$\ddot{Y}_1 + Y_1 = (2\omega_1 - 1)Y_0 \qquad Y_1(0) = \dot{Y}_1(0) = 0 \tag{3.27}$$

etc. With $Y_0 = a \cos \tau$ from the first problem inserted into (3.27), the inhomogeneous term of the ODE for Y_1 is itself a complementary solution of the ODE and therefore gives rise to a nonperiodic particular solution $\frac{1}{2}(2\omega_1 - 1)a\tau \sin \tau$ (similar to the situation in Section 3.3).

An inhomogeneous term which gives rise to an unbounded particular solution (and hence a perturbation solution which is not uniformly valid) is called a *secular term* of the differential equation. We may

3E: Suppress the secular term in (3.27) by choosing the yet unspecified parameter ω_1 to be $\frac{1}{2}$.

This choice of ω_1 eliminates the only inhomogeneous term in (3.27) and the resulting homogeneous IVP gives $Y_1 \equiv 0$. Except for terms of order ϵ^2, we now have

$$y = a \cos \tau + O(\epsilon^2) = a \cos \{ [1 + \tfrac{1}{2}\epsilon + O(\epsilon^2)]t \} + O(\epsilon^2) \qquad (3.28)$$

We may continue the process, if we wish, and

3F: Choose ω_k to eliminate secular terms in the IVP for Y_k.

The resulting problems are then solved for Y_k, $k = 1, 2, \ldots$. However, such calculations do not add to our understanding of the Poincaré method and will not be pursued here. Instead, we return to the IVP for planetary motion including the relativistic effect.

It is a straightforward calculation to show that the perturbation solution of the IVP for w,

$$w'' + w = 1 + \epsilon w^2 \qquad w(\theta_0) = 1 + e \qquad w'(\theta_0) = 0 \qquad (3.29)$$

is not uniformly valid. In particular, the right-hand side of the ODE for w_1 (see Equation 3.10) contains a secular term. A uniformly valid perturbation series for w in the form

$$w(\theta) = W_0(\psi) + \epsilon W_1(\psi) + \epsilon^2 W_2(\psi) + \cdots \qquad (3.30)$$

$$\psi = \omega(\epsilon)\theta = (1 + \epsilon\omega_1 + \epsilon^2\omega_2 + \cdots)\theta \qquad (3.31)$$

can be obtained formally by Poincaré's technique without any conceptual or computational difficulty. The determination of W_k and ω_j will be left as an exercise at the end of this chapter. On the other hand, it is not so easy to show that the formal solution, (3.30) and (3.31), is, in fact, uniformly valid. We must show that all the W_k's remain bounded for all ψ. For a general IVP, this is decidedly a nontrivial task. A number of rigorous results of this type can be found in Bellman (1966), O'Malley (1974), and D. R. Smith (1985). Rigorous results for uniformly valid perturbation solutions are still not available for many important classes of problems in science and engineering.

3.5   The Modeling Cycle

As a problem in applied mathematics, the development of the theory of planetary motion outlined in Chapters 2 and 3 may be separated into several distinct stages. We began with the *formulation* of a simple, highly idealized mathematical model which is expected to adequately govern the behavior of the observed phenomenon. We then performed an *analysis* of the mathematical problems associated with this model. An *evaluation* of the mathematical results obtained from the analysis of the simple model by comparing them with known data showed that the idealized model is extremely reliable and may be used to predict future motion of the planets. To account for a very small discrepancy between theoretical predictions and actual observations for the planet Mercury, the higher-order relativistic effect was incorporated into an improved *formulation*. A new *analysis* was then performed on the mathematical problems associated with the improved model. An *evaluation* of the mathematical results showed that the small but nevertheless undesirable discrepancy in the Newtonian theory does not appear in the improved theory. We pointed out in Section 2.4 that the stages of formulation, analysis, and evaluation form a cycle in the mathematical modeling process. Each stage of the cycle is equally important to a successful investigation. The cycle is to be repeated until the phenomenon under investigation is reasonably well understood as in the case of the orbital motion of Mercury, or until the analysis becomes too difficult to be continued.

Modern teaching of applied mathematics tends to focus on the "analysis" phase of the cycle (and often only on the technical aspects of it) because it is more systematic, more universal, and more suitable for classroom discussion. In contrast, there are no fixed rules for the formulation of mathematical models; the method of construction varies with scientific disciplines and often from phenomenon to phenomenon within a given discipline. Nevertheless, it is important that students of applied mathematics be exposed to all the stages of mathematical modeling somewhere in their curriculum. Through this book, we intend to offer such an exposure. In order to discuss the formulation of mathematical models and the evaluation of these models, we must look at specific phenomena. It is natural to start with problems in classical physics because the construction of mathematical models is relatively straightforward and the results are often spectacular. Interest in any discipline, however, cannot be sustained solely by the successes of the distant past. Students of applied mathematics must also be shown the recent progress, current activities, and future challenges of applied mathematics in more than one area of mathematical science. The content of this book must reflect the exciting developments in applied mathematics both in the phenomena studied and the mathematical ideas involved in the investigation of these phenomena. We will try to structure the book to meet these objectives and constraints. It has taken a few iterations to arrive at the present approximation of the ideal.

We noted also in Section 2.4 that the study of new mathematical models often stimulates the development of new mathematics, rejuvenates an existing branch of mathematics or fusing several branches of mathematics. For example, interest in the relativistic theory of planetary motion has led to two different asymptotic techniques for approximate solutions of ordinary differential equations which are not encountered in conventional methods courses. As we shall see, these techniques also have many important applications in different branches of mathematical science, from mechanics to economics.

While we are on the subject of approximate solutions for differential equations, it should be pointed out that the ODE for w in (3.29) is of the type $d^2y/dx^2 = f(y)$ and can therefore be solved exactly. A first integral of the ODE gives

$$(w')^2 + w^2 = c_0 + 2\left(w + \frac{\epsilon}{3} w^3 \right) \tag{3.32}$$

where c_0 is determined by the initial conditions in (3.29) to be

$$c_0 = -2(1 + e)\left[1 - \frac{1}{2}(1 + e) + \frac{\epsilon}{2}(1 + e)^2 \right]$$

The first-order ODE (3.32) is separable and its solution is

$$\int_{1+e}^{w} \frac{dw}{\sqrt{c_0 + 2w - w^2 + (2\epsilon/3)w^3}} = \pm(\theta - \theta_0) \tag{3.33}$$

The integral on the left (and other similar integrals) is called an *elliptic integral* and is not generally expressible in terms of elementary functions. [The special case $\epsilon = 0$ can be evaluated to give $w_0(\theta)$.] A number of properties of elliptic integrals have been obtained over the years. Unless we are willing to spend time on the mathematics of elliptic integrals, it would be difficult for us to extract from (3.33) the simplest properties of $w(\theta)$ such as periodicity and boundedness. In contrast, the approximate solutions by perturbation series and by Poincaré's method are simple and informative. Thus, we have a situation where it is preferable to have an adequate approximate solution even when the exact solution is available. This situation is not unique to planetary motion:

3G: Informative approximate solutions are often preferred over the corresponding exact solution for many problems in science and engineering.

We will see the evolution of other dynamical systems in this book. In the next few chapters, however, we will turn our attention to a different scientific process theme: the *stability* of phenomena and configurations. A phenomenon or configuration may be in static equilibrium such as a marble at rest at the bottom of a soup bowl or a simple pendulum at rest in a downward position. It may be in a state of evolution such as a growing fish population or an orbiting planet. Even in motion, a phenomenon may be thought to be in equilibrium with a fictitious inertia force ($= -$mass $\times$ acceleration) associated with the motion, a convenient device first introduced by D'Alembert. We will henceforth speak about the stability of an equilibrium configuration whether or not the phenomenon is in static equilibrium. As some of the stability problems to be considered in this book are about phenomena in motion, we will continue to work with models involving evolution of other dynamical systems. However, the focus of our attention will no longer be the process of evolution itself and the analysis we undertake will require a different set of mathematical techniques from those considered previously.

EXERCISES

1. Find the first *two* terms of the perturbation series solution (in powers of ϵ) for the IVP

$$y'' + (1 + \epsilon)y = 0 \qquad y(0) = 1 + \epsilon^2 \qquad y^{\cdot}(0) = \epsilon$$

2. $$\frac{d^2y}{dx^2} + \epsilon^2 y = 0 \quad (0 < x < 1) \qquad y(0) = a \qquad y(1) = 0$$

   (a) Obtain the first two terms of a regular perturbation solution for this boundary-value problem. [*Note:* The secular term notwithstanding, this perturbation solution can be used as an approximate solution for a fixed (and sufficiently small) ϵ since x is restricted to a bounded interval. This is a major difference between BVP and IVP.]

   (b) Obtain the exact solution of the same problem ($\epsilon \neq n/\pi$) and show that the first two terms of its Taylor series (in powers of ϵ) agree with the result of part (a). (*Hint:* You may need careful applications of l'Hôpital's rule here.)

3. $$y'' + (1 + \epsilon)y = 0 \qquad y(0) = 0 \qquad y^{\cdot}(0) = b$$

   (a) Obtain the first two terms of a regular perturbation solution in powers of ϵ and identify the secular term(s) in your result.

   (b) Obtain an approximate *periodic* solution by Poincaré's method, neglecting all terms of order ϵ^2 or smaller.

   (c) Check your result with the exact periodic solution.

4. The so-called Duffing's equation $y^{\cdot\cdot} + y + \epsilon y^3 = 0$ appears frequently in nonlinear vibration and other areas. With $y = Y_0(\tau) + \epsilon Y_1(\tau) + \epsilon^2 Y_2(\tau) + \cdots$ and $\tau = t(1 + \epsilon\omega_1 + \epsilon^2\omega_2 + \cdots)$, find ω_1 and ω_2 in terms of the amplitude a of the zeroth-order solution, $Y_0(\tau) = a\cos(\tau + \phi)$, to get a two-term approximate *periodic* solution $y \simeq Y_0(\tau) + \epsilon Y_1(\tau) + O(\epsilon^2)$. (*Note:* You don't have to include the complementary solutions for Y_1 or to concern yourself with initial conditions.)

5. The fundamental IVP of the relativistic theory of planetary motion is

$$w'' + w = 1 + \epsilon w^2 \qquad w(\theta_0) = 1 + e \qquad w'(\theta_0) = 0$$

Show that a two-term perturbation solution of this problem is

$$w = [1 + e\cos\bar\theta]$$

$$+ \epsilon\left[\left(1 + \frac{e^2}{2}\right) - \left(1 + \frac{e^2}{3}\right)\cos\bar\theta - \frac{e^2}{6}\cos 2\bar\theta + e\bar\theta\sin\bar\theta\right] + O(\epsilon^2)$$

where $\bar\theta = \theta - \theta_0$.

6. Apply Poincaré's method to get the following uniformly valid perturbation solution for the above IVP with $\theta_0 = 0$:

$$w = 1 + e\cos\psi + \epsilon\left[\left(1 + \frac{e^2}{2}\right) - \left(1 + \frac{e^2}{3}\right)\cos\psi - \frac{e^2}{6}\cos 2\psi\right] + O(\epsilon^2)$$

where $\psi = [1 - \epsilon + O(\epsilon^2)]\theta$. (*Note:* Setting $\theta_0 = 0$ is equivalent to rotating the cartesian reference frame by θ_0.)

7. Find a two-term perturbation solution for

$$my^{\cdot\cdot} + \frac{k_0}{(1 + \epsilon t)^2}y = 0 \qquad y(0) = 1 \qquad y^{\cdot}(0) = 0$$

8. Find ω_2 and $Y_2(\tau)$ (as in Exercise 4) for $y^{\cdot\cdot} + (1 + \epsilon)y = 0$, $y(0) = a$, and $y^{\cdot}(0) = 0$.

II

Stability of Equilibrium Configuration

When a simple pendulum is at rest in its downward vertical position, a small displacement from that position will result in a small-amplitude oscillation about the rest position. The amplitude of the oscillation will be of the order of the perturbation. In the presence of damping, the oscillation will decay and the pendulum eventually returns to rest. If the pendulum (with a rigid and weightless arm) is at rest in its upward vertical position, any small displacement will cause the pendulum mass to tumble downward and to oscillate about the downward vertical position. It does not return or pass the original equilibrium position. With damping, the pendulum will eventually come to rest at the downward position. The upward equilibrium position of the pendulum is intuitively precarious and is shown to be unstable by a critical point analysis in most elementary courses on differential equations, at the level of the text by Boyce and DiPrima (1976). For many phenomena and situations in science and engineering, their *stability* with respect to a small perturbation is a very important consideration. A poorly designed satellite may tumble and wobble in its orbit and thereby fail to perform its intended function. A small jolt from turbulent air may cause a poorly designed airplane wing to flutter.

There are countless interesting and important problems available for the illustration of instability in action. The illustrative problems in the following chapters were chosen because they are mathematically accessible for the level assumed in this book and for their relation to other topics of the book. Having just completed a discussion of planetary motion, it seems natural to start with a discussion of the stability of a planet's orbit.

This problem offers us an opportunity to review and apply the critical point analysis and phase-plane technique acquired in an elementary course on ODE. We will then take up Eulerian wobble [see also Wan (1979) and S. J. Colley (1987)] both because it is fun and because it is important in the design of satellites or any rotating bodies. The problem allows a rare visual display of trajectories in phase space for a system of more than two coupled differential equations.

Having had our fill of problems in particle or rigid-body dynamics, we will enter the world of deformable bodies in Chapter 5 with a discussion of the Euler column. The buckling of a slender column, so fundamental to structural engineering, can be demonstrated without very complicated or expensive equipment. Mathematically, a linear buckling analysis shows how eigenvalue problems in ODE arise naturally and directly. We will take the opportunity to introduce some rudiments of the calculus of variations and discuss a variational formulation of the buckling problem. In so doing, we will have touched upon another scientific process theme, namely *optimization,* to which we will return in the latter part of this book. We will also illustrate, by working out a Rayleigh-Ritz solution, numerical methods based on variational principles. However, we will not venture into a finite-element solution of our buckling problem.

Stability of another type will be discussed in Chapter 6. Modeling of automobiles on a highway by particles with their special dynamics leads to dynamical systems with the extra feature of delayed reaction. The analysis of ODE with delays offers a novelty not often encountered in conventional undergraduate courses.

4

Swing Low

The Stability of Periodic Orbits

Familiarity with phase-plane stability analysis for nonlinear autonomous systems is assumed. The derivation of one set of model equations involves differentiating vectors in a rotating frame. There is a brief discussion of numerical methods for initial-value problems but a prior exposure to such methods is not required. The phenomenon of parametric excitation in nonautonomous systems is illustrated by Mathieu's equation.

4.1 Planetary Orbits and Their Stability

As long as we do not have a relatively simple exact solution to confirm the adequacy of an approximate solution for a problem, it is always desirable to learn as many qualitative features of the exact solution as possible by independent methods (see Section 3.3). For the problem of relativistic planetary motion, we would like to know whether the exact solution for w is a periodic function of θ and whether it is bounded. In the next two sections, we will see how the method of critical point analysis and phase portraits for autonomous systems of ordinary differential equations may be used to obtain answers to these questions and other qualitative behavior of the solution. Critical point analysis and phase diagrams also provide us with another type of information, namely, whether a particular orbit is stable: that is, does a small perturbation of a planet's periodic orbit result in only a small orbital change for all later time or in a quantitatively and qualitatively significant (perhaps even catastrophic) change?

A *critical point* of a second-order autonomous system

$$x^{\boldsymbol{\cdot}} = f(x, y) \qquad y^{\boldsymbol{\cdot}} = g(x, y) \tag{4.1}$$

is a point $(\bar{x}, \bar{y})$ for which $f(\bar{x}, \bar{y}) = g(\bar{x}, \bar{y}) = 0$. The behavior of solutions of the dynamical system (4.1) initially near $(\bar{x}, \bar{y})$ may be found by looking at the system linearized about $(\bar{x}, \bar{y})$:

$$u^{\cdot} = au + bv \qquad v^{\cdot} = cu + dv \tag{4.2}$$

where $u = x - \bar{x}$
$\qquad\quad v = y - \bar{y}$
$\qquad\quad a = f_x(\bar{x}, \bar{y})$
$\qquad\quad b = f_y(\bar{x}, \bar{y})$
$\qquad\quad c = g_x(\bar{x}, \bar{y})$
$\qquad\quad d = g_y(\bar{x}, \bar{y})$

with $f_x = \partial f / \partial x$, etc. The critical point $(\bar{x}, \bar{y})$ is said to be *unstable* if for some initial conditions near the origin the solution of (4.2) becomes unbounded as $t \to \infty$. Otherwise, it is *stable*. In particular, $(\bar{x}, \bar{y})$ is *asymptotically stable* if the solution (4.2) tends to $(0, 0)$ for any nearby initial conditions so that any trajectory of (4.1) near $(\bar{x}, \bar{y})$ not only stays near the critical point but actually approaches it after awhile.

The local stability of $(\bar{x}, \bar{y})$ depends only on the eigenvalues λ_1 and λ_2 of the coefficient matrix

$$A = \begin{bmatrix} a & b \\ c & d \end{bmatrix} \tag{4.3}$$

If the real parts of λ_1 and λ_2 are both negative, then $(\bar{x}, \bar{y})$ is asymptotically stable. If either one is positive, then $(\bar{x}, \bar{y})$ is unstable. If λ_1 and λ_2 both have no real parts, $(\bar{x}, \bar{y})$ is called a *center*. A *center* is locally stable but not asymptotically stable; the trajectories of the linearized system are closed ellipses and are therefore periodic in time. When λ_1 and λ_2 are real and of opposite signs, $\lambda_1 < 0 < \lambda_2$, then the unstable critical point $(\bar{x}, \bar{y})$ is called a *saddle point* and the nearby trajectories are hyperbolas. The behavior of local trajectories for these and other combinations of λ_1 and λ_2 are discussed in detail in Boyce and DiPrima (1976).

The ODE (3.1) governing planetary motion in the Newtonian theory may be written as a first-order system:

$$w_0' = v_0 \qquad v_0' = 1 - w_0 \tag{4.4}$$

where $(\quad)' \equiv d(\quad)/d\theta$. This system has only one critical point $(1, 0)$ in the finite (w, v) plane. The corresponding linearized system about $(1, 0)$ is

$$u' = v \qquad v' = -u \tag{4.5}$$

Its coefficient matrix has eigenvalues $\pm i$ so that the local trajectories are periodic and the critical point $(1, 0)$ itself is stable but not asymptotically stable. Since the original system (4.4) is linear, the global behavior of trajectories in the (w_0, v_0) plane is the same. From these mathematical results, we see that it is possible for the planet to be in a circular orbit of radius $r_0 = p_0^2/GMm^2$. But any small perturbation would change it into a nearby elliptical periodic orbit which remains close (but does not return) to the original orbit. A small perturbation of an elliptical orbit throws it into another nearby elliptical orbit for all later time. In the next two sections, the critical point analysis will be applied to the fundamental equation of planetary motion in a relativistic theory to show how the relativistic effect qualitatively changes the above conclusions based on the classical theory.

4.2   A Critical Point Analysis for Relativistic Planetary Motion

Set

$$w' = v \tag{4.6}$$

and write the ODE (3.3) as

$$v' = 1 - w + \epsilon w^2 \tag{4.7}$$

By definition, a *critical point* $(\bar{w}, \bar{v})$ of the system, (4.6) and (4.7), is a solution of the system which is independent of θ, that is, $w' = v' = 0$. Upon setting the right sides of (4.6) and (4.7) equal to zero, the system of two equations gives two critical points, $(\bar{w}_1, 0)$ and $(\bar{w}_2, 0)$ where

$$\bar{w}_1 = \frac{1}{2\epsilon}\,(1 - \sqrt{1 - 4\epsilon}) = 1 + \epsilon + O(\epsilon^2)$$

$$\tag{4.8}$$

$$\bar{w}_2 = \frac{1}{2\epsilon}\,(1 + \sqrt{1 - 4\epsilon}) = \frac{1}{\epsilon}\,[1 - \epsilon - \epsilon^2 + O(\epsilon^3)]$$

For $0 < 4\epsilon < 1$, we have $\bar{w}_2 > \bar{w}_1$. Each of the two critical point solutions of the system corresponds to a circular orbit with a radius equal to $p_0^2/GMm^2\bar{w}_k$.

To see the behavior of other *trajectories,* that is, other solutions of the ODE (3.3), we look first at those in the vicinity of the critical point

$(\bar{w}_2, 0)$. We perform a *linear stability analysis* at this critical point. Let $u \equiv w - \bar{w}_2$ with $|u| \ll 1$ initially. We then linearize the ODE for u and v:

$$u' = v \qquad v' = \sqrt{1 - 4\epsilon}\, u + \epsilon u^2 \simeq \sqrt{1 - 4\epsilon}\, u \qquad (4.9)$$

The eigenvalues λ_1 and λ_2 of the coefficient matrix of the linearized system are the solutions of $\lambda^2 = \sqrt{1 - 4\epsilon}$ or

$$\lambda_1 = -(1 - 4\epsilon)^{1/4} \qquad \lambda_2 = (1 - 4\epsilon)^{1/4} \qquad (4.10)$$

Hence,

4A: The critical point $(\bar{w}_2, 0)$ is a *saddle point* and nearby trajectories of the linearized system (4.9) are hyperbolas with $v = \pm\sqrt{1 - 4\epsilon}\, u$ as local asymptotes (see Figure 4.1).

A similar analysis shows that

4B: The critical point $(\bar{w}_1, 0)$ is a *center* and nearby trajectories of the linearized system are ellipses (see Figure 4.1).

From these results for the linearized system, we would like to draw similar conclusions about the trajectories of the original nonlinear system. A second-order autonomous system is *almost linear* in the vicinity of a critical point if the coefficient matrix of the corresponding linearized system is not the zero matrix and the neglected terms tend to zero faster than $\sqrt{u^2 + v^2}$. The system of equations 4.6 and 4.7 is *almost linear* in the neighborhood of the critical points. We know (from Minorsky, 1962) that the trajectories of our almost linear system near $(\bar{w}_2, 0)$ are in fact like hyperbolas and, except along a separatrix (one of the asymptotes), the critical point is *unstable*. However, nothing can be said about the trajectories of (4.6) and (4.7) near

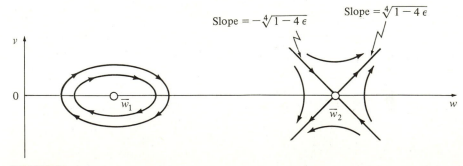

FIGURE 4.1

a center without further investigation. The results in Chapter 2 indicate that the solution for $w(\theta)$ is closer to $(\bar{w}_1, 0)$ than $(\bar{w}_2, 0)$; therefore, it is the trajectories near $(\bar{w}_1, 0)$ which are of interest here. We must work a little harder to find the behavior of those trajectories.

4.3   Phase Diagram for Relativistic Planetary Motion

The key to the behavior of trajectories near $(\bar{w}_1, 0)$ is the *separatrices* which correspond to the asymptotes of the saddle point $(\bar{w}_2, 0)$, that is, the trajectories which go into or come out of that critical point. We know from Equation 3.32 that a first integral of the ODE (3.3) is

$$v^2 = c_0 + 2w - w^2 + \frac{2\epsilon}{3}w^3 \tag{4.11}$$

where c_0 is determined by the initial conditions. This first integral gives different trajectories in the (w, v) plane for different values of c_0. Each separatrix corresponds to a trajectory which passes through $(\bar{w}_2, 0)$, that is, $v = 0$ at $w = \bar{w}_2$ for that trajectory, so that

$$\frac{2\epsilon}{3}\left(\bar{w}_2^3 - \frac{3}{2\epsilon}\bar{w}_2^2 + \frac{3}{\epsilon}\bar{w}_2 + \frac{3}{2\epsilon}c_0 \right) = 0$$

or

$$\bar{w}_2\left(\bar{w}_2^2 - \frac{1}{\epsilon}\bar{w}_2 + \frac{1}{\epsilon} \right) - \frac{1}{2\epsilon}\left(\bar{w}_2^2 - 4\bar{w}_2 - 3c_0 \right) = 0$$

The first term vanishes because $\bar{w}_2$ is a root of $v' = 0$ (see Equation 4.7); therefore, we have

$$c_0 = \tfrac{1}{3}\bar{w}_2(\bar{w}_2 - 4) \tag{4.12}$$

and (4.11) becomes

$$v^2 = \tfrac{1}{3}\bar{w}_2(\bar{w}_2 - 4) + 2w - w^2 + \frac{2\epsilon}{3}w^3$$

$$= \frac{2\epsilon}{3}(w - \bar{w}_2)\left[w^2 - \left(\bar{w}_1 + \frac{1}{2\epsilon} \right)w - \frac{1}{2\epsilon}\left(\bar{w}_2 - 4 \right) \right]$$

$$= \frac{2\epsilon}{3}(w - \bar{w}_2)^2(w - \bar{w}_3)$$

where

$$\bar{w}_3 = -\frac{1}{2\epsilon}\left(2\sqrt{1-4\epsilon}-1\right) \simeq -\frac{1-4\epsilon}{2\epsilon}$$

which is negative for $0 < 4\epsilon < \frac{3}{4}$. Thus, along a separatrix, we have

$$v = \pm \sqrt{\frac{2\epsilon}{3}}\,|w-\bar{w}_2|\sqrt{w-\bar{w}_3} \qquad\qquad (4.13)$$

The two branches of (4.13) make up a curve in the *phase plane* in the form of a figure $\propto$ (see the heavy curve in Figure 4.2), with the $+$ sign giving the part of the curve above the w axis and the $-$ sign the part below. All local asymptotes of the saddle point are merely different parts of the same continuous curve.

Since trajectories do not cross except possibly at critical points, the separatrix *separates* the trajectories of the nonlinear system, (4.6) and (4.7), into three groups. We are only interested in those trajectories which were at one time (say for $\theta = \theta_0$) near the critical point $(\bar{w}_1, 0)$. These trajectories

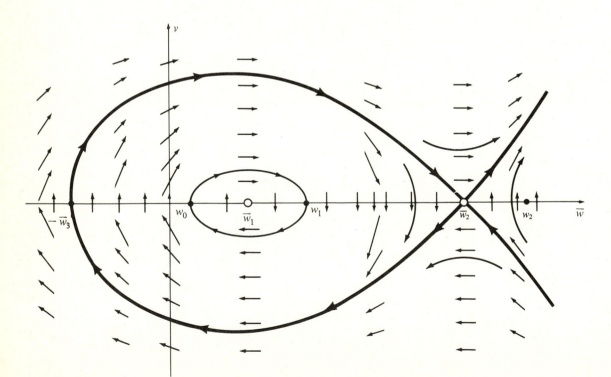

FIGURE 4.2

must forever remain inside the loop formed by the separatrix between $(\bar{w}_3, 0)$ and $(\bar{w}_2, 0)$. With the help of the direction field at some choice locations in the phase plane (such as along (1) $v = 0$, (2) $w = 0$, (3) $w = \bar{w}_1$, etc.), it is not difficult to see that a trajectory passing through a point on the w axis to the right of $\bar{w}_1$, say $(W_1, 0)$ with $\bar{w}_1 < W_1 < \bar{w}_2$, at one time must cross the w axis again between the critical points at a later time (having crossed the w axis somewhere to the left of $\bar{w}_1$ before that). What we want to know is whether or not it returns to the same point $(W_1, 0)$ to form a closed curve so that w is a periodic function of θ.

To find out whether we have a closed trajectory and therefore a periodic solution, we specialize the first integral (4.11) to the trajectory which passes through $(W_1, 0)$ at one time by choosing

$$c_0 = -\frac{2\epsilon}{3}\left(W_1^3 - \frac{3}{2\epsilon}W_1^2 + \frac{3}{\epsilon}W_1\right) \equiv c_0(W_1) \tag{4.14}$$

so that $v(W_1) = 0$. With (4.14), we can rearrange (4.11) to read

$$v^2 = \frac{2\epsilon}{3}(w - W_1)\left[w^2 - \left(\frac{3}{2\epsilon} - W_1\right)w - \frac{3}{2\epsilon}\frac{c_0(W_1)}{W_1}\right]$$

$$= \frac{2\epsilon}{3}(w - W_0)(w - W_1)(w - W_2) \tag{4.15}$$

where W_0 and W_2 are two other roots of the cubic equation $v^2(w) = 0$, that is, W_0 and W_2 are the two roots of the quadratic equation

$$w^2 - \left(\frac{3}{2\epsilon} - W_1\right)w - \frac{3}{2\epsilon}\frac{c_0(W_1)}{W_1} = 0 \tag{4.16}$$

We do not need the explicit expression for W_0 and W_2 here; all we need is the following relation among the three roots which can be deduced from these explicit expressions:

$$W_0 < W_1 < W_2 \tag{4.17}$$

To get a feel for the order of magnitude of the two other roots relative to W_1, we set $W_1 = \bar{w}_1 + \delta$ and obtain from the exact solution for W_0 and W_1 (not listed here)

$$W_0 = 1 - \delta + O(\epsilon) \quad \text{and} \quad W_2 = \frac{3}{2\epsilon}[1 + O(\epsilon)] > \bar{w}_2 \tag{4.18}$$

for $0 < 4\epsilon \ll \delta \ll 1/\epsilon$.

With these locations for the roots of $v^2(w) = 0$, we see that the trajectory passing through $(W_1, 0)$ is limited to a range of values for w between W_0 and W_1, that is, $W_0 \leq w \leq W_1$; otherwise, we have $v^2(w) < 0$ and v would be complex-valued. From the point $(W_1, 0)$, the trajectory follows the negative branch of $v(w)$ while both w and v decrease with θ until $w = \bar{w}_1$. At that point, v begins to increase while w continues to decrease with θ until the trajectory reaches the w axis at $(W_0, 0)$ [which is to the left of $(\bar{w}_1, 0)$]. The trajectory then goes along the positive branch of $v(w)$ as θ increases, both v and w increasing. It begins to head toward the w axis again after $w = \bar{w}_1$ and returns to $(W_1, 0)$ (which it had passed through earlier) as required by (4.15). The trajectory is therefore closed. Since $(W_1, 0)$ is arbitrary (as long as it is between $\bar{w}_1$ and $\bar{w}_2$),

4C: w is, in general, a periodic function of θ

which is what we wanted to know. On the basis of this result, we expect any uniformly valid approximate solution of the fundamental IVP of relativistic planetary motion to be periodic in θ. If we wish, we can obtain such an approximate solution by Poincaré's method as outlined in Section 3.4.

4.4 Approximate Solutions by Numerical Methods

Exact solutions in terms of elementary or special functions for initial-value problems in ordinary differential equations in science and engineering are few and far between. Even back in Newton's days, approximate solutions for many such problems were sought by numerical methods. One of the earliest numerical schemes still used today is the so-called *simple Euler* method. The method approximates a first derivative $y^\bullet$ at $t_n = nh$ by $(y_{n+1} - y_n)/h$ where $y_k \equiv y(kh) = y(t_k)$ and converts

$$y^\bullet = f(t, y) \qquad y(0) = a \qquad (4.19)$$

into its finite-difference analog

$$y_{n+1} = y_n + h f(nh, y_n) \qquad (n = 0, 1, 2, 3, \ldots) \qquad (4.20)$$

Starting with the known value $y_0 = y(0) = a$, the finite-difference equation (4.20) gives successively the values $y_1, y_2, \ldots$ and y_N, where $y_N = y(Nh) = y(T)$ for some terminal time T of interest. The *truncation error e_n* associated with the difference analog (4.20) at each step is of the order of the square of the time step h, that is, $e_n = O(h^2)$, since we have in effect approximated y_{n+1} by a two-term Taylor expansion at h. The accumulated truncation error after N steps is $E_N = O(Ne_n) = O(Nh^2) = O(h)$ since $Nh = T$.

Numerical schemes with higher-order accuracies than the simple Euler method have also been developed. One of the more popular higher-order schemes today is the fourth-order Runge-Kutta method with

$$y_{n+1} = y_n + \frac{h}{6}(k_{n1} + 2k_{n2} + 2k_{n3} + k_{n4}) \qquad (4.21a)$$

$$k_{n1} = f(x_n, y_n) \qquad\qquad k_{n2} = f(x_n + \tfrac{1}{2}h, y_n + \tfrac{1}{2}hk_{n1})$$

$$k_{n3} = f(x_n + \tfrac{1}{2}h, y_n + \tfrac{1}{2}hk_{n2}) \qquad k_{n4} = f(x_n + h, y_n + hk_{n3}) \qquad (4.21b)$$

The accumulation error for this scheme is $O(h^4)$.

Numerical methods for IVP in ODE have become more and more useful over the years. But it was the advent of high-speed computers which made them powerful tools for obtaining the long-time behavior of complex dynamical systems. From the orbital flight of Sputnik to the Apollo moon landing, computer-generated numerical solutions of IVP similar to (but more sophisticated than) those mentioned here were used to track and guide the flight paths of many space vehicles. These and other technological successes brought about by the use of numerical methods have led some of the uninitiated to believe that "we can just put it on the computer!" The problems in planetary motion discussed in these notes demonstrate that there are many things you just cannot "put on the computer" without some thought and finesse. Given that scientific computations will never be absolutely accurate because of inherent truncation errors and machine round-offs, there is no way for any numerical method to determine many of the qualitative features of the solution of an IVP. It is not enough to know that a function is approximately periodic. Either it is, or it is not! Because of the tremendous time span involved, a numerical solution for relativistic planetary motion sufficiently accurate to settle the issue of perihelion precession for the planet Mercury poses a challenge of the first magnitude. It should be compared with the back-of-the-envelop calculation adequate for the analytic approaches discussed in this and the last chapter.

Computing time and accuracy are not the only constraints in machine computation which force us to think and analyze before we put a problem on the computer. There are many others. In the next section, we will describe, by way of a few simple examples, the analysis needed for the numerical solution of differential equations in the presence of solution singularities.

4.5   Locating Solution Singularities Numerically

For linear equations, we can look out for singularities in its solution by checking the coefficients of the ODE to see whether they have singularities.

We are guaranteed in the case of linear systems that any solution singularity must also be a singularity of the ODE (but not necessarily conversely). For a nonlinear equation, the situation is much more complex. Consider the IVP

$$z' = 1 + z^2 \qquad z(1) = 1 \tag{4.22a}$$

which determines $z(x)$ for $x > 1$. It can be solved exactly to get $z = \tan(x - 1 + \pi/4)$. This solution becomes unbounded at $x = 1 + \pi/4$ and is therefore meaningless beyond that point. Yet the differential equation itself gives no warning of this singular behavior of the solution as it contains no "singularity."

We are fortunate to have the exact solution of (4.22a) so that we can see the solution singularity at $x = 1 + \pi/4$ and thereby know not to compute the solution beyond that point. For the following minor variation of (4.22a)

$$y' = x^2 + y^2 \qquad y(1) = 1 \tag{4.22b}$$

we are not so fortunate. An exact solution of this problem is not available in terms of simple functions or their integrals. Anyone who blindly computes the solution numerically would get a solution beyond the singularity. However, an astute applied mathematician would expect the solution y to have a singularity by comparing (4.22b) with (4.22a). They both start at the same initial point (1, 1), but from the two ODEs, the slope of y is everywhere as steep or steeper than the slope of z. Since z has a singularity at $x = 1 + \pi/4$, y must have a singularity before that! We may formalize this *comparison method* for the IVP

$$y' = f(x, y) \qquad y(x_0) = y_0 \tag{4.22c}$$

by finding another IVP

$$z' = g(x, z) \qquad z(x_0) = y_0 \tag{4.22d}$$

which satisfies the following conditions:

1. $g(x, u) \le f(x, u)$ for all $x \ge x_0$ and $|u| < \infty$.
2. z is known to have a singularity at x_1.

It follows from the relative steepness argument that y has a singularity at $\bar{x} \le x_1$.

It is not always straightforward to construct the IVP for z. Suppose the initial condition for the IVP (4.22b) is changed to $y(0) = 1$. It is not appropriate to use the ODE in (4.22a) with the new initial condition $z(0) = 1$ as the comparison problem, because $g(x, u) = 1 + u^2$ is not less than or equal to $f(x, u) = x^2 + u^2$ for $0 \leq x < 1$ and $|u| < \infty$. An appropriate IVP comparison which still has a solution singularity at a finite x_1 would be

$$z' = \begin{cases} x^2 & (0 \leq x < 1) \\ 1 + z^2 & (1 < x < \infty) \end{cases} \qquad z(0) = 1 \qquad (4.23a)$$

with z stipulated to be continuous at $x = 1$. The solution of the IVP (4.23a) is

$$z = \begin{cases} 1 + \frac{1}{3}x^3 & (0 \leq x \leq 1) \\ \tan(x + c_0) & (1 \leq x < x_1) \end{cases} \qquad (4.23b)$$

where $c_0 = \tan^{-1}(4/3) - 1$ was chosen so that z is continuous at $x = 1$. Evidently, z has a singularity at $x_1 = 1 + \pi/2 - \tan^{-1}(4/3)$. Now, we have

$$g(x, u) = \begin{cases} x^2 \leq x^2 + u^2 = f(x, u) & (0 \leq x < 1) \\ 1 + u^2 \leq x^2 + u^2 = f(x, u) & (1 < x < \infty) \end{cases} \qquad (4.23c)$$

for all $|u| < \infty$. It follows that y has a singularity at $\bar{x} \leq x_1$.

Suppose we want to know the actual location of the solution singularity of (4.22b). A numerical solution of the problem would be necessary. However, there will be overflows long before we get close to the singular point of the solution. We simply cannot do it by obtaining a numerical solution for the IVP itself as we cannot compute infinity. So, we will have to do a little thinking and not just "put the problem on the computer"!

To avoid overflows when we get close to the singular point, we note that $u = 1/y$ is small when y is large. In particular $u = 0$ when y becomes unbounded. This suggests that we work with u instead of y, for we can then get as close to the singular point as we wish. We illustrate this technique with the IVP (4.22b) for which there is no exact solution. In terms of u, we have $y' = (1/u)' = -u'/u^2$, so that (4.22b) may be written as

$$u' = -u^2\left(x^2 + \frac{1}{u^2}\right) = -x^2u^2 - 1 \qquad u(1) = 1 \qquad (4.24a)$$

From the ODE for u, we know that u is a decreasing function of x, vanishes at some (singular) point $\bar{x}$, and becomes negative for larger x. We start with $u(1) = 1$ and integrate the ODE numerically forward for larger and larger values of x (by a fourth-order Runge-Kutta scheme say) until u crosses the

x axis giving an interval $(x_n, x_n + h)$ which contains $\bar{x}$. We may then pin down more narrowly the location of the singularity $\bar{x}$ by using a smaller step size $\bar{h}$ for the numerical solution.

For a very accurate determination of $\bar{x}$, we have a more efficient method. Instead of solving for y or u as a function of x, suppose we try to solve for x as a function of u. The ODE for u in $(4.24a)$ may be written as

$$\frac{dx}{du} = \frac{-1}{x^2 u^2 + 1} \qquad (4.24b)$$

With the initial condition $x(u = 1) = 1$ [which is just another way of saying $u(x = 1) = 1$], we (numerically) integrate the equation for $x(u)$ backward from $u = 1$ to $u = 0$. The value of x at $u = 0$ is $\bar{x}$! In principle, we can get $\bar{x}$ as accurately as we wish by this method.

The comparison method for estimating solution singularities and the techniques for locating them numerically can be extended to more general problems. However, the simple examples of this section serve the purpose of reinforcing our view on computer simulations as stated in Section 1.8. Think and work for a qualitative understanding of a problem before getting on the computer. Ideally, *machine computation is done mainly to confirm what we already know qualitatively from an analysis of the problem.*

4.6   Trajectories on a Phase Sphere

Take an (unopened) cereal box and throw it up in the air with a rotary motion about an axis perpendicular to any one of its faces. (Use a textbook if you can put a rubber band around it to keep the pages from flying open.) You will find that the rotary motion is extremely smooth if it is about an axis perpendicular to the faces with the largest area (see Figure 4.3a). The same is true for the faces with the smallest area (Figure 4.3b). But if the rotary motion is about the third set of two faces with outward unit normals $\pm\mathbf{i}_2$, the box wobbles, often so badly that the direction of the vectors $\pm\mathbf{i}_2$ (which are fixed to the faces of the box) change from the original starting positions by 180° when the box lands in your hands again. For example, suppose the starting position of the box is as in Figure 4.3c, the box may end up in the position shown in Figure 4.3d. While the ending position may vary, there will always be severe wobbling about the $\pm\mathbf{i}_2$ axis as long as all six faces are not identical, that is, as long as the box is not a cube.

Just as the motion of the planets is a consequence of the fundamental laws of particle mechanics, the above phenomenon is a direct consequence of the fundamental laws of rigid-body mechanics. We saw in Chapter 2 that the motion of a particle in rectilinear motion is governed by Newton's

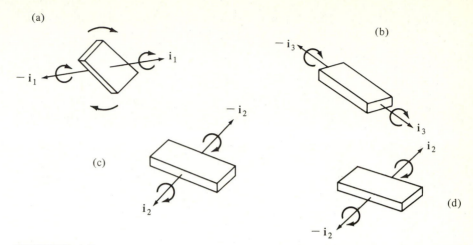

FIGURE 4.3

second law: The time rate of change of linear momentum $m\mathbf{v}$ is equal to the applied force. The rotary motion of a rigid body is governed by a corresponding law which stipulates that the time rate of change of *angular momentum* $\mathbf{p}$ is equal to the applied torque $\mathbf{T}$ so that $\mathbf{p}^{\bullet} = \mathbf{T}$. To develop the consequences of this law, we make the following three observations:

1. For our problem, we have $\mathbf{T} = \mathbf{0}$ since the box is free from external torques once thrown.

2. As the cartesian frame $(\mathbf{i}_1, \mathbf{i}_2, \mathbf{i}_3)$ is fixed to the box and therefore rotates in space with angular velocity ω, the direction of $\mathbf{i}_k$ changes with time and we have from vector calculus [see Greenspan and Benney (1973), p. 417] $\mathbf{i}_k^{\bullet} = \omega \times \mathbf{i}_k$ so that

$$\mathbf{p}^{\bullet} = (p_1\mathbf{i}_1 + p_2\mathbf{i}_2 + p_3\mathbf{i}_3)^{\bullet} = (p_1^{\bullet}\mathbf{i}_1 + p_2^{\bullet}\mathbf{i}_2 + p_3^{\bullet}\mathbf{i}_3) + \omega \times \mathbf{p}$$

$$= (\mathbf{p}^{\bullet})\,|_{\text{in rotating frame}} + \omega \times \mathbf{p}$$

3. Just as linear momentum is proportional to the rectilinear velocity, the angular momentum is "proportional" to the angular velocity. For the orthogonal cartesian frame $(\mathbf{i}_1, \mathbf{i}_2, \mathbf{i}_3)$, we have the simple scalar equations $p_k = I_k\omega_k$, $k = 1, 2, 3$, where $\omega = \omega_1\mathbf{i}_1 + \omega_2\mathbf{i}_2 + \omega_3\mathbf{i}_3$ and where I_k is the *moment of inertia* about the $\pm\mathbf{i}_k$ direction [see Greenspan and Benney (1973), p. 680].

We combine these three observations to get from

$$(\mathbf{p}^{\bullet})\,|_{\substack{\text{relative to a} \\ \text{fixed frame}}} = \mathbf{0}$$

the following vector equation for the components of $\mathbf{p}$:

$$(p_1^{\bullet}\mathbf{i}_1 + p_2^{\bullet}\mathbf{i}_2 + p_3^{\bullet}\mathbf{i}_3) + \left(\frac{p_1}{I_1}\mathbf{i}_1 + \frac{p_2}{I_2}\mathbf{i}_2 + \frac{p_3}{I_3}\mathbf{i}_3\right) \times \mathbf{p} = \mathbf{0}$$

This vector equation is equivalent to three scalar first-order ODE. For a box with $I_k = 1/k$, these three equations take the form

$$p_1^{\bullet} = p_2 p_3 \qquad p_2^{\bullet} = -2p_1 p_3 \qquad p_3^{\bullet} = p_1 p_2 \qquad\qquad (4.25)$$

Given the angular momentum vector at some initial time (say the moment you throw the box up), the system (4.25) determines $\mathbf{p}(t)$ for all time thereafter. The problem of interest here is one with $\mathbf{p}(0) \simeq p_0 \mathbf{i}_2$. The solution of this problem will give the angular velocity components as functions of time and therefore the motion of the cartesian frame ($\mathbf{i}_1$, $\mathbf{i}_2$, $\mathbf{i}_3$). In turn, this motion tells us how badly the box wobbles.

An exact solution of the IVP for the third-order nonlinear system (4.25) is possible. However, it is difficult to extract useful information from this exact solution, just as in the case of the relativistic theory of planetary motion. Fortunately, for the question we have posed for ourselves, we do not need detailed information on the evolution of the dynamical system provided by this exact solution. Limited qualitative results obtained through a critical point analysis suffice.

We begin our limited analysis of the system (4.25) by forming the following combination of the three ODE:

$$p_1 p_1^{\bullet} + p_2 p_2^{\bullet} + p_3 p_3^{\bullet} = \tfrac{1}{2}(p_1^2 + p_2^2 + p_3^2)^{\bullet} = 0$$

Upon integrating both sides of the above relation, we get $|\mathbf{p}|^2 = c_0$. With no loss in generality, we take the magnitude of the initial angular momentum vector to be one so that $c_0 = 1$ and therewith

$$p_1^2 + p_2^2 + p_3^2 = 1 \qquad\qquad (4.26)$$

Equation 4.26 is merely a restatement of the fact that

4D: In the absence of external torques, the magnitude of the angular momentum vector is conserved.

Geometrically, (4.26) says that the solution of the IVP for the system of ODE (4.25) is a curve in $\mathbf{p}$ space *lying on the surface of a unit sphere centered at the origin*. This curve is traced out by the tip of the angular momentum vector as it moves with time.

Next, we observe that the system (4.25) [with the constraint (4.26)] has six critical points located at $(\pm 1, 0, 0)$, $(0, \pm 1, 0)$, and $(0, 0, \pm 1)$, respectively. If the system is at one of these "equilibrium" states initially, it will stay there forever thereafter. For example, if $\mathbf{p}(0) = \mathbf{i}_2$, then we have $\mathbf{p}(t) = \mathbf{i}_2$ and $\omega(t) = \mathbf{i}_2 / I_2 = 2\mathbf{i}_2$ for all $t > 0$. In other words, the box will rotate about the $\mathbf{i}_2$ axis (with no wobbling) for all time if it started out that way *perfectly*. Unfortunately, we cannot be all that perfect with our hands. Though we might have aimed for $\mathbf{p}(0) = \mathbf{i}_k$, the box usually starts off with $\mathbf{p}(0) = \mathbf{i}_k + \mathbf{q}$ with $|\mathbf{q}| \ll 1$ instead. What we want to know is: What happens to the rotary motion of the box thereafter? Does it stay close to $\mathbf{p}(0)$? That is, are the critical points stable? This last question calls for a critical point analysis.

For the critical point $(1, 0, 0)$, we set $p_1 = 1 + u$, $p_2 = v$, and $p_3 = w$ and linearize the system (4.25) to get

$$u^{\boldsymbol{\cdot}} = vw = 0 \qquad v^{\boldsymbol{\cdot}} = -2(1 + u)w \simeq -2w \qquad w^{\boldsymbol{\cdot}} = (1 + u)v \simeq v$$

$$(4.27)$$

To a first approximation, $\mathbf{p}$ does not change in the $\mathbf{i}_1$ direction so that trajectories near $(1, 0, 0)$ lie in a (v, w) plane normal to the $\mathbf{i}_1$ axis. The usual eigenvalue calculation shows that the origin of this (v, w) plane is a *center* since the eigenvalues of the coefficient matrix of the two linear ODE for v and w are $\sqrt{-2}$. Similar calculations show that the critical points $(-1, 0, 0)$, $(0, 0, 1)$, and $(0, 0, -1)$ are also *centers*.

For the critical point $(0, 1, 0)$, we set $p_1 = \hat{u}$, $p_2 = 1 + \hat{v}$, and $p_3 = \hat{w}$ and linearize the system (4.25) to get

$$\hat{u}^{\boldsymbol{\cdot}} = (1 + \hat{v})\hat{w} \simeq \hat{w} \qquad \hat{v}^{\boldsymbol{\cdot}} = -2\hat{u}\hat{w} \simeq 0 \qquad \hat{w}^{\boldsymbol{\cdot}} = (1 + \hat{v})\hat{u} \simeq \hat{u}$$

$$(4.28)$$

Therefore, $\mathbf{p}$ does not change in the $\mathbf{i}_2$ direction, at least to a first approximation, so that the trajectories near the critical point lie in a $(\hat{u}, \hat{w})$ plane normal to the $\mathbf{i}_2$ axis. The eigenvalues of the coefficient matrix of the two linearized ODE for $\hat{u}$ and $\hat{w}$ are ± 1; the critical point $(0, 1, 0)$ is therefore a *saddle point* and unstable. By symmetry $(0, -1, 0)$ is also a saddle point and unstable. Any small initial deviation from either of the unstable equilibrium states $(0, \pm 1, 0)$ will evolve with time into a substantial deviation from that equilibrium state. In plain English,

4E: The rotating box wobbles about the $\pm \mathbf{i}_2$ axes unless it starts out rotating perfectly about $\pm \mathbf{i}_2$.

There is no (positive or negative) damping in our model; the dynamical system is *conservative*. We might expect the stability for the linearized system about the other four critical points (characterized by a center) to hold for the original nonlinear system as well. This is confirmed by the following exact first integrals of (4.25) (to be obtained in an exercise at the end of the chapter):

$$2p_1^2 + p_2^2 = c_3^2 \qquad 2p_3^2 + p_2^2 = c_1^2 \qquad (4.29)$$

with $c_1^2 + c_3^2 = 2$. Thus,

> **4F:** The projections of the trajectories in the neighborhood of (±1, 0, 0) onto a (p_2, p_3) plane are ellipses as are the projections near (0, 0, ±1) onto a (p_1, p_2) plane.

These four sets of trajectories near the four centers are separated by four trajectories the projections of which correspond to the four half-ellipses with $c_1^2 = c_3^2 = 1$. These four separatrices are great semicircles on the sphere going from one saddle point to the other, connecting up the asymptotes of the hyperbolas in the local portrait for a small neighborhood of the two saddle points as shown in Figure 4.4. (This figure was generated by Alar Toomre when he and the author were lecturing on the Eulerian wobble to M.I.T. students in a first course on ODE in the early seventies.)

Even without the solution **p** as a function of time, we now have a complete qualitative picture of the trajectories of our problem, as sketched in Figure 4.4. The sketch provides us all the information we need to understand the phenomenon of Eulerian wobble. In particular, we see that a trajectory will go from the neighborhood of (0, 1, 0) to the neighborhood of (0, −1, 0) if we wait long enough. In other words,

> **4G:** If you throw the box up high enough (and, of course, rotate it imperfectly about the i_2 axis at the same time), the $\pm i_2$ faces will have reversed their positions when the box lands in your hands again!

It should be stressed that our rather informal presentation of Eulerian mutations may not have done justice to this important result in classical mechanics. A knowledge of this rather unexpected phenomenon has been found useful in the design of space satellites, in the analysis of polar wandering (the shifting of the earth's axis of rotation) and in the study of continental drifts, just to mention a few of its modern applications.

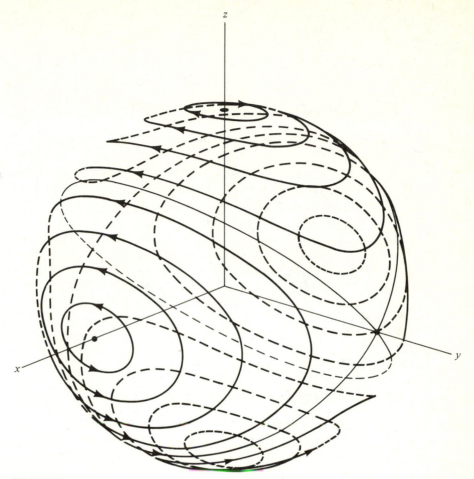

FIGURE 4.4

4.7   Parametric Excitation

Up to now, we have been concerned only with possible instability of the motion of a dynamical system caused by a small perturbation. In the worst scenario, the motion after perturbation runs away from the unperturbed motion without bound, at least until the system becomes overtaxed and cannot continue to function normally. It is important to note, however, that dynamical systems can become unstable without experiencing any external disturbance or perturbation. Most of us have the personal experience of getting a swing to go up higher and higher without any assistance from another person. The phenomenon is similar to a simple pendulum with the length of the pendulum arm varying periodically with time. Suppose we model the associated small-amplitude motion by the dimensionless equation

$$y^{..} + (1 + \epsilon \cos 2t)y = 0 \qquad (4.30)$$

with the initial condition (rescaled to read)

$$y(0) = \cos \phi \qquad y^{.}(0) = \sin \phi \qquad (4.31)$$

Equation 4.30, known as *Mathieu's equation,* is a linear equation with periodic coefficients. Linear ODE with periodic coefficients can have periodic or nonperiodic solutions. For example, the general solution of the first-order ODE with periodic coefficients $y^{.} = (1 + \sin t)y$ is the nonperiodic function $y(t) = c_0 \exp(t - \cos t)$. The second-order equation $y^{..} - y = 0$ also has periodic coefficients; it is well known that the general solution for this equation is nonperiodic.

For $0 \leq \epsilon \ll 1$, we seek a solution of (4.30) and (4.31) in the form of a perturbation series

$$y(t; \epsilon) = \sum_{n=0}^{\infty} y_n(t)\epsilon^n \qquad (4.32)$$

The leading term coefficient $y_0(t)$ is governed by the IVP $y_0^{..} + y_0 = 0$, $y_0(0) = \cos \phi$ and $y_0^{.}(0) = \sin \phi$. The exact solution for this problem is $y_0(t) = \cos(t - \phi)$.

For $\epsilon > 0$, the coefficient $y_1(t)$ is determined by

$$y_1^{..} + y_1 + \cos(2t)y_0 = 0 \qquad y_1(0) = y_1^{.}(0) = 0 \qquad (4.33)$$

The exact solution of the above IVP is

$$y_1 = \frac{1}{16}[\cos(3t - \phi) - \cos(t + \phi) - 4t \sin(t + \phi)] \qquad (4.34)$$

Hence, we have

$$y(t) = \cos(t - \phi)$$

$$+ \frac{\epsilon}{16}[\cos(3t - \phi) - \cos(t + \phi) - 4t \sin(t + \phi)] + O(\epsilon^2) \qquad (4.35)$$

It is not difficult to verify that to order ϵ^2, the "energy" of the system is given by

$$\tfrac{1}{2}[y^2 + (y')^2] = \frac{1}{2} + \frac{\epsilon}{16}\left[\cos 2\phi - \cos(4t - 2\phi) - 4t \sin 2\phi\right] + O(\epsilon^2)$$

(4.36)

which grows without bound unless the pendulum starts off at one of its equilibrium positions $2\phi = n\pi$. Thus,

4H: The motion of a pendulum is unstable even when there is no external forcing or disturbance.

The growth of energy with time is maximal if $\phi = \pm\pi/4$. This result agrees with our experience in how to get the swing up quickly.

Given our experience with the perturbation solution (3.20) for the IVP (3.13), it may be argued that the presence of a secular term in (4.36) does not prove anything about instability. However, Floquet's theory for ODEs with periodic coefficients (Minorsky, 1962) assures us that there is an unbounded solution for the Mathieu equation (4.30). Perturbation methods can also be applied to the more general equation

$$y^{\cdot\cdot} + (a + \epsilon \cos t)y = 0$$

(4.37)

to determine the regions of the (a, ϵ) plane where y is stable and the regions where it is unstable.

EXERCISES

1. Do a phase-plane analysis of Duffing's equation in the form of a system of first-order ODE: $y^{\cdot} = z$, $z^{\cdot} = -y - \epsilon y^3$.
   (a) Locate all the critical points.
   (b) Classify each critical point by a linear stability analysis.
   (c) Sketch a global portrait of the trajectories and give some justification (such as the tangent field, a first integral, etc.) for your sketch. Remember, the fact that a critical point is a center by a linear stability analysis does not say much about the trajectories near the critical point.

2. Repeat (a), (b), and (c) of the above problem for the equation of motion for an over-the-top pendulum: $\theta^{\cdot\cdot} + \sin\theta = 0$.

3. The equations of motion for our rectangular box are $L_1^{\cdot} = L_2 L_3$, $L_2^{\cdot} = -2L_1 L_3$, and $L_3^{\cdot} = L_1 L_2$. (**L** is the conventional notation for angular momentum.)

  (a) Show that $2L_1^2 + L_2^2 = C_3$, $2L_3^2 + L_2^2 = C_1$, and $L_1^2 - L_3^2 = \frac{1}{2}(C_3 - C_1)$.

  (b) Interpret these results in terms of the trajectories.

  (c) For $C_3 = C_1 = 1$, solve for $L_2(t)$ with $L_2(0) = 1 - \epsilon$. What is $L_2(\infty)$?

4. *Rabbits and Foxes.* In a forest, there is a rabbit population which feeds on the local vegetation and a fox population which preys on the rabbits. Let $r(t)$ and $f(t)$ be the size (measured in units of biomass) of the two populations, respectively. Suppose the growth of the rabbits and foxes is governed by the system

$$r^{\cdot} = r(1 - r) - rf \qquad f^{\cdot} = -f + 2rf$$

  (a) Discuss the intuitive meaning and effect of each of the terms in the two equations.

  (b) Locate all the critical points.

  (c) Classify each critical point by a linear stability analysis.

  (d) Sketch a global portrait of the trajectories.

  (e) What will happen to the rabbits and the foxes eventually? Be sure to analyze all the possibilities.

5. Do a phase-plane analysis of the equation: $y^{\cdot\cdot} + y - \epsilon y^3 = 0$.

  (a) Set $v = y^{\cdot}$ and convert the equation into a pair of first-order equations for y and v.

  (b) Locate all critical points of the system and classify each critical point by a linear stability analysis.

  (c) Sketch a global portrait of the trajectories and give some justification for your sketch. (*Remember:* The conclusion that a critical point is a center by a linear stability analysis is not conclusive!)

6. Show that the solution of $y' = e^{xy}$, $y(0) = 1$ has a singularity for some finite value $\bar{x} > 0$. Give an accurate estimate of $\bar{x}$.

7. To obtain a relation between a and ϵ for the solution of (4.37) to be periodic, we use the perturbation series (4.32) and

$$a = \sum_{n=0}^{\infty} a_n \epsilon^n$$

Choose a_n to eliminate the secular terms in the ODE for the coefficient $y_n(t)$ for $n = 0, 1, 2, \ldots$. Carry out this solution procedure for each of the following:

(a) Determine a_1 and a_2 given $a_0 = 0$.
(b) Determine a_1 given $a_0 = \frac{1}{4}$.
(c) Determine a_1 given $a_0 = 1$.

5

Hair

Euler Buckling and Elastic Stability

An eigenvalue problem for a simple ODE is solved without assuming any knowledge of the solution technique. Elements of calculus of variations are developed. The direct method for an approximate solution is illustrated. A boundary-value problem for a nonlinear ODE is formulated and its solution is obtained in terms of an elliptic integral.

5.1 The Legend of Samson and the Euler Column

As legend has it, strong man Samson lost his superior strength when seven locks were shaved off his head. He thus became a prisoner of the Philistines. On a later festive occasion, he was brought to the house of Dagon, the God of the Philistines, and chained between pillars. By then his hair had grown back and, along with it, his strength. He grasped the two middle pillars upon which the house rested and leaned his weight upon them. He then bowed with all his might; the house fell upon the Philistine lords and all the people that were in it.

Now, if you are a doubting Thomas and have not seen the heroics performed on the movie screen, be assured that Samson's feat is humanly possible. We prove it in the next few pages. All you need is a little knowledge of eigenvalue problems, not matrix eigenvalue problems as in Chapter 4, but eigenvalue problems in ODE.

To analyze the feasibility of Samson's feat, we begin by stripping away the unessential elements of the problem. The weight of the temple roof can be represented by a downward vertical force acting on the top end of the pillar. To maintain static equilibrium, the pillar is supported by the temple floor which exerts an equal and opposite force on its bottom. As a first approximation, we consider a pillar of uniform cross section and with a straight central axis (through the centroid of the cross sections and taken to be the x axis). Technically, a perfectly straight slender cylindrical body subject to equal and opposite axially directed compressive forces at its two ends is called a *column*.

Suppose the column is elastically deformable and the compressive forces are applied at the centroid of the cross section. Then the column will simply shorten until it reaches an equilibrium configuration and will return to its original length when the forces are removed (Figure 5.1). Now if a small lateral disturbance is applied to the compressed column (through a small lateral force exerted by Samson, say), it will cause the column to bow out as in Figure 5.2. If the axial force P is sufficiently small, our experience tells us that the column will return to its straight form when the lateral disturbance is removed. The column is said to be in a state of *stable equilibrium* under the compressive forces. But for some higher value of P, say P_{cr}, our intuition about a very long and slender column would suggest (and suitable experiments would show) that the column will remain bowed out even after the lateral disturbance is removed. The column is said to be in *neutral equilibrium* before we apply the small lateral disturbance. Finally, if P is sufficiently large, we expect that the column will pop out at the slightest disturbance (sometimes until it collapses). The column is said to be in a state of *unstable equilibrium* under the compressive axial forces. For obvious reasons, building and structural engineers want to know how large P can be for a given column to remain in stable equilibrium. Acting on a suggestion from D. Bernoulli, this question in a highly idealized form was answered by Euler in 1744. Many related but more complex problems have also been and are still being investigated.

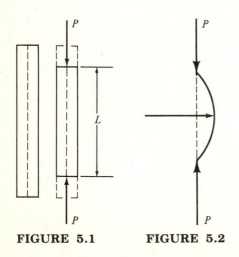

FIGURE 5.1   **FIGURE 5.2**

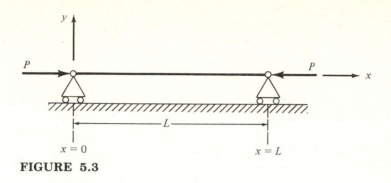

FIGURE 5.3

We will confine ourselves in the next few pages to a brief discussion of a special case of the problem studied by Euler. In Euler's investigation, the column is idealized by its central axis (Figure 5.3) with a compressive force at each end, equal in magnitude but opposite in direction. We will study the stability of such an Euler column, hinged at both ends so that the ends are free to rotate but are restrained from all but axial displacement. Similar to the stability of an orbiting mass point and the rotating box, the main question is again whether or not a small perturbation of the existing equilibrium configuration will lead to a substantial (possibly catastrophic) change.

5.2 Determination of the Buckling Load

If we are interested only in the critical load beyond which a given column is no longer in stable equilibrium, the problem can be treated as an eigenvalue problem in second-order linear ODE. In this case, we simply ask: Is there another equilibrium configuration for the column in the neighborhood of the straight form for the given compressive axial forces?

Suppose there is such a configuration in the form of a plane curve characterized by a displacement function $y(x)$ with the central axis of the column taken to be the portion of the x axis between the origin and L as shown in Figure 5.4. For a slightly deformed new configuration with $|dy/dx| \ll 1$ and $|y| \ll L$, the curvature at a point of the *deformed* central axis is (to a good approximation) proportional to the moment at that point. The proportionality factor was shown by J. Bernoulli in 1705 to be EI where E is *Young's modulus* and I is the *moment of inertia* of the *cross section* of the column [see Greenspan and Benney (1973)] to be specified more carefully later. Young's modulus E characterizes the stretching stiffness of the column material; it takes an axial force per unit cross-sectional area of magnitude E/L to stretch (or shorten) the column one unit length in the axial direction.

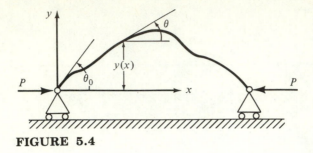

FIGURE 5.4

Unless stated otherwise, we will henceforth assume that the column material is homogeneous and that the cross section of the column is uniform over its length so that both E and I are constants. As the column is subject only to equal and opposite axial forces P at the ends, the moment at a point which was at x before deformation is simply Py (see Figure 5.5) and we therefore have

$$-\frac{EI}{R} = Py \qquad (5.1)$$

where $R(x)$ is the radius of curvature of the deformed central axis at x. The minus sign is introduced so that the Euler-Bernoulli relation (5.1) is consistent with the expression for R in what follows.

With $|dy/dx| \ll 1$, we have

$$\frac{1}{R} \simeq \frac{d^2y}{dx^2} \qquad (5.2)$$

Combining (5.1) and (5.2) gives us

$$\frac{d^2y}{dx^2} + \mu^2 y = 0 \qquad \mu^2 = \frac{P}{EI} \qquad (5.3)$$

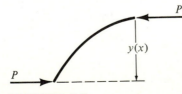

FIGURE 5.5

Now, the column is restrained in the y direction at the ends; so we have as boundary conditions for the above ODE[1]

$$y(0) = y(L) = 0 \qquad (5.4)$$

The ODE (5.3) has as its general solution

$$y(x) = c_1 \sin \mu x + c_2 \cos \mu x$$

The first boundary condition in (5.4) requires $y(0) = c_2 = 0$. The second condition requires $y(L) = c_1 \sin \mu L = 0$. For a nontrivial solution, we want $c_1 \neq 0$. It follows that we must have $\mu L = n\pi$, where n is any nonzero integer. Thus, the homogeneous boundary-value problem defined by (5.3) and (5.4) has a nontrivial solution if and only if

$$\mu^2 = \mu_n^2 \equiv \frac{n^2 \pi^2}{L^2} \qquad n = 1, 2, 3, \ldots \qquad (5.5)$$

or

$$P = P_n \equiv \frac{n^2 \pi^2 EI}{L^2} \qquad n = 1, 2, 3, \ldots \qquad (5.6)$$

The quantities μ_n^2, $n = 1, 2, \ldots$, are the *eigenvalues* of this homogeneous boundary-value problem. Thus, if $P = P_n$, the column has another equilibrium position described by

$$y_n = C_n \sin \mu_n x = C_n \sin \frac{n\pi x}{L} \qquad (5.7)$$

beside the straight form given by $y = 0$, the trivial solution. The solutions $\sin(n\pi x/L)$, $n = 1, 2, \ldots$, are the *eigenfunctions;* the constants $\{C_n\}$ are undetermined which is characteristic of all linear eigenvalue problems. We designate the lowest nonzero P_n as the *critical load* (also *buckling load* or *Euler's load*)

$$P_{\text{cr}} = \frac{\pi^2 EI}{L^2} \qquad (5.8)$$

---

[1] In the bent form, the two ends of the column are now separated by a distance slightly shorter than the original length L. But it is consistent with the approximation (5.2) to ignore this difference and prescribe $y = 0$ at $x = 0$ and $x = L$.

Thus we learn from the eigenvalue problem that

5A: The column may no longer be in stable equilibrium for $P \geq P_{cr}$.

The above analysis is not complete in that it only warns us of the possible danger if $P \geq P_{cr}$, but does not provide us with other crucial information. For instance, it does not tell us whether the bent equilibrium configuration is preferred over the straight configuration. One may also be led to the false conclusion that the column is in stable equilibrium if $P \neq P_n$. The proper interpretation of the above results can only be obtained from a more sophisticated analysis of the nonlinear problem associated with Equation 5.1. If the linear stability result is all you have (which is often the case), take no chances and design your column with a P_{cr} well above the load it is to carry. As the legend has it, there were about three thousand men and women on the roof who looked on while Samson made sport. The pillars apparently had not been designed for that, safety factor included.

5.3   The Nonlinear Elastic Stability Problem

To obtain more complete information about the stability of the Euler column, let us return once again to the Euler-Bernoulli formula (5.1) but now let us use the exact expression for the curvature of the bent column:

$$\frac{1}{R} = \frac{d\theta}{ds} \tag{5.9}$$

where s is the arc length of the column measured from the left end and θ is the angle made by the tangent to the deformed column with the x axis. Combining (5.9) and (5.1), we get

$$EI \frac{d\theta}{ds} + Py = 0 \tag{5.10}$$

To obtain a differential equation for θ alone, we differentiate (5.10) with respect to s to get (for a column of uniform properties and cross section)

$$EI \frac{d^2\theta}{ds^2} + P \frac{dy}{ds} = 0 \tag{5.11}$$

With $dy/ds = \sin\theta$, (5.11) becomes an ODE for θ,

$$\frac{d^2\theta}{ds^2} + \mu^2 \sin\theta = 0 \tag{5.12}$$

where $\mu^2 = P/EI$ as before.

The nonlinear differential equation (5.12) characterizes the deformation of the elastic line which serves as the classical model for the Euler column and is often called the *Elastica*. After we multiply (5.12) through by $d\theta/ds$ and integrate, we get

$$\frac{1}{2}\left(\frac{d\theta}{ds}\right)^2 - \mu^2 \cos\theta = C \tag{5.13}$$

where C is a constant of integration. The ends of the column are free to rotate, so we take $\theta = \theta_0 \neq 0$ at $x = 0$, θ_0 being an unknown at this point. With $\theta(x = 0) = \theta_0 \neq 0$ and $y(0) = 0$, we have from (5.10)

$$\frac{d\theta}{ds}\bigg|_{\theta=\theta_0} = 0 \tag{5.14}$$

which determines the constant of integration to be

$$C = -\mu^2 \cos\theta_0 \tag{5.15}$$

Equation 5.13 may now be written as

$$\frac{1}{2}\left(\frac{d\theta}{ds}\right)^2 = \mu^2(\cos\theta - \cos\theta_0) \tag{5.16}$$

From (5.4), we also have $y = 0$ at the right end of the column so that $d\theta/ds$ is also zero there [again because of (5.10)]. Equation 5.16 in turn requires that $\theta = \pm\theta_0 + 2n\pi$ at the right end, $0 \leq |\theta_0| \leq \pi$.

We will restrict our discussion to the special case where θ is a decreasing function of s and $\theta = -\theta_0$ at the right end so that

$$ds = -\frac{1}{\sqrt{2}\mu}\frac{d\theta}{\sqrt{\cos\theta - \cos\theta_0}} \tag{5.17}$$

[See Love (1944) for other possible configurations.] Upon integrating (5.17), we obtain

$$s = -\frac{1}{\sqrt{2}\mu} \int_{\theta_0}^{\theta} \frac{d\theta}{\sqrt{\cos\theta - \cos\theta_0}} \tag{5.18}$$

and hence

$$L = -\frac{1}{\sqrt{2}\mu} \int_{\theta_0}^{-\theta_0} \frac{d\theta}{\sqrt{\cos\theta - \cos\theta_0}} \tag{5.19}$$

In (5.19), we have made use of a result deduced in the general three-dimensional theory of elasticity: *To a first approximation, the central axis of our column does not change its length after deforming from the shortened straight form to the bent form.*

To write (5.19) in a more convenient form, we set

$$k = \sin\left(\tfrac{1}{2}\theta_0\right) \qquad \sin\left(\tfrac{1}{2}\theta\right) = k\sin\phi \tag{5.20}$$

We can then rewrite (5.19) as

$$L = \frac{2}{\mu} \int_0^{\pi/2} \frac{d\phi}{\sqrt{1 - k^2\sin^2\phi}} \equiv \frac{2}{\mu} J(k) \tag{5.21}$$

where

$$J(k) = \int_0^{\pi/2} \frac{d\phi}{\sqrt{1 - k^2\sin^2\phi}} \tag{5.22}$$

is called the *complete elliptic integral of the first kind.* For various values of k, the values of $J(k)$ can be found in tables [e.g., in Abramowitz and Stegun (1965)]. From (5.21), we have

$$\mu^2 = \frac{4}{L^2}[J(k)]^2 \quad \text{or} \quad P = \frac{4EI}{L^2}[J(k)]^2 \tag{5.23}$$

We note in passing that for small values of θ_0, we have $k^2 = \sin^2\left(\tfrac{1}{2}\theta_0\right) \ll 1$ so that

$$J \simeq \int_0^{\pi/2} d\phi = \frac{\pi}{2} \tag{5.24}$$

and therewith

$$P \simeq \frac{\pi^2 EI}{L^2} \equiv P_{cr} \tag{5.25}$$

which is the critical load for the column established by the linear eigenvalue problem approach. We note also

$$\int_0^\alpha \frac{d\phi}{\sqrt{1 - \sin^2 \phi}} = \frac{1}{2} \ln \frac{1 + \sin \alpha}{1 - \sin \alpha} \tag{5.26}$$

therefore, we have $J(k) \to \infty$ as $k \to 1$.

We will now use the above expression for P_{cr} to rewrite (5.23) in the form

$$\frac{P}{P_{cr}} = \frac{4}{\pi^2} [J(k)]^2 \quad \text{where} \quad k = \sin(\tfrac{1}{2}\theta_0) \tag{5.27}$$

The shape of the bent column may be calculated from

$$y = \int_0^s \sin \theta \, ds = -\frac{1}{2\mu} \int_{\theta_0}^\theta \frac{\sin \theta \, d\theta}{\sqrt{\sin^2(\tfrac{1}{2}\theta_0) - \sin^2(\tfrac{1}{2}\theta)}} \tag{5.28}$$

For our purpose, it suffices to determine the value $\delta = |y|_{max}$ which is attained at $\theta = 0$:

$$\delta = \frac{1}{2\mu} \int_0^{\theta_0} \frac{\sin \theta \, d\theta}{\sqrt{\sin^2(\tfrac{1}{2}\theta_0) - \sin^2(\tfrac{1}{2}\theta)}} = \frac{2k}{\mu} \int_0^{\pi/2} \sin \phi \, d\phi = \frac{2k}{\mu} \tag{5.29}$$

We may use (5.21) to write (5.29) in dimensionless form:

$$\frac{\delta}{L} = \frac{k}{J(k)} \tag{5.30}$$

The above expression for δ/L and the expression (5.27) for P contain much of the important information about the stability of a hinged column. For example, Equation 5.27 gives the magnitude of P needed to maintain a bent form corresponding to an angle θ_0 at the end $x = 0$. For $\theta_0 = 0+$, we have $P = P_{cr}$ corresponding to the lowest eigenvalue of the linear stability analysis and P increases monotonically with θ_0, tending to infinity as $\theta_0 \to$

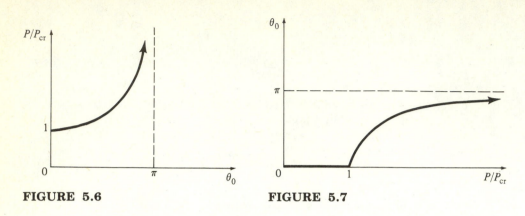

FIGURE 5.6

FIGURE 5.7

π, that is, as $k \to 1$. Figure 5.6 gives a plot of $P(\theta_0)$ which may be inverted to get $\theta_0(P)$ as shown in Figure 5.7. This last figure tells us the following about the *Elastica:*

> **5B:** The column can only be straight (after the removal of the lateral disturbance) if $P < P_{cr}$. For $P > P_{cr}$, the column may be either straight or bent; furthermore there is a unique bent form for each $P > P_{cr}$.

The question whether the column prefers the straight form or the bent form must still be analyzed [see Wang (1953), for example]. However, given the legend of Samson and other experimental evidence, engineers usually take no chances and design their columns well below P_{cr}.

5.4   The Potential Energy in a Hinged Column

The buckling analysis in the last section typifies one of the two broad approaches to the mechanics of flexible (or deformable) bodies. That approach postulates the validity of the laws of mechanics (and materials) and views physical phenomena as the consequences of these laws. The deduction of these consequences usually involves the solution of some initial-boundary value problem. For example, the critical load of a hinged column was determined in Section 5.2 by the linear eigenvalue problem

$$y'' + \frac{P}{EI}y = 0 \qquad y(0) = y(L) = 0 \tag{5.31}$$

(which is just a restatement of Equations 5.3 and 5.4). The alternative approach to the mechanics of deformable continuous media (or *continuum mechanics* for brevity) views nature as a very efficient machine which tends to "optimize" its effort. This approach considers mechanical phenomena involving deformable media to be the result of this optimization (and the material properties of the body). Within the optimization framework, the problem is usually to search for some function(s) which extremizes some quantity summarizing the effort exerted. In mechanical phenomena such as the Euler column, this quantity is usually the total energy in the mechanical system. Historically, this was Euler's approach to the problem of the buckling of a column.

If both approaches are appropriate for the analysis of problems in continuum mechanics, then there must be a bridge between them. This bridge is the *calculus of variations*. For the purpose of bringing out the basic issues involved, we will discuss the equilibrium configurations of a hinged column again, this time using the optimization approach. It is possible to show by way of this calculus that the two approaches are equivalent. We will also demonstrate how the new approach can be used to get approximate solutions for more complex problems by the direct method of calculus of variations. This method for obtaining approximate solutions of problems in differential equations has led to a host of modern numerical techniques such as the finite-element methods which are of current interest in both research and applications.

To arrive at the proper Eulerian optimization formulation, we take from the linear theory of elastic solids (Fung, 1965) that the total energy in the bent column is

$$J_h \equiv \frac{1}{2} \int_0^L [EI(y'')^2 - P(y')^2] dx \qquad (5.32)$$

The second term in J_h is the loss of *potential energy* associated with the decrease in distance between the two ends of the column when the straight column is bent. This energy loss is equal to the negative of the work done by P to produce this decrease in distance:

$$-\int_0^L P\left[1 - \frac{1}{\sqrt{1 + (y')^2}}\right]dx \simeq -\frac{1}{2}\int_0^L P(y')^2\, dx \qquad (5.33)$$

The first term in J_h is the *bending strain energy* stored in the column as it is kept in the bent form by the end forces. It is equal to the negative of the work required to straighten out the bent column by suitable bending moments. For an elemental arc of the bent column, this negative work is

$$-\tfrac{1}{2} M\, d\theta \simeq -\tfrac{1}{2} M y''\, dx = \tfrac{1}{2} EI(y'')^2\, dx \qquad (5.34)$$

where the factor $\frac{1}{2}$ comes from the fact that the work is equal to the area of the triangle in an $(M, d\theta)$ diagram between the $M = -EI\,d\theta$ line and the $d\theta$ axis. The bending strain energy is another source of potential energy not encountered in particle or rigid-body mechanics. However, we can almost feel the release of this energy when the bent column straightens itself out with the removal of the end forces.

5.5 The Variational Problem for a Hinged Column

The Eulerian optimization formulation for the determination of equilibrium states of a hinged column postulates the following:

> *When a hinged column is at rest, its shape, characterized by* y(x), *must be one which minimizes the total energy of the column, given by* J_h *when* $|y'|^2 \ll 1$.

The quantity J_h as defined by (5.32) is, of course, just a number; but its numerical value varies with the choice of y. For the solution of the buckling problem, we must find a shape function which minimizes J_h. There may be more than one such choice.

Evidently, the mathematical problem associated with the Eulerian minimum potential energy approach to the equilibrium configurations of a column is quite distinct from that associated with the Newtonian mechanics approach to the same problem. If correct, both approaches must lead to the same results; hence, there must be a link between the two corresponding types of mathematical problems. This link lies in the following mathematical fact:

5C: Among all four-times-continuously differentiable functions $y(x)$, $0 < x < L$, satisfying the conditions,

$$y(0) = y(L) = 0 \qquad (5.35)$$

those which make the value of J_h stationary also satisfy the differential equation

$$\frac{d^2}{dx^2}\left(EI\,\frac{d^2y}{dx^2}\right) + \frac{d}{dx}\left(P\,\frac{dy}{dx}\right) = 0 \qquad (5.36)$$

and the boundary conditions

$$x = 0,\, L: \qquad EIy'' = 0 \qquad (5.37)$$

Before we prove this assertion, note that the boundary conditions on $-EIy''$ are in fact implied by (5.31) if we take the ODE to hold also at the end points. Physically, these new boundary conditions simply express the fact that we allow the end points to rotate freely and hence the column is free from any bending moments there. Mathematically, they are needed to complement $y(0) = y(L) = 0$ for the fourth-order ODE (5.36). In order for the fourth-order ODE to be meaningful, $y(x)$ must be four times continuously differentiable. We will call the class of four-times-continuously differentiable functions which satisfy (5.35) the *admissible class of comparison functions* for our problem. For a constant P, we can integrate (5.36) twice to get the ODE in (5.31); the two constants of integration must be set to zero because of (5.37) and (5.35). The variational formulation leading to (5.36) and (5.37) is therefore more general than (5.31). The fourth-order ODE (5.36) not only allows for an axial force P which varies along the column (when the weight of the column is not negligible, for example), but also the possibility of a nonvanishing transverse distributed load along the column (such as a steady wind pressure, for example). Of course, we can also include the same effects in the Newtonian formulation and get the same results if we wish to do so.

To prove theorem **5C**, we let $\bar{y}$ be a function which makes J_h stationary and let $y = \bar{y} + \epsilon z$ where $z(x)$ is another member of the admissible class of comparison functions [in particular, $z(0) = z(L) = 0$], and ϵ is a parameter independent of x. We have then

$$J_h(\epsilon) = \frac{1}{2} \int_0^L [EI(\bar{y}'' + \epsilon z'')^2 - P(\bar{y}' + \epsilon z')^2] \, dx \qquad (5.38)$$

Since $\bar{y}$ makes J_h stationary, it follows that

$$\left. \frac{dJ_h}{d\epsilon} \right|_{\epsilon=0} = 0 \qquad (5.39)$$

for an arbitrary admissible z, or

$$\int_0^L (EI\bar{y}''z'' - P\bar{y}'z') \, dx = 0 \qquad (5.40)$$

Upon integrating by parts, we get

$$\left[EI\bar{y}''z' - (EI\bar{y}'')'z - P\bar{y}'z \right]_0^L + \int_0^L [(EI\bar{y}'')'' + (P\bar{y}')']z \, dx = 0 \qquad (5.41)$$

Since z satisfies the boundary conditions $z(0) = z(L) = 0$, the last two terms inside the brackets vanish and we are left with

$$\left[EI\bar{y}''z' \right]_0^L + \int_0^L [(EI\bar{y}'')'' + (P\bar{y}')'] z \, dx = 0 \qquad (5.42)$$

Equation 5.42 must be true for arbitrary admissible $z(x)$. We conclude that $\bar{y}$ must satisfy the ODE (5.36) and the boundary conditions (5.37). Equation 5.36 is called the *Euler differential equation* for J_h and the boundary conditions (5.37) are called *Euler* (or *natural*) *boundary conditions*. So $\bar{y}$, which makes J_h stationary, is in fact the solution of the differential equation (5.36) and the boundary conditions (5.37) *and* (by stipulation) (5.35).

The step from (5.42) to (5.36) and (5.37) needs proving. While an applied mathematician often bypasses a rigorous deduction to get to the heart of a problem, he or she must be able to justify all the deductions upon request (or to make clear that the results are of the nature of conjectures). Since steps similar to that from (5.42) to (5.36) and (5.37) occur quite frequently in the variational models in applied mathematics, we will give a proof of the following general mathematical result here for future reference:

5D: Let f be continuous in $[a, b]$. If

$$\int_a^b f(x)z(x)dx = 0 \qquad (5.43)$$

for all four-times-continuously differentiable z in $[a, b]$ with $z(a) = z(b) = 0$, then $f(x) = 0$ for all x in $[a, b]$.

Note that our choice of $z(x)$ ensures that it belongs to the admissible class of comparison functions for the Euler column.

To prove the above theorem, suppose that $f(x_0) \neq 0$ for some x_0 in $[a, b]$. With no loss in generality, we can take $f(x_0) > 0$ and x_0 as an interior point of $[a, b]$. By continuity, we have $f(x) > 0$ for $c < x < d$ with $a \leq c < x_0 < d \leq b$. Consider the admissible comparison function

$$z(x) = \begin{cases} (x - c)^6(x - d)^6 & c \leq x \leq d \\ 0 & \text{elsewhere} \end{cases} \qquad (5.44)$$

For this $z(x)$, we have $f(x)z(x) > 0$ for $c < x < d$ and therefore

$$\int_a^b f(x)z(x)dx = \int_c^d f(x)z(x)dx > 0$$

This contradicts the other part of the hypothesis that the left side of (5.43) must vanish for all admissible comparison functions.

The theorem is still true if we restrict z to have any number of derivatives vanishing at the end points. The $z(x)$ in (5.44) used to induce the contradiction allows for that. We can therefore apply this theorem to (5.42) first for admissible comparison functions with $z'(0) = z'(L) = 0$ to get the ODE (5.36) leaving just

$$\left[EI\bar{y}''z' \right]_0^L = 0 \tag{5.45}$$

By choosing two admissible comparison functions $y_k = \bar{y} + \epsilon z_k (k = 1, 2)$ with $z_1'(0) = 0$, $z_2'(L) = 0$ but $z_2'(0)z_1'(L) \neq 0$, we get (5.37) as well.

Taking the optimization approach for the determination of the equilibrium states of the column literally, we would have to search for a $\bar{y}$ among the admissible class of functions to minimize J_h. This is humanly impossible, given the infinity of comparison functions available, unless the blind search is replaced by a more systematic scheme. We formulate this more systematic scheme through the use of the calculus of variations and the scheme itself is to solve the eigenvalue problem, (5.35) to (5.37), which reduces to (5.31) since P is a constant. This might seem circular at first, but you must remember that in arriving at (5.31) via this "minimum energy" approach to continuum mechanics we never had to appeal to the law of mechanics. [We did use the Euler-Bernoulli relation (5.1) which is a law obeyed by a certain class of materials and not a part of Newtonian mechanics.]

We have not yet shown that the stationary value $J_h(\bar{y})$ is actually a minimum. The nontrivial argument which proves this assertion will be carried out in the next section for an analogous but more familiar problem. Also, we note here for later reference that the variational problem of minimizing (5.32) applies to a nonuniform column (e.g., tapered column, Young's modulus varying along the length of the column, etc.) as well.

5.6   The Spring-Mass Analogy

For a uniform column (for which EI is a constant) and constant axial force P, the Euler differential equation (5.36) of the integral J_h can be written as

$$EI\psi'' + P\psi = 0 \tag{5.46}$$

with $\psi = y''$. Both (5.46) and its integrated form in (5.31) have the same form as the differential equation of motion for the more familiar linear spring-mass system:

$$m\ddot{y} + ky = 0 \tag{5.47}$$

with m and k taking the place of EI and P, respectively. It is not difficult to see that (5.47) is the Euler differential equation of the integral

$$H = \frac{1}{2T} \int_{t_1}^{t_2} [m(y^{\cdot})^2 - ky^2] dt \qquad (5.48)$$

where $T \equiv t_2 - t_1$. Now $\frac{1}{2}m(y^{\cdot})^2$ is the kinetic energy of the system while

$$-W = -\int_0^y (-ky)\, dy = \tfrac{1}{2}ky^2$$

is the potential energy (with ky being the restoring spring force). The integrand $\frac{1}{2}[m(y^{\cdot})^2 - ky^2]$ is therefore the energy differential between kinetic and potential in the spring-mass system and H is the average energy differential in the system over the time interval $T \equiv t_2 - t_1$. The variational problem here is a special case of the Hamilton principle; it postulates that

Among all time paths y(t) *between the same two end points* y(t$_1$) = Y$_1$ *and* y(t$_2$) = Y$_2$, *the one actually realized minimizes the energy differential* TH [*as given in* (5.48)] *over the time interval* (t$_1$, t$_2$).

It is not difficult to repeat the calculation of Section 5.5 and show that among all admissible comparison functions [twice-continuously-differentiable $y(t)$ with the same end values Y_1 and Y_2], those which make H a stationary value also satisfy the equation of motion (5.47). Suppose $\bar{y}(t)$ renders H stationary, then for any admissible comparison function $y = \bar{y} + \epsilon z$, we have $z(t_1) = z(t_2) = 0$ since $y(t_k) = \bar{y}(t_k) = Y_k$. It is much less obvious that $H(\bar{y})$ is actually a minimum relative to all other admissible comparison functions in the neighborhood of $\bar{y}(t)$. To see the nature of the difficulty and its eventual resolution, consider

$$H(\epsilon) \equiv H(\bar{y} + \epsilon z) = H_0 + \epsilon H_1 + \frac{1}{2!} H_2 \epsilon^2 + \frac{1}{3!} H_3 \epsilon^3 + \cdots$$

where $H_k = d^k H/d\epsilon^k |_{\epsilon=0}$. Since $H(\bar{y})$ is stationary, we have $H_1 = 0$. For sufficiently small ϵ, $H(\bar{y})$ is a minimum if

$$H_2 = \frac{1}{T} \int_{t_1}^{t_2} [m(z^{\cdot})^2 - kz^2] dt > 0 \qquad (5.49)$$

for all $z(t)$. [$H(\bar{y})$ is a maximum if $H_2 < 0$. If $H_2 = 0$, then we must investigate H_3, H_4, etc which all vanish for the expression (5.48).] Now the integrand of H_2 is the difference of two positive terms; it appears that H_2 may be made positive, negative, or zero by an appropriate choice of $z(t)$. As it turns out, the integral H_2 can be put into a form which is always nonnegative by the following ingenious artifice of Jacobi.

First, we integrate the $(z^{\cdot})^2$ term in (5.49) by parts and use $z(t_1) = z(t_2) = 0$ to get

$$TH_2 = -\int_{t_1}^{t_2} (mz^{\cdot\cdot} + kz)z \, dt \qquad (5.50)$$

Next, consider any nontrivial solution $u(t)$ of the equation of motion (5.47) of the spring-mass system which does not vanish at both end points. We can write $mu^{\cdot\cdot} + ku = 0$ as $k = -mu^{\cdot\cdot}/u$ and use it to eliminate k from (5.50) to get

$$TH_2 = m \int_{t_1}^{t_2} (u^{\cdot\cdot}z - z^{\cdot\cdot}u)\left(\frac{z}{u}\right)dt \qquad (5.51)$$

With $u^{\cdot\cdot}z - uz^{\cdot\cdot} = (u^{\cdot}z)^{\cdot} - (uz^{\cdot})^{\cdot}$, we integrate (5.51) by parts once more to get

$$TH_2 = m \int_{t_1}^{t_2} (z^{\cdot}u - zu^{\cdot})\left(\frac{z}{u}\right)^{\cdot} dt = m \int_{t_1}^{t_2} \left(\frac{z^{\cdot}u - u^{\cdot}z}{u}\right)^2 dt \qquad (5.52)$$

where we have used $z(t_1) = z(t_2) = 0$ to eliminate the integrated terms. In this new form, the integrand of H_2 is nonnegative so that $H_2 > 0$ unless $z^{\cdot}u - u^{\cdot}z \equiv 0$ or $z = c_0 u$ for some constant c_0 [in which case the expression (5.51) for TH_2 vanishes]. As u does not vanish at both arbitrary end points t_1 and t_2 but z does, we must have $c_0 = 0$ or $z \equiv 0$. Therefore, $H(\bar{y})$ is a minimum except for the special case where t_1 and t_2 are *conjugate points,* so that $u(t_1) = u(t_2) = 0$.

5.7   The Direct Method of Calculus of Variations

While the solution of the boundary-value problem (5.31) or (5.35) to (5.37) was very straightforward for columns whose geometrical and material properties are uniform along the central axis and P is a constant, an exact solution in terms of elementary or special functions may be difficult or impossible if any one of *E, I,* or *P* is a function of *x*. We will show presently that the variational formulation offers a straightforward way to obtain the solution in the more complicated cases. The general procedure, which we will illustrate with the uniform hinged column case, is applicable to a very wide range of problems in continuum mechanics.

Instead of searching for a function $\bar{y}(x)$ which makes J_h a minimum value among all admissible comparison functions for the problem, we will only look for a $\bar{y}$ among a restricted class of admissible comparison functions. One such subclass is

$$y_1(x) = C_1 x(L - x) \tag{5.53}$$

The problem is to choose the yet unspecified constant C_1 so that the corresponding J_h is at least stationary. Note that $y_1(0) = y_1(L) = 0$ and y_1 is at least four times continuously differentiable; hence, for any fixed C_1, it is admissible.

To determine the critical load of the column, we insert (5.53) into (5.32). After carrying out the integration, we get

$$J_h(y_1) = \frac{1}{2} \int_0^L \left\{ EI(-2C_1)^2 - P[C_1(L - 2x)]^2 \right\} \equiv J_h^{(1)}(C_1) \tag{5.54}$$

For a stationary value of J_h, we choose the yet unspecified constant C_1 so that $dJ_h^{(1)}/dC_1 = 0$ or

$$4C_1 \left(1 - \frac{\mu^2}{12} \right) = 0 \tag{5.55}$$

For a nontrivial solution for $y_1(x)$, we must have $\mu^2 = \bar{\mu}_1^2 \equiv 12$ which gives an approximate value of the exact critical load $\mu_1^2 = \pi^2 = 9.8696 \ldots$. As in the case of the exact solution for the linear theory of Section 5.2, the approximate buckled shape of the column is determined up to an amplitude factor C_1.

Evidently, $\bar{\mu}_1^2$ is not a very accurate approximation of μ_1^2. More accurate solutions may be obtained by using more complicated admissible comparison functions. For the admissible comparison functions

$$y_2(x) = C_1 x(L - x) + C_2 x^2(L - x)^2 \tag{5.56}$$

we have

$$J_h = \frac{1}{2} \int_0^L \left\{ EI[C_1(-2) + C_2(2L^2 - 12xL + 12x^2)]^2 \right.$$

$$\left. - P[C_1(L - 2x) + C_2(2xL^2 - 6x^2L + 4x^3)]^2 \right\} dx$$

$$\equiv J_h^{(2)}(C_1, C_2) \tag{5.57}$$

For $J_h^{(2)}$ to be a stationary value, we must choose C_1 and C_2 so that $\partial J_h^{(2)}/\partial C_k = 0$, $k = 1, 2$. These two conditions give two linear equations for C_1 and C_2:

$$\left(1 - \frac{\mu^2}{12}\right)C_1 - \frac{\mu^2}{60}(C_2 L^2) = 0$$

$$-\frac{\mu^2}{12}C_1 + \left(1 - \frac{\mu^2}{42}\right)(C_2 L^2) = 0$$

(5.58)

For this homogeneous system to have a nontrivial solution, the determinant of the coefficient matrix must vanish. This condition gives a quadratic equation for μ^2:

$$\mu^4 - 180\mu^2 + 1680 = 0$$ (5.59)

The smaller root is $\tilde{\mu}_1^2 = 9.8751$ which is considerably closer to μ_1^2 than $\bar{\mu}_1$. For this choice of μ^2, we can use one of the equations in (5.58) to express $C_2 L^2$ in terms of C_1 and $\tilde{\mu}_1^2$ in (5.56). The (approximate) buckled shape is again determined up to an amplitude factor.

More complex admissible functions do not always give better results. For example, it can be verified that

$$y_3 = C_1 x(L - x) + C_2 x^2(L - x)$$ (5.60)

also gives an approximate solution of 12 for μ_1^2. It is not better than the simpler function y_1; but it cannot be worse. (Why not?) On the other hand, a function such as

$$y_4 = C_1 x^3(L - x)^3$$ (5.61)

is expected to give an even better result than $y_1(x)$. This function not only satisfies the end constraints $y(0) = y(L) = 0$, but also the Euler boundary conditions $y''(0) = y''(L) = 0$.

5.8   The Rayleigh Quotient

Instead of minimizing J_h, a more efficient optimization procedure for the buckling load is through the Rayleigh quotient. For the case of a constant P, we can multiply the ODE in (5.31) through by y and integrate over $(0, L)$. After integration by parts, we have

$$0 = \int_0^L (EIy'' + Py)y\,dx = \left[EIyy'\right]_0^L + \int_0^L [Py^2 - (EIy)'y']\,dx \qquad (5.62)$$

With $y(0) = y(L) = 0$, we may rewrite the above relation as

$$P = \frac{\int_0^L (EIy)'y'\,dx}{\int_0^L y^2\,dx} \qquad (5.63)$$

It follows that

5E: The critical load of the Euler column is the minimum value of the **Rayleigh quotient** (5.63).

In the case of a uniform column for which EI is a constant, (5.63) simplifies to

$$\frac{P}{EI} \equiv \frac{\mu^2}{L^2} = \frac{\int_0^L (y')^2\,dx}{\int_0^L y^2\,dx} \qquad (5.64)$$

To determine the critical (buckling) load of our column, the problem is to find the shape function $\bar{y}(x)$ which makes the quotient a minimum value. Of course, an admissible comparison function in this case must also satisfy the boundary conditions $y(0) = y(L) = 0$.

To illustrate, consider a function

$$y_0(x) = \begin{cases} x & (x < L/2) \\ L - x & (x > L/2) \end{cases} \qquad (5.65)$$

For this function, we also get $\mu^2 = \mu_1^2 \equiv 12$ obtained previously by working with y_1 of (5.53) and J_h. The simpler expression $y_0(x)$ is only piecewise differentiable; so it does not even qualify as an admissible comparison function for J_h.

If we use the legitimate comparison function (5.53) in the Rayleigh quotient, we get

$$\mu^2 = \frac{L^2 \int_0^L (L - 2x)^2\,dx}{\int_0^L x^2(L - x)^2\,dx} = 10 \qquad (5.66)$$

Evidently, the same trial function y_1 gives a much more accurate approximate critical load than what we got from $J_h[y_1]$.

EXERCISES

1. For the following problems, obtain the form of the eigenfunctions and the (transcendental) equation which determines the eigenvalues:
   - **(a)** $y'' + \lambda y = 0$, $y'(0) = y(1) + y'(1) = 0$
   - **(b)** $y'' + \lambda y = 0$, $y(0) = y(\pi) + y'(\pi) = 0$
   - **(c)** $y'' - \lambda y = 0$, $y(0) + y'(0) = y(1) = 0$
   - **(d)** $y'' + \lambda y = 0$, $y(0) - y'(0) = y(1) + y'(1) = 0$
   - **(e)** $y'' + (1 + \lambda)y' + \lambda y = 0$, $y'(0) = y(1) = 0$

2. Use Newton's method for root finding to determine an approximate value for the smallest eigenvalue (in absolute value) for the problems in Exercise 1. (*Hint:* Get a first approximation from the graph and do one iteration.)

3. From the graph of the two sides of the relevant transcendental equation, find an approximate value for λ_n when $n \gg 1$ for the problems in Exercise 1. (*Note:* We can improve this approximate value by Newton's method if necessary.)

4. Determine the eigenvalues and eigenfunctions of
$$x^2 y'' + xy' - \lambda y = 0 \qquad y(1) = y(e) = 0$$
   (*Hint:* The ODE is equidimensional.)

5. Let L be a linear differential operator defined by $L[y] \equiv (py')' - qy$ where $p(x)$ and $q(x)$ are known functions continuous in the interval $[a, b]$. (We will also assume p to be continuously differentiable for simplicity.) Suppose $\alpha y + \beta y' = 0$ at $x = a$ and $\gamma y + \delta y' = 0$ at $x = b$ where α, β, γ, and δ are known constants with $\alpha\beta \neq 0$ and $\gamma\delta \neq 0$. Show that if both $u(x)$ and $v(x)$ satisfy the same boundary conditions as y, then they satisfy the *Lagrange identity*
$$\int_a^b (vL[u] - uL[v])dx = 0$$

6. **(a)** For $L[y]$ as defined in Exercise 5 and $r(x) > 0$ continuous in $[a, b]$, the eigenvalue problem
$$L[y] + \lambda r(x)y = 0 \qquad y(a) = y(b) = 0$$
   with $p(x) > 0$ and $q(x) \geq 0$ in $[a, b]$ is known to have an infinite sequence of distinct real nonnegative eigenvalues $\{\lambda_k\}$ and a corresponding sequence of eigenfunctions $\{\phi_k\}$. Use Lagrange's identity to show
$$\int_a^b r(x)\phi_n(x)\phi_m(x)dx = 0 \qquad (n \neq m)$$

(b) How would you use the nonnegativity of the eigenvalues for the eigenvalue problem in part (a) to conclude that the eigenvalues of Exercise 4 must be nonpositive?

7. *Brachistochrone or Shortest Time.* A bead, strung on a frictionless wire and acted upon only by gravity, is to slide down the wire between two fixed end points. Determine the shape of the wire which minimizes the time of descent. The bead is initially at rest. (You need only obtain the ODE and the auxiliary conditions for the determination of the shape of the wire.)

8. Consider a slender cylindrical body of length L, uniform cross section A, and homogeneous elastic material with a constant Young's modulus E. The bottom of this slender body is fixed to the ground so that its central axis cannot be displaced laterally or rotated from the x axis (positive upward). The top end of the central axis is free from any constraint so that it can be displaced laterally and rotated from the x axis. There is no axial force acting at the top end, and the only force acting on the body is its own weight.

   (a) Consider a segment of the deformed body from $x = x_1$ to $x = x_1 + \Delta x$. Show that moment equilibrium requires that, as the segment Δx approaches zero,

   $$\frac{dM}{dx} = Q$$

   where $M(x)$ is the bending moment at the point x and $Q(x)$ is the transverse force at x (Figure 5.8).

   (b) Show that the force in the x direction, $P(x)$, at location x is given by $P(x) = -\rho g A (L - s)$ where ρ is the mass density, A is the cross-sectional area, g is the gravitational acceleration, and

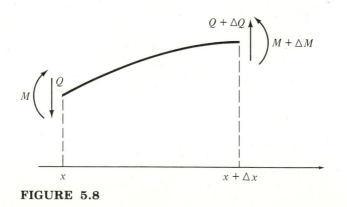

FIGURE 5.8

s is the arc length of the deformed central axis measured from the bottom end $x = 0$. (As long as we prescribed boundary condition at $x = 0$ and $x = L$, we will take $s = x$.)

(c) If $f(x)$ is the lateral distributed external force per unit length of the central axis acting on the body (e.g., wind pressure), show that $dQ/dx = f(x)$.

9. Obtain the Euler differential equation and the boundary conditions by minimizing the total energy J_g of the body of Exercise 8:

$$J_g \equiv \frac{1}{2} \int_0^L [EI(y'')^2 - P(y')^2 - f(x)y]\, dx$$

Specify the appropriate class of admissible comparison functions.

10. **(a)** Show that with $f(x) \equiv 0$ and $y' \equiv \phi$, the Euler differential equation in Exercise 9 may be written as $\ddot{\phi} + \alpha z \phi = 0$ where $(\ )^\bullet \equiv d(\ )/dz$ and $z = L - x$.

(b) Obtain the solution of the ODE for ϕ in terms of Airy functions (Bessel's function of order $\frac{1}{3}$).

(c) Determine the critical load of the body in terms of the first zero of the appropriate Bessel function.

11. **(a)** Obtain an approximate critical load of the Euler column by using $y_4(x) = x^3(L - x)^3$ in the Rayleigh quotient (5.64).

(b) Obtain an approximate value for the dimensionless critical load μ_1^2 by using $y_5 = C_1 x^2(L - x)^2 + C_2 x^3(L - x)^3$ in the Rayleigh quotient (5.64).

12.

$$J = \frac{1}{2} \int_0^L [EI(y'')^2 - P(y')^2 - \omega^2 \rho y^2]\, dx$$

$$y(0) = y'(0) = y(L) = y'(L) = 0$$

(a) Show that the $\bar{y}(x)$ which renders J stationary must be a solution of the Euler differential equation $(EIy'')'' + (Py')' - \omega^2 \rho y = 0$.

(b) Show that, for all four-times-continuously differentiable functions u and v which also satisfy the four boundary conditions stated above, we have

$$\int_0^L (vL[u] - uL[v])\, dx = 0$$

where $L[y] \equiv (EIy'')'' + (Py')'$.

13.

$$\frac{d^4 y}{dx^4} - \lambda^4 y = 0 \qquad y(0) = y'(0) = y(1) = y'(1) = 0$$

(a) Show that the eigenvalues $\{\lambda_n^4\}$ of this problem are the roots of $1 - \cosh \lambda \cos \lambda = 0$.

(b) Show that the eigenfunctions $\{y_k(x)\}$ satisfy the orthogonality relation

$$\int_0^1 y_n(x) y_m(x)\, dx = 0 \qquad (m \neq n)$$

(c) Obtain an approximate value for (the lowest) λ^4 by the direct method of calculus of variations with $y(x) = c_1 x^2 (1-x)^2$. You may leave your answer in the form of a ratio of two definite integrals.

14. (a) Obtain the Rayleigh quotient of the eigenvalue problem in Exercise 13.

(b) Repeat part (a) for the eigenvalue problem in Exercise 12.

15. With $y \simeq c_1 \sin \pi x + c_2 \sin 2\pi x$, find an approximate solution of the following problem by the direct method of calculus of variations:

$$y'' + \lambda y = 0 \qquad y(0) = y(1) = 0$$

16. The direct method of calculus of variations is useful for finding approximate solutions for inhomogeneous boundary-value problems as well. Consider

$$y'' + xy = -x \qquad y(0) = y(1) = 0$$

(a) Show that the above ODE is the Euler differential equation of the functional

$$J = \int_0^1 [\tfrac{1}{2}(y')^2 - \tfrac{1}{2}xy^2 - xy]\, dx$$

(b) Find an approximate solution of the BVP by the direct method, taking $y \simeq c_1 x(1-x)$.

(c) Repeat part (b) with $y \simeq c_1 x(1-x) + c_2 x^2(1-x)$.

17. For the eigenvalue problem

$$y'' + \lambda xy = 0 \qquad y(0) = y(1) = 0$$

the appropriate functional for a variational formulation is

$$J = \frac{1}{2} \int_0^1 [(y')^2 - \lambda xy^2]\, dx$$

(a) Find an approximate solution of the eigenvalue problem with $y \simeq c_1 x(1-x)$.

(b) Repeat part (a) with $y \simeq c_1 x(1-x) + c_2 x^2(1-x)$.

18. $\qquad (xy')' + y = x \qquad y(0) = 0 \qquad y(1) = 1$

    (a) Construct an appropriate functional J for a variational formulation of the above inhomogeneous BVP.

    (b) Find an approximate solution with $y \simeq x + c_1 x(1 - x)$.

    (c) Repeat part (b) with $y \simeq x + x(1 - x)(c_1 + c_2 x)$.

19. $\qquad [(1 + x)y']' + \lambda y = 0 \qquad y(0) = y(1) = 0$

    (a) Construct an appropriate functional J for a variational formulation.

    (b) Find an approximate solution with $y \simeq c_1 x(1 - x)$.

    (c) Repeat part (b) with $y \simeq c_1 x(1 - x) + c_2 x^2 (1 - x)$.

20. Show by the method of Jacobi that the stationary value of J in Exercises 9 and 11 is actually a minimum.

6

A Menace on Any Road

Car Following

Several methods of solution are developed for simple differential-difference equations.

6.1 Lagrangian and Eulerian Formulation of Traffic Flow

The daily activities of modern industrialized societies depend vitally on the efficient and safe flow of vehicular traffic. In the last few decades, there has been considerable effort to analyze the basic mechanisms which govern traffic movements. A better understanding of these mechanisms has led to better road development, traffic-light systems, and regulations which help to reduce traffic congestion and accidents. In this chapter, we discuss some elementary aspects of one of the two main approaches to traffic-flow problems: the microscopic analysis of road traffic initiated by a research group at General Motors in the mid-fifties. The other approach, known as the continuum theory of traffic flow, will be discussed in the next chapter. As we shall see, the two theories complement each other; each is useful in a different range of traffic density.

The microscopic analysis of traffic flow focuses on the movement of individual vehicles in traffic, each vehicle being treated as a single mass particle. Such a treatment is known as a *Lagrangian* formulation for the traffic flow problem. The general objective of this approach is to know the position, velocity, and acceleration of individual cars for all time. We will discuss only the simplest mathematical models in this area. All such models are concerned with unidirectional traffic in a single lane (idealized as a straight line) with no passing. The cars (or particles) move along the lane with constant speed u_0 for $t \leq 0$. At $t = 0+$, the lead car begins to deviate from this constant speed motion, at least for awhile (e.g., it may slow down to avoid a pedestrian and then speed up again to the previous pace). This deviation generates a disturbance which propagates to the rear as each of the successive cars tries to adjust for the disturbance created by the lead car. For this so-called *car-following problem,* we are particularly interested in two types of questions:

1. For a pair of cars, one following the other, under what circumstances does a maneuver by the front car cause an overcompensation by the follower? From a mathematical viewpoint, the question is concerned with the behavior of the single follower's motion as a function of time. In response to a change of the lead car's motion, does the follower repeatedly overshoot the needed adjustment and thereby execute an oscillatory motion (about the desired motion) which may or may not decay with time? This is the *local stability* (or instability) problem in car following.

2. For a long line of cars, under what circumstances does a small disturbance by the lead car get amplified as it travels (propagates) through the line of cars and thereby cause a collision among cars further back in the platoon? This is a problem of *asymptotic stability* (or instability) in car following.

Acceptable models of varying degrees of sophistication for the particle (Lagrangian) formulation of traffic-flow problems invariably involve a dynamical system (characterized by one or more ODE) whose response to external excitations lags behind the excitatory signals. As we attempt to answer the two types of questions posed above for some of these models, we will be exposed to several methods of solution for differential equations with a time lag (and more generally, differential-difference equations) not normally encountered in most undergraduate courses in ODE.

In contrast, the continuum theory does not focus on individual cars; rather it keeps a record of information such as the speed of cars passing through different points in space at different times. The approach is similar to the Eulerian formulation used for continuum (especially fluid) mechanics. With space and time both as independent variables, the analysis in an Eulerian formulation will involve partial differential equations.

6.2 Instantaneous Velocity Control

Let $x_n(t)$ be the position of the nth car along the road (taken as the x axis) at time t, and $v_n(t) \equiv x_n^{\cdot}(t)$ be its velocity. As before, $(\ )^{\cdot}$ indicates differentiation of $(\ )$ with respect to time. The simple models described in Section 6.1 stipulate $v_n(t) = u_0$ for $t \leq 0$ since all cars are going at the same steady speed u_0 up to $t = 0$ in these models. We can eliminate the appearance of u_0

from much of the analysis to follow by setting $u_n(t) \equiv v_n(t) - u_0$ so that $u_n(t) = 0$ for $t \leq 0$. In other words, $u_n(t)$ is the velocity of the nth car in a reference frame moving with a constant speed u_0 along the x axis.

As a first approximation, we expect the driver of the nth car to adjust its speed according to the relative speed between it and the car in front. In the simplest model, we take this dependence on the relative speed to be linear and write

$$u_2^{\cdot} = \lambda(u_1 - u_2) \qquad u_3^{\cdot} = \lambda(u_2 - u_3) \qquad \cdots \qquad (6.1)$$

or

$$u_{n+1}^{\cdot} = \lambda(u_n - u_{n+1}) \qquad n = 1, 2, 3, \ldots, N-1 \qquad (6.2)$$

where N is the number of cars on the road. For simplicity, we have taken all drivers to be identical so that the *sensitivity factor* λ is the same for all cars. The early experiments at General Motors found λ to be somewhere between 0.3 and 0.4 s^{-1} with 0.37 s^{-1} as a reasonable estimate (Chandler, Herman, and Montroll, 1958). For initial conditions, we have

$$u_n(0) = 0 \qquad n = 1, 2, \ldots, N-1 \qquad (6.3)$$

Once the speed of the lead car, that is, its deviation from the constant speed u_0, is prescribed, the first equation of (6.1) and $u_2(0) = 0$ determines $u_2(t)$ for $t > 0$. Having $u_2(t)$, we can use the second equation of (6.1) and $u_3(0) = 0$ to determine $u_3(t)$ for $t > 0$, etc. In this way, the motion of all N cars can be investigated by examining the solution $u_n(t)$, $n = 2, 3, \ldots, N$.

As indicated in Section 6.1, we are interested in the stability of the dynamical system characterized by (6.2). For the purpose of examining the amplification of the disturbance generated by the lead car, it suffices to consider the case $u_1(t) = \sin \omega t$ as we can decompose an arbitrary disturbance into its Fourier components. The solution of the IVP for $u_2(t)$ with such a $u_1(t)$ is

$$u_2(t) = U[e^{-\lambda t} \sin \phi_2 + \sin(\omega t - \phi_2)] \qquad (6.4)$$

where the amplitude U and phase angle ϕ_2 are given in terms of ω and λ by

$$U = \left(1 + \frac{\omega^2}{\lambda^2}\right)^{-1/2} \equiv \cos \phi_2 \qquad (6.5)$$

Note that as $t \to \infty$, we have $|u_2| \leq U$. With u_2 given by (6.4), we can solve the IVP for $u_3(t)$ to get

$$u_3(t) = U^2\{[\sin(2\phi_2) + \omega t]e^{-\lambda t} + \sin(\omega t - 2\phi_2)\} \qquad (6.6)$$

Now, we have $|u_3(t)| \le U^2$ as $t \to \infty$. Continuing the process, it can be shown that $u_k(t)$ is proportional to U^{k-1} with $|u_k| \le U^{k-1}$ as $t \to \infty$. Since $U < 1$, $u_k(t)$ is at most of the same order of magnitude as the forcing function $u_1(t)$ for all $t > 0$; it is considerably smaller than $u_1(t)$ for large k and large t. Thus, our seemingly reasonable model of car following yields a rather unreasonable prediction:

> **6A:** Any disturbance generated by the lead car would (according to the instantaneous velocity control model) never get amplified as it travels through the platoon of cars.

This is simply not realistic.

6.3   Velocity Control with Lag Time

One shortcoming of the above model which may have led to the unrealistic conclusion of unconditional stability is the assumption of instantaneous response on the part of the "follower" cars. There is necessarily a lapse of time between a change of motion of the lead car and a change of motion of the car immediately behind it. The same is true for any two consecutive cars. There are three different sources for the lag time: (1) the time lapse before the driver of the "follower" car observes the change of motion of the lead car, (2) the time lapse between the follower's observation and his or her response, that is, stepping on the gas pedal or brake, and (3) the time it takes the car to respond to the driver's action. This observation suggests that, for an improved model, we modify (6.2) by introducing a lag time:

$$\dot{u}_{n+1}(t) = \lambda [u_n(t - T) - u_{n+1}(t - T)] \qquad (t > T) \qquad (6.7)$$

$n = 1, 2, 3, \ldots, N - 1$, where the lag time T ranges from 1.0 to 2.2 s according to experiments carried out at General Motors. For about 50 percent of the drivers tested, T is slightly less than 1.5 s (Chandler, Herman, and Montroll, 1958). The auxiliary conditions for these equations are

$$u_n(t) = 0 \qquad [t \le (n - 1) T] \qquad (6.8)$$

$n = 1, 2, 3, \ldots, N - 1$, in which the effect of lag time has been included. In most problems, the motion of the lead car is completely known.

Linear IVP involving differential equations with constant coefficients and a lag time are most conveniently solved by the method of Laplace transforms. Define

$$U_{k+1}(s) \equiv \int_0^\infty e^{-st} u_{k+1}(t)\,dt = \int_{kT}^\infty e^{-st} u_{k+1}(t)\,dt \qquad (6.9)$$

where we have used (6.8) in the second equality. With $u_{k+1}(t) \equiv 0$ for $t \le kT$, we have

$$\int_0^\infty e^{-st} \dot{u}_{k+1}(t)\,dt = \left[u_{k+1}(t) e^{-st} \right]_{kT}^\infty + s \int_{kT}^\infty e^{-st} u_{k+1}(t)\,dt$$

$$= sU_{k+1}(s) \qquad (6.10)$$

and

$$\int_0^\infty e^{-st} u_k(t-T)\,dt = \int_{(k-1)T}^\infty e^{-st} u_k(t-T)\,dt$$

$$= \int_{(k-2)T}^\infty e^{-s(\tau+T)} u_k(\tau)\,d\tau = e^{-sT} U_k(s) \qquad (6.11)$$

as we must have $e^{-st} u_{k+1}(t) \to 0$ for the improper integral (6.9) to exist. Now, multiply (6.7) through by e^{-st} and integrate over $(0, \infty)$; we get in this way

$$sU_{n+1}(s) = \lambda [e^{-sT} U_n(s) - e^{-sT} U_{n+1}(s)]$$

or

$$U_{n+1}(s) = \frac{\lambda U_n(s)}{\lambda + se^{sT}} \qquad (6.12)$$

For the case $u_1(t) = \sin \omega t$, we have

$$\int_0^\infty e^{-st} \sin \omega t\, dt = \frac{\omega}{s^2 + \omega^2} \qquad (6.13)$$

and from (6.12)

$$U_2(s) = \frac{\lambda U_1(s)}{\lambda + se^{sT}} = \frac{\lambda}{\lambda + se^{sT}} \frac{\omega}{s^2 + \omega^2} \qquad (6.14)$$

where $U_k(s)$ is the Laplace transform of $u_k(t)$. It remains to invert $U_2(s)$ to get $u_2(t)$. The process can be continued to get $u_3(t)$, $u_4(t)$, etc. Unfortunately, the inverse transform of $U_2(s)$ cannot be found from the usual table of transforms or obtained from those listed in such a table by suitable manipulation. Indeed, the inversion requires an adroit use of contour integration of complex-valued functions. Moreover, the results obtained are too complicated to readily yield useful information for the car-following problem without further analyses [see Herman et al. (1959)]. In the next two sections, we will obtain, by elementary means, conditions for local and asymptotic stability of the dynamical system (6.7). In Section 6.8, we will describe an elementary method for constructing the solution of the initial-value problem (6.7) and (6.8).

6.4   Approximate Solution for a Short Lag Time

In this section, we obtain some approximate results for the IVP (6.7) and (6.8) by a Taylor series expansion of $u_k(t - T)$ about t:

$$u_k(t - T) = u_k(t) + (-T)u_k^{\cdot}(t) + \frac{1}{2!}(-T)^2 u_k^{\cdot\cdot}(t) + \cdots \qquad (6.15)$$

If T is not large, we may approximate $u_k(t - T)$ by retaining the first two terms in (6.15) and write (6.7) as

$$(1 - \lambda T)u_{n+1}^{\cdot} + \lambda u_{n+1} = \lambda[u_n(t) - Tu_n^{\cdot}(t)] \qquad (6.16)$$

For $u_1 = \sin \omega t$, the general solution of this first-order ODE for $n = 1$ is

$$u_2(t) = c_2 e^{-\lambda t/(1-\lambda T)} + A_2 \sin \omega t + B_2 \cos \omega t \qquad (6.17)$$

where

$$A_2 = \frac{\lambda^2(1 + \omega^2 T^2) - \omega^2 \lambda T}{\lambda^2(1 + \omega^2 T^2) - \omega^2(2\lambda T - 1)} \qquad B_2 = \cdots \qquad (6.18)$$

The solution for $n > 2$ can be obtained similarly. The structure of these solutions is similar to (6.17). It follows from the complementary solution that the amplitude of the motion of the cars does not increase with time if $\lambda T < 1$. The effect of the lead car's motion also does not get amplified as n increases if $\lambda T > 2\lambda T - 1$ (see the expression for A_2 and B_2) or (again) $\lambda T < 1$. We have then

6B: For the first approximation model (6.16), the motion of the cars is locally and asymptotically stable if $\lambda T < 1$.

A more accurate result can be obtained by retaining more terms in (6.15). For example, by keeping the quadratic term in T, we have from (6.7)

$$\frac{\lambda}{2} T^2 \ddot{u}_{n+1} + (1 - \lambda T) \dot{u}_{n+1} + \lambda u_{n+1} = \lambda [u_n(t) - T\dot{u}_n(t) + \tfrac{1}{2} T^2 \ddot{u}_n(t)]$$

$$(6.19)$$

As before, one of the factors influencing stability comes from the complementary solutions of (6.19). The characteristic exponents of the complementary solutions taken in the form $e^{\alpha t}$ are the roots of

$$(\tfrac{1}{2} \lambda T^2) \alpha^2 + (1 - \lambda T) \alpha + \lambda = 0 \qquad (6.20)$$

The two roots of (6.20) are

$$\binom{\alpha_1}{\alpha_2} = \frac{-(1 - \lambda T) \pm \sqrt{(1 - \lambda T)^2 - 2(\lambda T)^2}}{\lambda T^2} \qquad (6.21)$$

Evidently, $\lambda T > 1$ will again lead to instability. However, we now see that the motion for $\lambda T < 1$ is qualitatively different from the solution (6.17) depending on whether or not we have $\lambda T < \sqrt{2} - 1$. For $\sqrt{2} - 1 < \lambda T < 1$, the motion is a damped oscillation, which may or may not be troublesome.

For $n = 1$ and $u_1 = \sin \omega t$, the general solution of (6.19) for u_2 is

$$u_2(t) = \bar{c}_{21} e^{\alpha_1 t} + \bar{c}_{22} e^{\alpha_2 t} + \bar{A}_2 \sin \omega t + \bar{B}_2 \cos \omega t \qquad (6.22)$$

with

$$\bar{A}_2 = 1 + \frac{\omega^2}{\Delta_2} (\lambda T - 1) \qquad \bar{B}_2 = -\frac{\lambda \omega}{\Delta_2} (1 - \tfrac{1}{2} T^2 \omega^2)$$

$$(6.23)$$

$$\Delta_2 = \lambda^2 (1 - \tfrac{1}{2} T^2 \omega^2)^2 + \omega^2 (1 - \lambda T)^2$$

It follows from (6.23) that the effect of the lead car's motion will not be amplified only if $\lambda T < \tfrac{1}{2}$. This is a substantial (100 percent) reduction from the corresponding result for (6.16).

6C: The motion of cars, as governed by the approximate model (6.19), is locally and asymptotically stable if $\lambda T < \sqrt{2} - 1$.

The criterion on λT for the amplification of the effect of the lead car motion may be further modified by the effect of the higher-order terms in (6.15). The development shows how much more complex are differential equations with delays compared to the ordinary ones (without delays). We will not pursue an investigation of the effects of the higher-order terms in the series (6.15). Instead, we will return in the next section to the original model defined by (6.7) and (6.8) and analyze it directly.

6.5   Local and Asymptotic Stability

Guided by the case $T = 0$ [and the Laplace transform solution (6.14) for $T \neq 0$], we seek a complementary solution of the ODE with delay

$$\dot{u}_2(t) = \lambda[u_1(t - T) - u_2(t - T)] \qquad (t > T) \qquad (6.24)$$

in the form of an exponential function

$$u_{2c}(t) = Ve^{-\alpha t} \qquad (6.25)$$

for some constants V and α. The dynamical system (6.24) [as well as the more general system (6.7)] is *locally stable* if and only if all complementary solutions are of the form (6.25) with $\alpha > 0$. With $u_1(t - T) = 0$, (6.24) requires that α be a solution of the transcendental equation

$$(\lambda T)e^z = z \qquad (z \equiv \alpha T) \qquad (6.26)$$

From a graph of both sides of this equation (see Figure 6.1) we see that (6.26) has

1. Two real positive roots if $0 < \lambda T < 1/e$
2. Only one real root, $z = 1$, if $\lambda T = 1/e$
3. No real root if $\lambda T > 1/e$

If some of the roots of (6.26) are complex, a disturbance generated by the lead car will still be damped out as long as $\mathrm{Re}(\alpha) > 0$. To see what

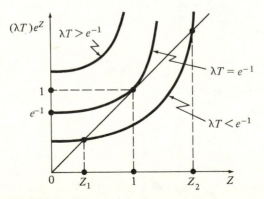

FIGURE 6.1

restriction this imposes on λT, we set $z = x + iy$ and obtain from (6.26) two equations for x and y:

$$x = (\lambda T)e^x \cos y \qquad y = (\lambda T)e^x \sin y \qquad (6.27)$$

If $y = 0$, these equations reduce to (6.26) for real z. For $y \neq 0$, the two equations in (6.27) are equivalent to

$$x = y \cot y \quad \text{and} \quad x = \ln \frac{y}{\sin y} - \ln \lambda T \qquad (6.28)$$

The graph of the first equation of (6.28) is shown as the dashed curve in Figure 6.2. The graphs of the second equation for different values of λT are given in the same figure by the solid curves.

In the range $|y| < \pi$, the solid and dashed curves intersect on or below $x = 0$ for $\lambda T > \pi/2$ so that (6.26) has roots with a negative real part in this range of λT. From the figure, it is also evident that for $|y| > \pi$, all roots of (6.28) with a negative real part, that is, all intersections between the dashed and solid curves below the line $x = 0$, correspond to values of λT greater than $\pi/2$. Therefore, we must have $\lambda T < \pi/2$ for all exponential complementary solutions of (6.24) to damp out eventually.

For $\lambda T < 1/e$, the dashed curves do not intersect the solid curve in $|y| < \pi$; they do for $|y| > \pi$ but only in the range $x > 0$. Therefore, Equation 6.26 has no complex roots in the strip $|y| < \pi$. [The figure does not apply to the $y = 0$ case as (6.28) is not equivalent to (6.27) when $y = 0$.] On the other hand, the existence of real roots of (6.26) has already been analyzed earlier. In particular, there are only two positive real roots of (6.26) for $\lambda T < 1/e$. We conclude that the solution (6.25) for (6.24) is damped oscillatory if $\lambda T < \pi/2$ and is nonoscillatory damped if $\lambda T < 1/e$.

6D: The system (6.7) is locally stable for $\lambda T < 1/e = 0.368 \ldots$.

Incidentally, the above analysis is equivalent to that for the position of the zeros of $\lambda + se^{sT}$ of the denominator of (6.14) in the complex s plane for Laplace transforms.

Turning now to the particular solution of (6.24), we again consider the case $u_1(t) = \sin \omega t$, $t > 0$. For this case, we expect a particular solution of the form $u_{2p}(t) = B \sin \omega t + C \cos \omega t$. The method of undetermined coefficients gives

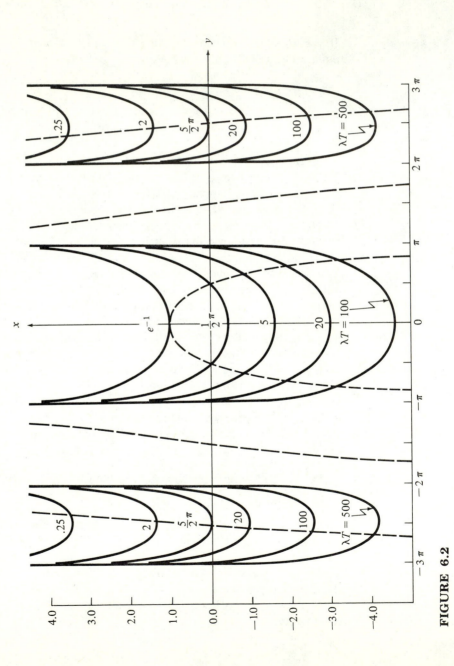

FIGURE 6.2

$$B = \frac{1}{D^2}\left(1 - \frac{\omega}{\lambda}\sin \omega T\right) \equiv \frac{1}{D}\cos \phi_2$$

$$C = -\frac{1}{D^2}\frac{\omega}{\lambda}\cos \omega T = -\frac{1}{D}\sin \phi_2$$

(6.29)

where

$$D^2 = 1 + \left(\frac{\omega}{\lambda}\right)^2 - 2\left(\frac{\omega}{\lambda}\right)\sin \omega T$$

(6.30)

so that

$$u_{2p}(t) = \frac{1}{D}\sin (\omega t - \phi_2) \qquad (t > T)$$

(6.31)

Similar calculations show that the corresponding particular solution for $u_n(t)$, $n = 3, 4, \ldots$, is of the form $D^{1-n}\sin(\omega t - \phi_n)$. Therefore, the effect of the oscillatory motion of the lead car will be amplified as it travels through the line of cars (each trying to adjust to the motion of the car in front of it) unless we have $D^2 > 1$ or

$$\lambda T < \frac{1}{2}\frac{\omega T}{\sin \omega T}$$

(6.32)

The dynamical system (6.7) is *asymptotically stable* if $D^2 > 1$ for all ω. Since $x/\sin x$ is monotone increasing for $0 < x \le \pi/2$ and

$$\lim_{x \to 0}\frac{x}{\sin x} = 1$$

(6.33)

we have that

> **6E:**   The dynamical system (6.7) is asymptotically stable if $0 < \lambda T \le \frac{1}{2}$.

Together with results already obtained on the type of transient solutions for different ranges of λT, we see that asymptotic stability also ensures a damped oscillatory transient solution of the type (6.25), though not local stability.

For $\lambda = 0.37$ s^{-1} and $T = 1.5$, we have $\lambda T = 0.555$ which is slightly above $\frac{1}{2}$. Given that the model (6.7) is a relatively crude approximation of reality, the result cannot be conclusive at such a borderline situation. A more sophisticated model of the car-following phenomenon, which includes some secondary effects, will be needed in this case. Our analysis does confirm that

> **6F:** Most of us are living closer to disaster on the highway than we would like to believe,

with the occasional spectacular pileup on our freeway serving as a reminder.

6.6 Solution of the Initial-Value Problem

In this section, we introduce the so-called *continuation process* to obtain the solution of (6.7) and (6.8) for a prescribed $u_1(t)$. To illustrate this elementary method of solution for differential equations with a lag time, we consider first the simpler IVP

$$v^{\cdot}(t) + \lambda v(t-1) = 0 \qquad (t > 0) \qquad\qquad (6.34)$$

$$v(t) = 1 \qquad (-1 < t \le 0) \qquad\qquad (6.35)$$

In the interval $0 < t \le 1$, we have $v^{\cdot}(t) = -\lambda$ so that $v(t) = V_1 - \lambda t$ since with $\tau = t - 1$, we have $v(\tau) = 1$ in the interval $-1 < \tau \le 0$. On physical grounds or otherwise, we stipulate that the solution $u(t)$ be continuous at the junction $t = 0$. With $v(0) = 1$, the continuity condition gives $V_1 = 1$ so that

$$v(t) = 1 - \lambda t \qquad (0 < t \le 1) \qquad\qquad (6.36)$$

Let $H(t)$ be the *Heaviside unit step function* defined to be 0 for $t < 0$ and 1 for $t > 0$. With the help of $H(t)$, we can write the solution of (6.34) and (6.35) in the interval $-1 < t \le 1$ more compactly as

$$v(t) = 1 - \lambda t H(t) \qquad (-1 < t \le 1) \qquad\qquad (6.37)$$

Next, we look at the interval $1 < t \le 2$. For this interval, (6.34) becomes

$$v^{\cdot}(t) = -\lambda v(t-1) = -\lambda[1 - \lambda(t-1)] \qquad\qquad (6.38)$$

Therefore, we get

$$v(t) = 1 - \lambda t + \tfrac{1}{2}\lambda^2(t-1)^2 \qquad (1 < t \le 2) \qquad\qquad (6.39)$$

where the continuity at $t = 1$ has been used to determine the constant of integration. Equation 6.39 can be combined with (6.37) to give

$$v(t) = 1 - \lambda t H(t) + \frac{1}{2!} [\lambda(t-1)]^2 H(t-1) \qquad (-1 < t \le 2)$$

$$(6.40)$$

A graph of $v(t)$ for $t \le 2$ already suggests that λT must be considerably less than unity if the dynamical system is to be locally stable. We can continue the above process to get

$$v(t) = 1 + \sum_{m=0}^{M} \frac{1}{(m+1)!} [-\lambda(t-m)]^{m+1} \qquad (M < t \le M+1)$$

$$(6.41)$$

or

$$v(t) = 1 + \sum_{m=0}^{M} \frac{1}{(m+1)!} [-\lambda(t-m)]^{m+1} H(t-m) \qquad (6.42)$$

for all $t > -1$.

Evidently, the above continuation process can be used to solve inhomogeneous equations as well, in particular the IVP

$$v^{\cdot}(t) + \lambda v(t-T) = f(t-T) \qquad (t > T) \qquad (6.43)$$

$$v(t) = g(t) \qquad (0 < t \le T) \qquad (6.44)$$

The sequence of IVP, defined by (6.7) and (6.8) with a prescribed $u_1(t)$, is of this type. For the special case where $f(t) = H(t)$ and $g(t) = 0$, this method gives as a solution for (6.43) and (6.44)

$$v(t) = - \sum_{m=1}^{\infty} \frac{1}{m!} [-\lambda(t-mT)]^m H(t-mT) \qquad (6.45)$$

corresponding to the solution for $u_2(t)$ of (6.7) and (6.8) with $u_1(t) = H(t)$ and $u_2(t) = 0$ for $t < T$.

6.7   Fluctuation of Car Spacings

The position of the nth car at time t is related to its velocity by

$$x_n(t) = x_n(0) + \int_0^t v_n(y)\,dy = x_n(0) + \int_0^t [u_n(y) + u_0]\,dy$$

$$= x_n(0) + u_0 t + \int_0^t u_n(y)\,dy \qquad (6.46)$$

The spacing $s_{n+1}(t)$ between the nth and $(n + 1)$st car at time t is therefore

$$s_{n+1}(t) \equiv x_n(t) - x_{n+1}(t)$$

$$= [x_n(0) - x_{n+1}(0)] + \int_0^t [u_n(y) - u_{n+1}(y)]\,dy \qquad (6.47)$$

Let $d_{n+1}(t)$ be the deviation from the initial spacing $s_{n+1}(0)$ between the nth and the $(n + 1)$st car. We have from (6.47)

$$d_{n+1}(t) \equiv [x_n(t) - x_{n+1}(t)] - [x_n(0) - x_{n+1}(0)]$$

$$= \int_0^t [u_n(y) - u_{n+1}(y)]\,dy \qquad (6.48)$$

By the equation of motion (6.7) and the initial condition (6.8), we have

$$d_{n+1}(t) = \frac{1}{\lambda} \int_0^t u^{\cdot}_{n+1}(y + T)\,dy = \left[\frac{1}{\lambda} u_{n+1}(y + T) \right]_0^t = \frac{1}{\lambda} u_{n+1}(t + T)$$

$$= \frac{1}{\lambda} [v_{n+1}(t + T) - u_0] \qquad (6.49)$$

To illustrate the use of this formula, suppose the lead car comes to a halt at $t = 0$ (instantaneously), so that $x_1^{\cdot} = v_1(t) = 0$ for all $t > 0$ and $u_1(t) = -u_0$ for all $t > 0$. After a delay of T units of time, the driver of the second car will slow the car down and eventually bring it to a stop (if a collision has not taken place before that) so that $u_2(t_f) = -u_0$ for some $t_f > T$. Therefore, we have $v_2(t_f) = 0$ and, from (6.49),

$$d_2(t_f) = \frac{-u_0}{\lambda} \qquad (6.50)$$

6G: If u_0/λ is less than the initial spacing, then the first two cars will still be separated by some distance less than the initial spacing. If not, then the second car will have plowed into the lead car before t_f.

It is important to realize that the conclusion of no collision for the case of small u_0/λ is valid only if we have local stability. Otherwise, the conclusion might not be valid as $|u_2(t + T)/\lambda|$ can be larger than the initial spacing for some $t < t_f$.

More extensive numerical results have been obtained for the more realistic case when the lead car slows down gradually (instead of stopping

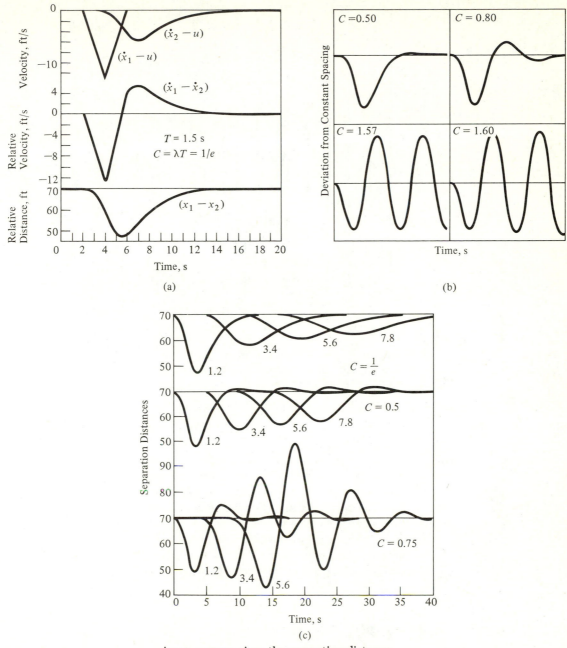

An *m-n* curve gives the separation distance
between the *m*th and *n*th cars.

FIGURE 6.3

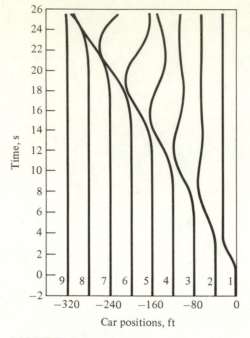

FIGURE 6.4
At $t = 0$, lead car decelerates from 50 mph
to 45 mph in 2 s and then accelerates back
up to 50 mph in the next 2 s. $T = 2$ s and C
$= 0.8$.

abruptly). For a lead car slowing down linearly from its original steady speed
of u_0 miles per hour (mph) for 2 s and then speeding up linearly back to u_0
in the next 2 s (and staying at that speed thereafter), numerical results for
the various quantities (such as the relative velocity, the separation distance,
etc.) characterizing the response of the second car, are given in Figure 6.3a
and b. The separation distance between successive cars of the same problem
is given in Figure 6.3c. The propagation of the disturbance generated by the
lead car for a similar problem (the differences involve only the input values
of the various parameters) is shown in Figure 6.4. Note that a collision takes
place between the seventh and eighth cars about 24 s later.

6.8 Density and Velocity Relation

For an application of the results of the present microscopic theory to the
macroscopic theory of the next chapter, we introduce the concept of a *velocity*

field $u(x, t)$ for a heavily traveled road. Among the many definitions of $u(x, t)$ used in the literature, we choose to define it as the velocity of the car which happens to be passing through point x on the road at time t (hours), that is, if $x_n(t) = x$, then

$$u(x, t) = u(x_n(t), t) = \dot{x}_n(t) \tag{6.51}$$

To gain a better understanding of the quantity $u(x, t)$, consider ten cars (points) at a unit (mile) distance apart from each other initially ($t = 0$). Each car travels at a constant speed (in the positive x direction) with the speed of the nth car being $[45 - \frac{3}{2}(n - 1)]$ miles per hour, so that the first car has a speed of 45 mph, the second, 43.5 mph, etc. If we take $x_1(0) = 0$, we have the graph in Figure 6.5 for the position of each car at a later time. At $t > 0$, the position of the nth car is evidently given by $x_n(t) = -(n - 1) + [45 - \frac{3}{2}(n - 1)]t$. In particular, we have $x_7(\frac{1}{4}) = 3$ miles, and correspondingly $u(3, \frac{1}{4}) = 36$ mph which is the slope [of the straight line emanating from $(-6, 0)$] at the point $(3, \frac{1}{4})$. The reader may want to work out $u(12, \frac{1}{2})$, $u(25, 1)$, and $u(5, \frac{1}{6})$.

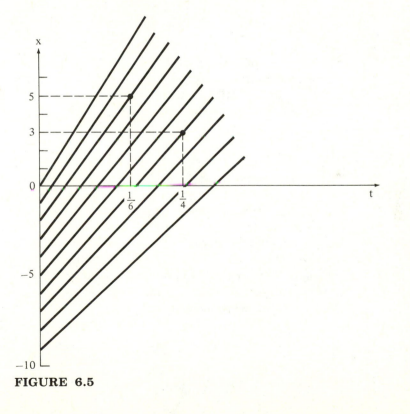

FIGURE 6.5

We know from our daily experience that traffic moves more slowly if it is heavy and faster if it is light. We now wish to relate $u(x, t)$, which characterizes the speed of the traffic (not individual cars), to the car density on the road. We do this by integrating (6.7) once to get

$$\dot{x}_{n+1}(t + T) = \lambda[x_n(t) - x_{n+1}(t)] + C_{n+1} \tag{6.52}$$

where C_{n+1} is a constant of integration. Now, consider a steady-state situation in which all cars are equal distances apart and moving at the same velocity. In that case, we have $\dot{x}_{n+1}(t) = \dot{x}_{n+1}(t + T)$ and $C_{n+1} = C_0$ for all n so that Equation 6.52 can be written as

$$\dot{x}_{n+1}(t) = \lambda[x_n(t) - x_{n+1}(t)] + C_0 \tag{6.53}$$

The distance between two consecutive cars $[x_n(t) - x_{n+1}(t)]$ is a reasonable measure of $(\text{density})^{-1} = (\text{number of cars per unit distance})^{-1}$. In the notation of the next chapter, we set $[x_n(t) - x_{n+1}(t)] = 1/k$. Equation 6.53 becomes

$$u = \frac{\lambda}{k} + C_0 \tag{6.54}$$

The constant of integration C_0 may be determined from the observation that, at maximum density k_j (j for "jammed" in bumper-to-bumper traffic), there should be (little or) no movement, giving $C_0 = -\lambda/k_j$ and

$$u = \lambda\left(\frac{1}{k} - \frac{1}{k_j}\right) \tag{6.55}$$

The velocity-density relation is unacceptable in the low-density range for $u \to \infty$ as $k \to 0$. For this reason (and others), improved models of the car-following problem have been proposed. Several linear models with delay can be studied through the exercises at the end of this chapter. We mention here one nonlinear model which improves upon the model of (6.7). The improvement consists of the observation that the sensitivity factor is not a constant. A driver tends to be more alert or sensitive if the car in front is close up and less so when it is far ahead. A simple way to incorporate this effect into a model would be to take

$$\lambda = \frac{\mu}{[x_n(t - T) - x_{n+1}(t - T)]^2} \tag{6.56}$$

for some constant μ. With $\dot{u}_n(t) = [\dot{x}_n(t) - u_0]^{\cdot}$, we have from (6.7)

$$\overset{..}{x}_{n+1}(t) = \frac{\mu[\overset{.}{x}_n(t-T) - \overset{.}{x}_{n+1}(t-T)]}{[x_n(t-T) - x_{n+1}(t-T)]^2} \tag{6.57}$$

Equation 6.57 can be integrated once immediately to get

$$\overset{.}{x}_{n+1}(t) = -\mu[x_n(t-T) - x_{n+1}(t-T)]^{-1} + C_{n+1} \tag{6.58}$$

Again, in a steady-state situation, Equation 6.58 becomes

$$u = -\mu k + C_0 \tag{6.59}$$

The constant C_0 is again determined by $u(k = k_j) = 0$ giving

$$u = \mu(k_j - k) \tag{6.60}$$

The nonlinear model (6.58) is more plausible than the linear model (6.7) with a constant λ as (6.60) gives $u(k = 0) = \mu k_j$. In reality, there may be a lower maximum speed u_{max} ($< \mu k_j$) imposed by law. In that case, we may set $u = u_{max}$ for $k \leq k_c$ with k_c determined by $u_{max} = \mu(k_j - k_c)$ for a continuous transition. It will be noted in the next chapter that the relation (6.60) is still not completely satisfactory with or without the cutoff speed for other reasons. Other car-following models have been studied with more adequate density-velocity relations. Some of these models will be studied in the exercises at the end of this chapter.

EXERCISES

1. Investigate the local stability of the constant-spacing control model characterized by $\overset{..}{x}_{n+1}(t) = \lambda[x_n(t-T) - x_{n+1}(t-T)] - [x_n(0) - x_{n+1}(0)]$. Show that the model is asymptotically unstable for all $\Lambda T > 0$.

2. Investigate the local stability of the acceleration control model with $\overset{.}{u}_{n+1}(t) = \lambda[\overset{.}{u}_n(t-T) - \overset{.}{u}_{n+1}(t-T)]$.

3. Consider the model given by

   $$\overset{..}{x}_{n+1}(t) = \lambda[x_n^{(m)}(t-T) - x_{n+1}^{(m)}(t-T)]$$

   where $x_j^{(m)} \equiv d^m x_j/dt^m$. Show that it is locally stable for some values of λT if m is odd and locally unstable if m is even.

4. Drivers do look ahead beyond the front car. Consider the model

$$u_{n+2}^{\cdot}(t) = \lambda_1 [u_{n+1}(t - T_1) - u_{n+2}(t - T_1)]$$

$$+ \lambda_2 [u_n(t - T_2) - u_{n+2}(t - T_2)]$$

for $n = 0, 1, 2, \ldots$. The motion of the first and second car in this line is assumed to be known. Discuss the stability of the car motion in this model. Is a local collision more or less likely to occur than the model of (6.7)?

5.   Drivers are also affected by the car behind. Consider the model

$$u_{n+1}^{\cdot}(t) = \lambda_1 [u_{n-1}(t - T) - u_n(t - T)]$$

$$+ \lambda_2 [u_{n+1}(t - T) - u_n(t - T)],$$

with $\lambda_1 > \lambda_2 > 0$. For simplicity, assume a disturbance induced by the lead car only. Discuss asymptotic stability.

6.   A lead car is traveling at 60 mph along a highway. A second car follows at the same steady speed. At $t = 0$, the lead car slows down to 55 mph and maintains that speed thereafter (perhaps sensing a radar trap). Describe the response of the follower car by the model of (6.7) and (6.8). Is the follower-car driver able to maintain control of the vehicle in a stable fashion if the lag time T is 1.8 s and sensitivity factor λ is 0.33 s^{-1}? What if T is 1.5 s instead?

7.   Use the continuation process to calculate the solution of

$$u^{\cdot}(t) = 2u(t - 1) \qquad (t > 1)$$

with

$$u(t) = t \qquad\qquad 0 < t \le 1$$

for $0 < t \le 5$.

8.   Find the solution of $u^{\cdot}(t - 1) = u(t)$, with $u(t) = t^2 (0 < t \le 1)$ for all $t > 1$. Can we require the solution to be continuous for all $t > 0$?

9.   **(a)**   Show that by setting $w(t) = u(t)e^{-bt}$, the equation

$$w^{\cdot}(t) + bw(t) + aw(t - T) = F(t) \qquad (t > T)$$

can be transformed into

$$u^{\cdot}(t) + Au(t - T) = f(t) \qquad (t > T)$$

where $A = ae^{bT}$ and $f(t) = F(t)e^{bt}$.

(b)   With $u(t) = g(t), 0 < t \le T$, show that in the interval $nT < t < (n + 1)T$, $u(t)$ is a polynomial of degree at most $(n + r)$ if f and g are polynomials of degree at most r.

10. Use the continuation process to get the velocity of the second car for $t \leq 3T$ in a linear velocity control model with lag time (6.7) and (6.8) if $u_1(t) = \sin \omega t$ for $t > 0$.

11. Another more complex car-following model is given by

$$\ddot{x}_{n+1}(t) = \lambda \dot{x}_{n+1}(t - T) \frac{\dot{x}_n(t - T) - \dot{x}_{n+1}(t - T)}{[x_n(t - T) - x_{n+1}(t - T)]^2}$$

Obtain the velocity-density relation implied by this model. Using (a) $u(k_{\max}) = 0$, and alternatively (b) $u(0) = u_{\max}$, compare the results.

12. In recent years, physiologists have studied the IVP

$$\dot{x}(t) = -x(t) + \frac{x(t - 1)}{1 + [x(t - 1)]^n} \qquad (t > 1)$$

with $x(t) = x_0$ (a constant) for all $t \leq 0$, as one possible mathematical model for the oscillation of white blood cell level in several kinds of leukemia patients. (Some parameters in the actual model have been set equal to 1 for simplicity.)

(a) Obtain the exact solution of this IVP for $0 < t \leq 1$.
(b) For $n = 1$, obtain the exact solution of this same IVP for $1 < t \leq 2$. (You may leave your answer in the form of an integral.)

13. An initially 2.5-mile-long queue of cars follows the Presidential limousine in a motorcade from City Hall to the airport. Owing to gradual thinning of crowds (and of enthusiasm), the lead car accelerates so that its speed u_P (in mph) = $2t$ (in minutes from the start). Assuming that the rest of the politicians are driven so that their u (in mph) = $(3000/k) - 15$, where k is the local density in cars/mile, how far behind will the third alderman of the 29th District be when the President reaches Air Force One 15 miles away? Also, how fast will the alderman be traveling then?

14. $$\ddot{x}_n(t) = -\lambda(t - T)[\dot{x}_n(t - T) - \dot{x}_{n-1}(t - T)]$$

$$\lambda(t) = \frac{\gamma}{x_{n-1}(t) - x_n(t)} \qquad (\gamma > 0)$$

(a) Obtain an expression for $\dot{x}_n(t)$ in terms of the car spacing.
(b) Translate the above result into a u versus k relation with $u = 0$ at $k = k_j$.

Wave Propagation

A pebble dropped in a still pond generates a circular ripple which spreads outward with time. It is a sight which titillates our visual sense and mind's eye. The phenomenon is even more amazing to those who know about the motion of individual water particles. By dropping a dye into the pond, we can trace its motion as the ripple propagates away from its source point. As the ripple passes by, we see that the dye moves in a closed-loop "circular" motion not far from its original at-rest position. It is rather amazing that something physical such as a ripple or "wave" can move outward without taking any fluid particle along with it. Is there anything which left with the wave? How does wave propagation work? A complete understanding of water waves turns out to be an intellectual and scientific challenge of the highest order. Understanding of any phenomenon usually leads to its mastery and applications. Our understanding of water waves has led to better ship design and navigation and many other engineering accomplishments.

Many other observed phenomena also involve wave propagation— not so obvious are sound, light, and electromagnetic events. A better understanding of these different types of wave propagation phenomena has led to a more sophisticated technology in communication, transportation, information processing, and health care, just to name a few. Mathematics has been most instrumental in gaining an understanding of wave phenomena in continuum mechanics, electromagnetics, optics, and acoustics. Conversely, these application areas have been responsible for the development of much of what we know in partial differential equations (PDE) today.

The important concept of wave propagation is introduced in Chapter 7 by way of the Lighthill-Whitham theory of traffic flow. While it is not as physically intuitive a traveling-wave phenomenon as water waves, the theory involves only a single first-order PDE which is a simple way to start a discussion. It provides a simple setting to introduce the method of characteristics. It also offers an alternative to the particle dynamics type treatment of vehicular traffic described in Chapter 6 and allows for a smooth transition from the stability theme to the wave propagation theme.

The method of characteristics and the formation of shocks first introduced in Chapter 7 are further developed in Chapter 8 through a discussion of water waves. Many of the standard results for the linear wave equation are introduced there along with a discussion of bores.

Standing waves associated with finite solution domains are discussed in terms of normal modes and natural frequencies of vibrating strings and membranes in Chapter 9. The method of eigenfunction expansions is used to solve both the forced and unforced problems. The Rayleigh quotient is introduced at the end of that chapter to offer a different way of looking at eigenvalues.

For traffic flow, shallow-water theory, vibrating strings, and vibrating membranes, the relevant mathematical models can be formulated without a great deal of background material. This is not the case for most wave phenomena, in electromagnetics and geophysics for example. However, one-dimensional problems of longitudinal wave motion, which provide some contrast to the transverse wave motion in Chapters 8 and 9, will be discussed through the exercises of the next three chapters.

7

The Shock of the Crash

Traffic Flow on a Long and Crowded Road

Concepts and solution techniques for a single first-order partial differential equation in two variables are introduced. The method of characteristics is developed in some detail. No prior knowledge of PDE is assumed.

7.1 Eulerian Formulation of Traffic Problems

In the theory of "car-following," attention was focused on the motion of individual cars. Each car was identified by its initial position at some reference time $t = 0$ as no two cars occupy the same position at the same time, at least initially. Certain relationships among the position, velocity, and acceleration of these vehicles were postulated, for example, the assumption of velocity control with a lag response time as in (6.7). Starting with their (known) initial position and velocity, the mathematical problem resulting from these postulates determines the evolution of these kinematic quantities with time and thereby predicts the motion of the individual vehicles for all later time. Such a microscopic approach to (or a Lagrangian formulation of) the problem of vehicular movement has generated a considerable amount of information concerning the mechanisms governing the traffic-flow phenomenon. However, for long roads with a large number of cars (particularly during rush hours), this microscopic approach becomes impractical since there are thousands of differential-difference equations of motion to be solved by the continuation process or other techniques in this case. For such a situation, a macroscopic approach similar to the Eulerian formulation of fluid mechanics would be more appropriate.

A macroscopic approach to traffic-flow problems does not keep track of the evolution of individual vehicular motion. Instead, it fixes our attention at different points along the road and describes what happens at each of these points as time goes by. It is like standing at the corner of Colorado Boulevard and Oak Street in Pasadena on New Year's Day to watch the Rose Parade with floats and marching bands passing by one at a time. [The reference to Pasadena seems particularly appropriate here since Professor Gerald Whitham, one of the two originators of this macroscopic theory of traffic flow (Lighthill and Whitham, 1955), is currently a member of the Department of Applied Mathematics at the California Institute of Technology located in that city.] By stationing yourself at one place, you will not be able to gather enough information to learn about the detailed movement of individual bands and floats in the parade. However, you can find out a great deal about the parade if you wish, by counting how many bands and floats pass by during consecutive half-hour intervals. This so-called *flow rate* (or *flux*) of bands and floats at one point in space contains much less information than the time history of individual cars but may be quite adequate for certain purpose such as getting an estimate of the pace of the parade. You will get a more accurate estimate if you station more people at different spots along the parade route all reporting their observed flow rate to you by walkie-talkie at half-hour intervals. In an *Eulerian formulation* of traffic flow, you will have, figuratively, people lining the entire road to observe the flow rate at every point of the road and reporting to you continuously for the duration of the whole parade. Now, you cannot ask for more than that!

In this chapter, we shall first make the above qualitative description of a macroscopic approach quantitative. We do so by formulating a *conservation law,* which in turn gives rise to an initial-value problem (IVP) in partial differential equations. The solution of this IVP can be obtained by the method of *characteristics*. The macroscopic theory will be used to analyze some typical problems associated with traffic movement on long crowded roads, for example, the effect of a bottleneck and the function of traffic lights. The phenomenon of *shock waves* will also be discussed. Applications of the theory developed here to other interesting problems can be found in Haberman (1977).

7.2 Conservation of Cars

Let us once more idealize the road as the x axis and introduce two quantities which characterize car movements on long and crowded roads: the *flow rate* (or *flux*) $q(x, t)$ and the *car density* (or *concentration*) $k(x, t)$.

The flow rate q(x, t) *at a point* x *and at an instant* t *is the number of cars passing through* x *per unit time at instant* t *with*

$$\int_{t_1}^{t_2} q(x, t)\,dt = \begin{array}{l} \textit{number of cars passing through } x \\ \textit{over the time interval } (t_1, t_2) \end{array} \qquad (7.1)$$

We take $q(x, t)$ to be positive if the flow is in the positive x direction. If q is continuous in $[t_1, t_2]$, we have, with $\Delta t \equiv t_2 - t_1$,

$$\int_{t_1}^{t_2} q(x, t)\,dt = q(x, \bar{t})\,\Delta t$$

for some $\bar{t},\ t_1 \leq \bar{t} \leq t_2$.

The car density *at point* x *and instant* t *is the number of cars per unit road length at time* t *and at point* x *on the road with*

$$\int_{x_1}^{x_2} k(x, t)\,dx = \begin{array}{l} \textit{number of cars in a stretch of the road} \\ (x_1, x_2) \textit{ at instant } t \end{array} \qquad (7.2)$$

If k is continuous in $[x_1, x_2]$, we have, with $\Delta x \equiv x_2 - x_1$,

$$\int_{x_1}^{x_2} k(x, t)\,dx = k(\bar{x}, t)\,\Delta x$$

for some $\bar{x},\ x_1 \leq \bar{x} \leq x_2$.

Since cars can neither be created nor destroyed along any stretch of the road without an entry or exit, we have the following *postulate* which states the fact that the number of cars must be conserved over a stretch of the road $[x_1, x_2]$:

$$\left\{ \begin{array}{l} \text{Number of cars entering the} \\ \text{stretch of road at } x_1 \text{ over} \\ \text{the period } (t_1, t_2) \end{array} \right\} - \left\{ \begin{array}{l} \text{number of cars leaving the} \\ \text{stretch of road at } x_2 \text{ over} \\ \text{the period } (t_1, t_2) \end{array} \right\}$$

$$= \left\{ \begin{array}{l} \text{number of cars in } (x_1, x_2) \\ \text{at time } t_2 \end{array} \right\} - \left\{ \begin{array}{l} \text{number of cars in } (x_1, x_2) \\ \text{at time } t_1 \end{array} \right\}$$

With the help of (7.1) and (7.2), the above *conservation law* can be expressed in terms of k and q as follows:

$$\int_{t_1}^{t_2} q(x_1, t)\,dt - \int_{t_1}^{t_2} q(x_2, t)\,dt = \int_{x_1}^{x_2} k(x, t_2)\,dx - \int_{x_1}^{x_2} k(x, t_1)\,dx$$

Under suitable differentiability conditions on k and q, we can write the above equation in the form

$$\int_{t_1}^{t_2} \int_{x_2}^{x_1} q_x(x,\,t)\,dx\,dt = \int_{x_1}^{x_2} \int_{t_1}^{t_2} k_t(x,\,t)\,dt\,dx$$

where a subscript on x or t indicates partial differentiation with respect to the subscripted variable. If k_t is continuous in both arguments, we can interchange the order of integration on the right to get

$$\int_{t_1}^{t_2} \int_{x_1}^{x_2} [k_t(x,\,t) + q_x(x,\,t)]\,dx\,dt = 0$$

The above equation must hold for any pair of $(t_1,\,t_2)$ and $(x_1,\,x_2)$ (provided that there is no entry or exit along the stretch of the road in question). The continuity of k_t and q_x requires that the integrand vanish identically so that

$$k_t(x,\,t) + q_x(x,\,t) = 0 \qquad (7.3)$$

Starting with an initial road configuration at a reference time $t = 0$, we are interested in the quantities k and q at all points of the road for later time. These quantities completely describe the macroscopic behavior of traffic on long and crowded roads. The theory developed so far has only one equation, namely (7.3), for the determination of these two quantities. We need a second equation to supplement (7.3), and we get it by postulating a definite relationship between q and k which can be found by controlled experiments. That is, there is an *equation of state* relating q and k, similar to the experimentally determined Boyle's law relating the volume, pressure, and temperature of a body of perfect gas. (Do you remember $pV = nRT$?) For many problems in traffic flow, a relation of the form

$$q = Q(k) \qquad (7.4)$$

provides an adequate first approximation of the exact equation of state. However, there is no universal agreement on the exact form of the functional $Q(\cdot)$. Possible forms of Q as well as equations of state more complex than (7.4) will be discussed in the next section.

For an equation of state of the form (7.4), we can eliminate q from (7.3) to get a single equation for $k(x,\,t)$ alone. With $q_x = Q'(k)k_x \equiv c(k)k_x$, we get from (7.3)

$$k_t + c(k)k_x = 0 \qquad c(k) = \frac{dQ}{dk} \qquad (7.5)$$

which is a PDE for k. It is a *first-order* PDE since no partial derivative of higher order than the first appears in the equation. If $c(k)$ is independent of k, then the PDE is *linear*, as the equation is linear in the unknown k and its derivatives. Otherwise, the PDE is said to be *nonlinear*. Equation 7.5 will be analyzed for specific forms of $c(k)$ in Sections 7.4–7.7.

For a complete determination of $k(x, t)$, the PDE (7.5) must be supplemented by some initial or boundary conditions, analogous to the ODE situation. For example, we often know the distribution of car concentration at some initial time $t = 0$ so that we have

$$k(x, 0) = k_0(x) \qquad (-\infty < x < \infty) \qquad (7.6)$$

The PDE (7.5) and the initial condition (7.6) define an initial-value problem. As we shall see later, such an IVP completely determines $k(x, t)$ once we have $c(\cdot)$ and $k_0(\cdot)$. If $k(x, 0)$ is known only for a finite or semi-infinite interval, k may have to be prescribed at one or more end points for all time in order that it be completely determined.

7.3   On the Flow-Density Relation

We postulated in Section 7.2 a definite relationship between the flow rate q and the traffic density k in the form $q = Q(k)$. Even without further theoretical or experimental study, we expect $Q(k)$ to vanish at $k = 0$ (no car, no flow!) and $k = k_j$ (j for "jam") with k_j being the density corresponding to a "bumper-to-bumper" traffic condition. Unlike $Q(0) = 0$ which follows from the definition of q and k, the property $Q(k_j) = 0$ is based on our observation of the behavior of human drivers. The flow rate at k_j is not exactly zero but is usually so small in practice that we might as well take it to be zero to simplify the mathematical problem. Between the two extremes, that is, $0 < k < k_j$, Q must be positive as we cannot have a negative flow rate for a unidirectional traffic flow. Therefore, we have $c(k) \equiv dQ/dk > 0$ near $k = 0$ and $c(k) < 0$ near $k = k_j$. Beyond these observations, there is some general agreement that we may take Q to be concave downward so that $d^2Q/dk^2 < 0$ for $0 < k < k_j$. It is felt that a more accurate description of $Q(k)$ may not be meaningful as long as we continue to ignore effects such as "looking more than one car ahead," etc. Putting all these observations together, we have the following rough sketch of $Q(k)$ (see Figure 7.1) where q reaches a maximum value q_m at $k = k_m$ with $q_m = Q(k_m)$.

It is easy to calculate k_j since we know the average car length and therefore how many cars a stretch of the road can hold, bumper-to-bumper. Roughly, we should allow about 25 ft/car (20 ft on the average for the car length in North America and another 5 ft of spacing at k_j) giving $k_j = 200$ cars/mile. In contrast to k_j, the numerical values for k_m and q_m and the qualitative features of the q versus k curve vary from road to road. As we

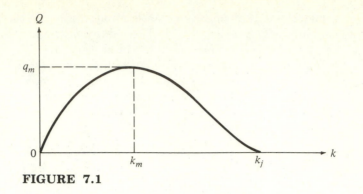

FIGURE 7.1

shall see later, the model developed in this chapter actually provides us with a relatively simple way of measuring k_m and q_m with the help of a traffic light. As far as the qualitative features of the q versus k curve are concerned, we shall make use of the results in Section 6.8 obtained from the microscopic type of analyses studied in Chapter 6 and gather data about a particular road to complete the picture.

To carry out the above program, we need the concept of a *velocity field* $u(x, t)$ for a heavily traveled road introduced in Section 6.8. With the car spacing $x_{n-1}(t) - x_n(t)$ taken to be the reciprocal of $k(x_n, t)$, the density at x_n and t, we obtained there a relation between $u(x, t)$ and $k(x, t)$ for several car-following models. Now, it is not difficult to see that

7A: Traffic flow equals traffic density times velocity so that

$$q = uk \tag{7.7}$$

To see this, consider the number of cars through point x in a small time interval $(t, t + \Delta t)$. In that small time interval, we have, to a good approximation, $u(x, t)\Delta t$ being the total length of cars passing by the observer. The actual number of cars passing by is approximately that length times the density or $[u(x, t)\Delta t]k(x, t)$. The number of cars passing through x per unit time, that is, the flow rate $q(x, t)$, is therefore $u(x, t)k(x, t)$.

It is evident from the above discussion that u and k must be smooth functions of x and t. We will assume this to be true in this chapter. It follows that the theory evolved from this chapter applies only to traffic which is heavy relative to the time scale involved.

Density-velocity relations were obtained for two car-following models in Section 6.8. One for the model (6.7) gives

$$u(x, t) = \lambda \left[\frac{1}{k(x, t)} - \frac{1}{k_j} \right] \tag{7.8}$$

Correspondingly, we have from (7.7)

$$q = uk = \lambda\left(1 - \frac{k}{k_j}\right) \tag{7.9}$$

where the sensitivity coefficient λ summarizes the special characteristics of a particular road and is to be determined by observing the behavior of drivers on the road. Unfortunately, the result based on the model (7.9) is not satisfactory because $q\ (k = 0) \neq 0$ and $u\ (k = 0) = \infty$. In reality, speed is limited by laws, by the power available in the car engine, and by the driver's interest in his/her own safety. A simple remedy for the fact $Q(0) \neq 0$ adopted by many investigators is to use (7.8) for $k > k_c$ and $u = u_{max}$ for $k < k_c$, keeping in mind that our macroscopic theory is not applicable for extremely low density flow anyway. The threshold value k_c may be set to 50 cars/mile and taken as the lower limit of applicability for a continuum theory. The resulting $Q(k)$ from such a modification agrees extremely well with some observations conducted by B. D. Greenshield in 1935. Unfortunately, Greenshield's data [see reference given in Lighthill and Whitham (1955)] are rather sparse and there is a large intermediate range of k where there are no observations.

A more fundamental remedy of the model (6.7) is to allow the sensitivity factor λ to vary with the spacing between cars. It is reasonable to assume that the driver will be more sensitive to the difference between his/her speed and that of the car in front if the two cars are close to each other. We may therefore take $\lambda = \mu/(x_{n-1} - x_n)^2$ giving us the model in Equation 6.57 with μ summarizing the characteristics of the road and the driver. For this model, we have

$$u(x, t) = \mu(k_j - k) \tag{7.10}$$

and from (7.7), we have

$$q = \mu k(k_j - k) \equiv Q(k) \tag{7.11}$$

which is the parabolic curve we sought for $Q(k)$ with $Q(0) = Q(k_j) = 0$ and $Q > 0$ for $0 < k < k_j$. It is not difficult to check that $k_m = k_j/2$ and the maximum value of q is $q_m = \mu k_j^2/4$ in this case. We shall see later how μ can be determined with the help of a traffic light.

The equation of state deduced from (6.57) fits available observational data such as those obtained by H. Greenberg for the Lincoln Tunnel, connecting New York and New Jersey, reasonably well (see p. 288 of Haberman, 1977). However, the data for u display some curvature while (7.10) is linear in k. Furthermore, $u(x, t)$ of our model does not level off at low density.

7.4   The Linear Initial-Value Problem

We consider in this section the simplest (and not particularly realistic) traffic-flow problem consisting of the conservation law with the linear flow-density relation

$$q = 3000 - 15k \tag{7.12}$$

so that $c = -15$ mph. Note that (7.12) is a special case of (7.9). Together, they give the linear PDE with constant coefficients

$$k_t - 15k_x = 0 \qquad (-\infty < x < \infty, t > 0) \tag{7.13}$$

Given a particular initial density distribution $k_0(x)$, $-\infty < x < \infty$, at a reference time $t = 0$, that is, $k(x, 0) = k_0(x)$, we have an initial-value problem in PDE.

For the solution of this IVP, we introduce (with the help of hindsight) a new set of two independent variables $\xi = x + 15t$ and $\eta = x - 15t$. By the chain rule, we have

$$k_t = k_\xi \xi_t + k_\eta \eta_t = 15k_\xi - 15k_\eta$$

$$k_x = k_\xi \xi_x + k_\eta \eta_x = k_\xi + k_\eta \tag{7.14}$$

With (7.14), the conservation law (7.13) can be written as

$$k_t - 15k_x = -30k_\eta = 0 \tag{7.15}$$

In the form (7.15), the conservation law implies that k is independent of η and hence

$$k = f(\xi) = f(x + 15t) \tag{7.16}$$

In other words, the general solution of the PDE (7.13) is any differentiable function of $x + 15t$. The solution of the IVP is obtained by noting that $k(x, 0) = f(x) = k_0(x)$ so that $f(x + 15t) = k_0(x + 15t)$ and

$$k(x, t) = k_0(x + 15t) \tag{7.17}$$

For example, if

$$k_0(x) = \begin{cases} 1 + \epsilon(1 - |x|) & (|x| \le 1) \\ 1 & (|x| > 1) \end{cases} \tag{7.18}$$

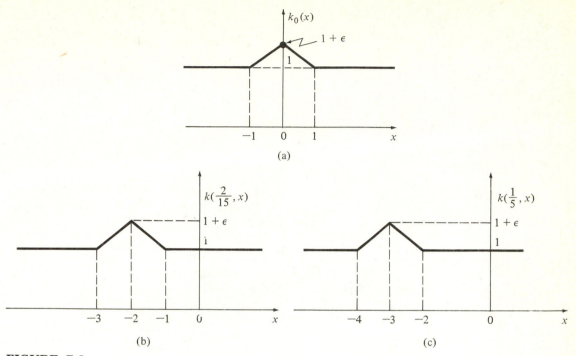

FIGURE 7.2

as shown in Figure 7.2a, we then get

$$k(x, t) = \begin{cases} 1 + \epsilon(1 - |x + 15t|) & (|x + 15t| \le 1) \\ 1 & (|x + 15t| > 1) \end{cases} \tag{7.19}$$

Note that for a later time t, the graph of $k(x, t)$ is simply $k_0(x)$ translated without changes by $15t$ units of distance to the left. In particular, the traffic hump moves to the left with a constant speed of 15 mph as shown in parts (b) and (c) of Figure 7.2. An important observation for later references is that $k(x,t)$ remains constant along any straight line with slope $-\frac{1}{15}$ in the t versus x plane (see Figure 7.3). For example, we have $k(x,t) = k(x_0, 0) = k_0(x_0)$ along $x + 15t = x_0$. These straight lines are called the *characteristics* or, more accurately, the *characteristic projections* of the PDE (7.13). More generally,

7B: The straight lines $x - c_0t = x_0$ are characteristic projections of the first-order linear PDE with constant coefficients $k_t + c_0k_x = 0$, and k_0 remains constant along these lines. The general solution of this

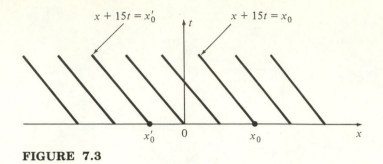

FIGURE 7.3

> equation is $k(x, t) = f(x - c_0 t)$ with the arbitrary function $f(\cdot)$ determined by the initial condition. For $c_0 > 0$, an initial traffic hump will travel to the right (without distortion) with a *wave speed* equal to c_0.

The development above suggests that we seek the solution of a higher-order linear PDE with constant coefficients in the form $f(x + \alpha t)$. This observation will be useful in the next chapter.

7.5   The Nonlinear Initial-Value Problem

We now turn to the more realistic (and more difficult) case of a nonconstant wave speed c. Typically, we have $c = c(k)$ for traffic flow with (7.11) as the equation of state relating flow to density. To be concrete, we consider the IVP:

$$k_t + c(k) k_x = 0 \qquad k(x, t = 0) = k_0(x) \qquad (7.20)$$

where $c(k) = 3(100 - k)/5$. This particular wave speed corresponds to $q = 3k(200 - k)/10$ with $k_j = 200$ cars/mile and $q_m = 3000$ cars/h.

To develop a method of solution for this type of nonlinear IVP, we will be guided by the results for the linear case and look for the locus in the x,t plane along which k remains constant. These plane curves are also called characteristic projections or the *characteristic base curves* of the given PDE. Suppose there is such a characteristic base curve, given by $x(t)$. Along this curve, we have $k(x, t) = k(x(t), t)$, and

$$\frac{dk}{dt} = k_x \frac{dx}{dt} + k_t \qquad (7.21)$$

Now, k will be constant along the characteristic base curve if $x^{\cdot} = c(k(x, t))$ for then the right side of (7.21) is equal to zero by the PDE in (7.20). In other words, we have along a characteristic base curve

$$\frac{dx}{dt} = c(k) \quad \text{and} \quad \frac{dk}{dt} = 0 \qquad (7.22)$$

with

$$x(0) = x_0 \quad \text{and} \quad k(x(0), 0) = k(x_0, 0) = k_0(x_0) \qquad (7.23)$$

Evidently, (7.22) and (7.23) define an IVP for a system of two ODE with x_0 as a parameter. The solution of this problem is immediate. The second equation in (7.22) gives

$$k = k_0(x_0) \qquad (7.24)$$

where we have used the fact $k = k_0(x_0)$ at $t = 0$. The first equation then gives

$$x = c(k_0(x_0))t + x_0 \qquad (7.25)$$

We get the solution of the original IVP in PDE by eliminating x_0 from the two equations (7.24) and (7.25).

For the special case where c is a constant ($c = c_0$ corresponding to the linear IVP), we have from (7.25) $x_0 = x - c_0 t$. Upon substituting this expression into (7.24) for x_0, we get again $k = k_0(x - c_0 t)$ as in (7.17) where $c_0 = -15$.

To illustrate the nonlinear case, consider a problem with $k_0(x) = 1 - x$ [and $c(k) = 3(100 - k)/5$]. For this problem, (7.25) and (7.24) become

$$x = [60 - \tfrac{3}{5}(1 - x_0)]t + x_0 \qquad k = 1 - x_0 \qquad (7.26)$$

The first equation in (7.26) can be solved for x_0 in terms of x and t to get

$$x_0 = \frac{x - 3(99t)/5}{1 + 3t/5} \qquad (7.27)$$

This result is then used to eliminate x_0 from the second to get

$$k = \frac{1 - x + 60t}{1 + 3t/5} \qquad (7.28)$$

For another $k_0(x)$, it is often not possible to eliminate x_0 from (7.24) and (7.25) to get $k(x, t)$. In this sense, the solution cannot be given by a single formula as in the linear case. But this deficiency is merely cosmetic as (7.24) and (7.25) give a parametric representation of the exact solution; the value of k at any point in the x,t plane can be calculated from them without any difficulty. In particular, we see that for each value of x_0, (7.25) gives a straight line in the x,t plane so that

7C: The characteristic base curves are straight lines (as in the linear case) along which the traffic density remains unchanged.

Equation (7.24) gives the value of k along each of these lines.

Since k is constant along a characteristic base curve, the locus of k above the characteristic base curves is also a straight line parallel to the x,t plane. Such a locus is called a *characteristic* of the original PDE in (7.20). There is one such characteristic for each x_0; together, these characteristics form a surface in the (x, t, k) space which is the solution surface $k(x, t)$ of the IVP (7.20).

The above concept of characteristics for PDE is very important as it allows us to develop a method for solving IVP involving more general PDE. For more complex cases, k will not be constant along a characteristic and the characteristic base curves will not be straight lines as in the traffic-flow problem. A description of the general method of characteristics is given in the appendix of this chapter. It should be noted that the characteristic space curve $k(x(t), t)$ and its projection in the x,t plane, that is, the characteristic base curve $t(x)$, are both called a characteristic of the PDE in many references. While this practice occasionally causes minor confusion, the meaning is usually clear enough from the context for us to do the same at times.

7.6 The Green Light Problem

A traffic light at $x = 0$ is red for $t < 0$ so that $k = k_j$ for $x < 0$ and $k = 0$ for $x > 0$. Of course, $q = 0$ for $-\infty < x < \infty$ in that case. Suppose the light turns green at $t = 0$; we know from our daily experience that the traffic begins to move across $x = 0$ (from left to right) and the car density begins to thin out near $x = 0$. As time increases, more and more of the car platoon will be in motion so that the head of that part of the car queue which remains motionless recedes further and further to the left. Now, if the Lighthill-Whitham theory is at all useful, it should at least be able to predict this phenomenon. In other words, the solution of the appropriate IVP in that theory should contain all the features of the flow of traffic described above.

The relevant IVP problem of course involves the PDE in (7.20) for $-\infty < x < \infty$ and $t > 0$. To be concrete, we will again use a quadratic equation of state $q = 3k(200 - k)/10$ so that $c(k) = 3(100 - k)/5$. At $t = 0$, when the light turns green, we have (see Figure 7.4a)

$$k(x,\,0) = k_j H(-x) = 200 H(-x) = \begin{cases} 200 & (x < 0) \\ 0 & (x > 0) \end{cases} \tag{7.29}$$

To solve the IVP defined by (7.20) and (7.29) with the prescribed $c(k)$, we set up the characteristic IVP:

$$x^{\cdot} = c(k) \qquad k^{\cdot} = 0 \tag{7.30}$$

with

$$x(0) = \eta \quad \text{and} \quad k(x(0), 0) = k(\eta, 0) = k_0(\eta) = 200 H(-\eta) \tag{7.31}$$

$-\infty < \eta < \infty$, where we have used η (instead of x_0) for a point along the road at $t = 0$. The solution of (7.30) and (7.31) is

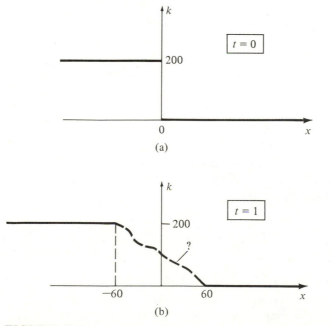

(a)

(b)

FIGURE 7.4

$$x = c(k_0)t + \eta = \tfrac{3}{5}[100 - k_0(\eta)]t + \eta$$
$$\qquad\qquad\qquad\qquad\qquad\qquad (-\infty < \eta < \infty) \qquad (7.32)$$
$$k = k_0(\eta) = 200H(-\eta)$$

While we cannot use the first relation in (7.32) to eliminate η from the second to get an explicit solution in the form of $k = K(x, t)$, we do have the solution in parametric form. We note in particular that

1. The solution (7.32) satisfies the initial condition (7.29) at $t = 0$.
2. The characteristic base curves are straight lines and k is constant along these curves whatever form the initial data may be.

The second feature is the same one we found earlier in the linear case when $c = c_0$ (a constant). The main difference between the linear and the nonlinear problem lies in the slope of the characteristic base curves in the x,t plane. For the nonlinear case, the slope may vary from characteristic to characteristic.

To see the actual distribution of car density along the road for $t > 0$, consider the instant $t = 1$ for which

$$x = 60 - 120H(-\eta) + \eta = \begin{cases} -60 - |\eta| & (\eta < 0) \\ 60 + \eta & (\eta > 0) \end{cases} \qquad (7.33)$$

$$k = 200H(-\eta) = \begin{cases} 200 & (\eta < 0) \\ 0 & (\eta > 0) \end{cases}$$

The two equations in (7.33) may be combined to give

$$k(x, t = 1) = \begin{cases} 200 & (x < -60) \\ 0 & (x > 60) \end{cases} \qquad (7.34)$$

Nothing is said about the interval $-60 < x < 60$ (see Figure 7.4b)!
It is not difficult to see that for any $t > 0$, we have from a similar analysis

$$k(x, t) = \begin{cases} 200 & (x < -60t) \\ 0 & (x > 60t) \end{cases} \qquad (7.35)$$

and nothing is said about the solution in the interval $-60t < x < 60t$! A little reflection suggests that our difficulty lies in the fact that the initial condition $k_0(\eta)$ is discontinuous at $\eta = 0$ while the use of the conservation law in the form (7.3), (7.5), or (7.20) requires that k be differentiable.

To remove this difficulty, let us approximate the actual discontinuous initial distribution of k by a slightly distorted but now continuous distribution, namely (see Figure 7.5),

$$k(x, 0) = k_0(x) = \begin{cases} 200 & (x < -\epsilon) \\ 100(1 - x/\epsilon) & (-\epsilon < x < \epsilon) \\ 0 & (x > \epsilon) \end{cases} \tag{7.36}$$

We shall see that the corresponding solution for the IVP (7.20) and (7.36) is physically acceptable even in the limiting case with $\epsilon = 0$. Furthermore, any other monotone decreasing behavior in the initial distribution of k from k_j to 0 in a narrow interval near $x = 0$ gives the same solution for $k(x, t)$.

For the specific modified initial condition (7.36), we have as the solution of the IVP

$$x = [60 - \tfrac{3}{5}k_0(\eta)]t + \eta = \begin{cases} -60t + \eta & (\eta < -\epsilon) \\ (60t + \epsilon)(\eta/\epsilon) & (|\eta| < \epsilon) \\ 60t + \eta & (\eta > \epsilon) \end{cases}$$

$$k = k_0(\eta) = \begin{cases} 200 & (\eta < -\epsilon) \\ 100(1 - \eta/\epsilon) & (|\eta| < \epsilon) \\ 0 & (\eta > \epsilon) \end{cases} \tag{7.37}$$

We can now express the solution (7.37) in the form $k = K(x, t)$:

$$k(x, t) = \begin{cases} 200 & (x < -60t - \epsilon) \\ 100[1 - x/(60t + \epsilon)] & (-60t - \epsilon < x < 60t + \epsilon) \\ 0 & (x > 60t + \epsilon) \end{cases} \tag{7.38}$$

and the solution is continuous for $-\infty < x < \infty$ (and $t > 0$) even in the limiting case $\epsilon = 0$. This solution has all the expected features described in

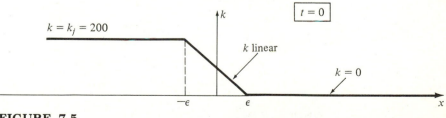

FIGURE 7.5

the introductory remarks of this section. Moreover, this solution predicts the particular (linear) way in which car density thins out with time to the left of the origin (Figure 7.6). It also shows how cars move to the right into the previously empty section of the road.

Note that the car density at the traffic light remains equal to $k_j/2 = 100$ for all times corresponding to the maximum flow for a quadratic equation of state with $q_m = 3000$ cars/h in our case. This particular observation is used in the experimental determination of q_m [and hence μ in (7.11)] in the equation of state.

> **7D:** To determine q_m, we simply measure the flow rate at the traffic light when the traffic is flowing smoothly.

To see that the solution (7.38) for $\epsilon = 0$ is independent of the particular form of the modified (continuous) initial condition (7.36), consider an arbitrary monotone decreasing k from k_j to 0 in the interval $-\epsilon < x < \epsilon$ (see Figure 7.7a). We still have Equations 7.30 which give

$$k = k_0(x_0) \quad \text{and} \quad x = c(k_0(x_0))t + x_0 \tag{7.39}$$

in which we have used x_0 instead of η as our parameter. The characteristic curves of the PDE in (7.20) in the x,t plane are therefore straight lines with slope $c(k_0(x_0))$ which depends on the initial location x_0. For $x_0 < -\epsilon$, we have $c(k_0(x_0)) = c(k_j) = -60$; so all the lines in that range of x_0 have the same slope, slanting to the left. For $x_0 > \epsilon$, we have $c(k_0(x_0)) = c(0) = 60$; so all the lines there have the same slope, slanting to the right. For $-\epsilon < x_0 < \epsilon$, $c(k_0(x_0))$ increases monotonically from -60 to 60; so the slope of the characteristic projections changes continuously from slanting to the left to slanting to the right. [The exact slope at each x_0 would depend on the shape of the initial data $k_0(x_0)$ but is not important in this development.] Altogether,

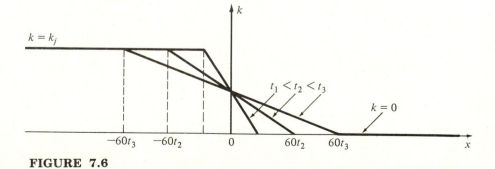

FIGURE 7.6

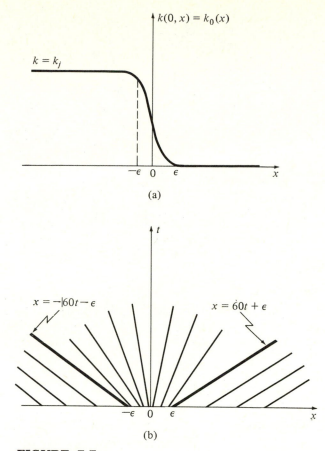

FIGURE 7.7

we have the picture of Figure 7.7b in the x,t plane. As $\epsilon \to 0$, the picture becomes as shown in Figure 7.8. In other words, we have

$$x = \begin{cases} x_0 - 60t & (x_0 < 0) \\ x_0 + 60t & (x_0 > 0) \end{cases} \qquad (7.40)$$

and between these two sets of parallel lines, there is a pencil of straight lines, emanating from the origin $x_0 = 0$ with slopes varying from -60 to 60, that is,

$$\frac{x}{t} = c = 60 - \tfrac{3}{5}k \qquad (-60 < c < 60) \qquad (7.41)$$

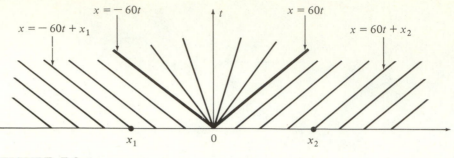

FIGURE 7.8

Now, we have k = constant along each of the lines (the characteristic projections or characteristic base curves of the PDE) in Figure 7.8. In particular, we have $k = k_j = 200$ along the lines from $x_0 < 0$ and $k = 0$ along the lines from $x_0 > 0$. In other words, we have

$$k = \begin{cases} 200 & (x < -60t) \\ 0 & (x > 60t) \end{cases} \qquad (7.42)$$

For the pencil of lines from the origin ($x_0 = 0$), we have from (7.41)

$$k = 100\left(1 - \frac{x}{60t}\right) \qquad (-60t < x < 60t) \qquad (7.43)$$

Equations 7.42 and 7.43, for which the monotone decrease of k from k_j to 0 is arbitrary in the initial condition, may be combined to give the same solution as (7.38) with $\epsilon = 0$.

Before leaving this problem, we should note the difference between the qualitative features of our solution and those of the corresponding solution for the linear IVP with $c = -15$. The latter solution (to be worked out as an exercise) simply shifts the entire initial distribution of traffic density by a distance equal to $15t$ to the left without distorting the shape of the distribution. This is not the case in the solution of the nonlinear problem.

7.7   A Traffic Hump and Shock Formation

Suppose the governing equations remain the same as those of Section 7.6, that is,

$$k_t + c(k)k_x = 0 \qquad c(k) = 60 - \tfrac{3}{5}k \qquad (7.44)$$

But now the initial density distribution at $t = 0$ is

$$k(x, 0) = k_0(x) = \begin{cases} 50 & (|x| > 1) \\ 25(3 - |x|) & (|x| \leq 1) \end{cases} \tag{7.45}$$

instead, so that we have a triangular traffic hump centered at $x = 0$. The solution for this problem is

$$k = k_0(x_0) = \begin{cases} 50 & (|x_0| \geq 1) \\ 25(3 - |x_0|) & (|x_0| \leq 1) \end{cases} \tag{7.46}$$

$$x = \begin{cases} 30t + x_0 & (|x_0| \geq 1) \\ 15t(1 + |x_0|) + x_0 & (|x_0| \leq 1) \end{cases} \tag{7.47}$$

obtained by the method of Section 7.5 without any difficulty. To see how the density profile evolves for $t > 0$, we will look at three instants of time, $t = (1 - \epsilon)/15$, $t = \frac{1}{15}$, and $t = (1 + \epsilon)/15$ (for some $0 < \epsilon < 1$) which bring out significantly different qualitative features of our solution.

For $t < \frac{1}{15}$, we may use (7.47) to eliminate x_0 from (7.46) to get

$$k(t, x) = \begin{cases} 50 & (|x - 30t| \geq 1) \\ 25\left(3 - \dfrac{x - 15t}{1 + 15t}\right) & (15t \leq x \leq 1 + 30t) \\ 25\left(3 + \dfrac{x - 15t}{1 - 15t}\right) & (-1 + 30t \leq x \leq 15t) \end{cases} \tag{7.48}$$

In particular, we have for $t = (1 - \epsilon)/15$

$$k = \begin{cases} 50 & (x \geq 3 - 2\epsilon) \\ 25(7 - 4\epsilon - x)/(2 - \epsilon) & (1 - \epsilon \leq x \leq 3 - 2\epsilon) \\ 25(x - 1 + 4\epsilon)/\epsilon & (1 - 2\epsilon \leq x \leq 1 - \epsilon) \\ 50 & (x \leq 1 - 2\epsilon) \end{cases} \tag{7.49}$$

with its graph shown in Figure 7.9.

The solution (7.48) breaks down for $t = \frac{1}{15}$ and we must return to the parametric representations (7.46) and (7.47) which now take the form

$$k = \begin{cases} 50 & (|x_0| > 1) \\ 25(3 - |x_0|) & (|x_0| \leq 1) \end{cases} \tag{7.50}$$

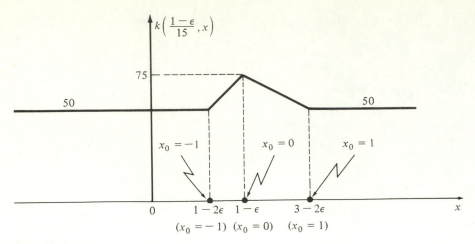

FIGURE 7.9

$$x = \begin{cases} 2 + x_0 & (x_0 > 1) \\ 1 + 2x_0 & (0 \le x_0 \le 1) \\ 1 & (-1 \le x_0 \le 0) \\ 2 + x_0 & (x_0 < -1) \end{cases} \tag{7.51}$$

The graph of $k(x, t = \frac{1}{15})$, as shown in Figure 7.10, may also be gotten from the graph for $t = (1 - \epsilon)/15$ in Figure 7.9 as ϵ tends to zero. Finally, we have from (7.47)

$$x = \begin{cases} 2(1 + \epsilon) + x_0 & (|x_0| > 1) \\ (1 + \epsilon) + (2 + \epsilon)x_0 & (0 \le x_0 \le 1) \\ (1 + \epsilon) - \epsilon x_0 & (-1 \le x_0 \le 0) \end{cases} \tag{7.52}$$

for $t = (1 + \epsilon)/15$. The graph for k from (7.52) and (7.46) is as shown in Figure 7.11 with $x = 1 + \epsilon$, $1 + 2\epsilon$, and $3 + 2\epsilon$ corresponding to $x_0 = 0$, -1, and 1, respectively! For a fixed $\epsilon > 0$, the density distribution given by (7.46) and (7.52) is multivalued in x between $1 + \epsilon$ and $1 + 2\epsilon$. More specifically, there are three different values for the car density at any point on the road between $x = 1 + \epsilon$ and $1 + 2\epsilon$. This is not possible in actual traffic-flow situations since the car density at any point on the road is a well-defined physical quantity and must have a unique value! Something is wrong with the theory; it must have left out some factors which influence the traffic-flow phenomenon in a significant way for the traffic hump problem. We

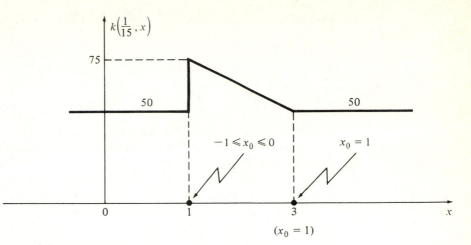

FIGURE 7.10

should look for an improved mathematical model and hence a reformulation of the problem.

To track down the deficiency of the Lighthill-Whitham theory for the traffic hump type problems, let us examine the solution a little more closely. For the present problem, the velocity field is

$$u = \frac{q}{k} = 60 - \tfrac{3}{10}k \qquad (7.53)$$

which says the car speed is slower at points on the road with higher car density, a very reasonable result. As high-speed cars catch up with low-speed

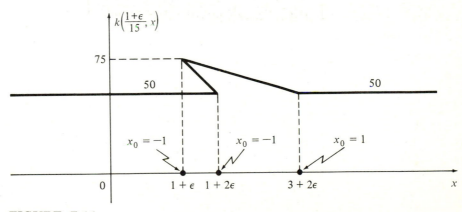

FIGURE 7.11

cars at high-density locations, more and more cars accumulate behind the location of $k_{\max}$, and, in time, shift the location of $k_{\max}$ further to the left (in a moving frame of constant speed 30 mph) with a steeper rise from 50 cars/mile to 75 cars/mile. This in turn leaves less time for a car coming into the (new) traffic hump to slow down; change in speed becomes more abrupt and a higher deceleration is required of these cars. At $t = \frac{1}{15}$, cars arriving at the traffic hump must make a finite speed change instantaneously (or infinite deceleration), giving rise to a vertical slope in the density profile at the hind side of the traffic hump. Such a vertical rise in the (density) profile is referred to as a *shock wave* or simply a *shock* and the development of a multivalued profile for $t > \frac{1}{15}$ is called *shock breaking* (similar to the breaking of a wave as it runs up a sloping beach). Thus, the traffic hump problem shows that

> **7E:**   When the wave speed c depends on k, shocks may form in the traffic density profile and, if allowed to continue in time, eventually break.

Since infinite deceleration is not possible, the mathematical results cannot be meaningful at and beyond the instant of shock formation (at $t = \frac{1}{15}$ for our problem). Indeed, the solution given by (7.46) and (7.47) is well defined only for $t < \frac{1}{15}$.

With the above interpretation of our mathematical results, we now see that the unrealistic phenomenon of *shock formation* and then its *breaking* (with the multivalued profile being similar to the breaking of a wave) are definitely connected with the equation of state employed in our model. By postulating $q = Q(k)$ and therefore $u = Q(k)/k \equiv U(k)$, we have, in effect, assumed that car speed at any position along the road depends only on the car density at the same point in space and time. In real life, drivers look out for traffic conditions ahead and slow down if the traffic ahead is heavy and slow (even if the traffic nearby is neither heavy nor slow). Such a driving habit eliminates the need for an "abrupt" change in speed at the last minute and must be included in the equation of state if we want to eliminate the occurrence of the shock formation and breaking phenomenon which is unrealistic for the traffic flow problem.

In Chapter 10 we will formulate a more adequate theory of traffic flow, taking into account the "looking ahead" effect which is an important factor, at least when there is a "large" change in k over a "short" distance. The mathematical problem associated with this improved theory will be more difficult than the one in this chapter. It is therefore important to note that the present theory should be completely adequate for many problems (e.g., the green light problem) and adequate for other problems for an interval of time (e.g., $0 \le t < \frac{1}{15}$ for the traffic hump problem). As long as sensitivity to heavy traffic ahead is associated only with a significant change over a short distance, we may continue to use the simple Lighthill-Whitham theory of the present chapter and avoid the more complicated improved theory. It is understood that we do not take the results literally when shocks begin to form but should know how to spot the development of shocks (by **7H** below) and obtain information about the traffic flow after shock formation (see Chapter 10).

It is possible to determine whether a shock develops in time without working out the distribution of $k(x, t)$ for different times. Let us look at the characteristic base curves in the x,t plane given by

$$x = \begin{cases} 30t + x_0 & (|x_0| \geq 1) \\ 15t(1 + |x_0|) + x_0 & (|x_0| \leq 1) \end{cases} \qquad (7.54)$$

The graphs of these characteristic base curves for this traffic-flow problem are straight lines, as shown in Figure 7.12. Since the slope of all the straight lines starting at $x \leq -1$ (when $t = 0$) are less steep than those starting in the interval $(-1, 1)$, the two sets of lines will intersect at some $t > 0$. In particular, the earliest instant for an intersection of two lines is at $t = \frac{1}{15}$; in fact, all lines starting at x in the interval $(-1, 0)$ intersect at the same time $(t = \frac{1}{15})$ and at the same position $(x = 1)$, and all other intersections occur at later times. But we know k is constant along any characteristic and is equal to its value at the corresponding initial position. It follows that $k(\frac{1}{15},1)$ must be simultaneously $k_0(x_0) = 25(3 - |x_0|)$ for $-1 \leq x_0 \leq 0$, and is therefore multivalued there. In other words, a shock is formed at $x = 1$ and $t = \frac{1}{15}$, the earliest instant when the characteristic base curves intersect. It is not difficult to see now that

7F: The intersection of two or more characteristic base curves is generally a warning of possible shock formation at a point in the space-time plane given by the position of the intersection.

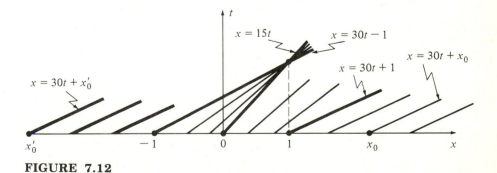

FIGURE 7.12

If a shock does form, we should stop using the Lighthill-Whitham theory at the earliest instant of intersection. How to obtain information about the traffic flow beyond that point will be discussed in Chapter 10.

EXERCISES

1. Show by the methods described below that any (differentiable) solution of the equation $k_t + c_0 k_x = 0$, where c_0 is a known constant, must be of the form $k(x, t) = f(x - c_0 t)$:

    (a) Let $u = x - c_0 t$ and $v = x + c_0 t$. Show that $k_t + c_0 k_x = 0$ is equivalent to $k_v = 0$ so that $k = f(u) = f(x - c_0 t)$.
    (b) Repeat part (a) but now with $v = x$.
    (c) Would you expect the result to be different if you had set $v = t$ instead?

2. (a) Show that any solution of the linear PDE with constant coefficients

    $$k_t + c_0 k_x = ak$$

    where a and c_0 are known constants, must be of the form $k(x, t) = e^{at} f(x - c_0 t)$. (*Hint:* Let $u = x - c_0 t$ and $v = t$ be two independent variables and proceed as in Exercise 1.)
    (b) Find the solution of

    $$k_t + c_0 k_x = -b^2 k \qquad (-\infty < x < \infty, t > 0)$$

    with

    $$k(x, 0) = \sin x \qquad (-\infty < x < \infty)$$

    where c_0 and b are known constants.

3. With the help of a few graphs, describe the behavior of the solution for $t > 0$ of the initial-value problem:

    $$k_t + c_0 k_x = 0 \qquad (-\infty < x < \infty, t > 0)$$

    $$k(x, 0) = k_0(x) = \begin{cases} 1 - |x| & (-1 \le x \le 1) \\ 0 & (|x| > 1) \end{cases}$$

    (*Note:* Graph the two cases $c_0 < 0$ and $c_0 > 0$ separately.)

4. Solve the following problems by the method of Lagrange:

    (a) $x^2 z_x + y^2 z_y = (x + y)z, \quad z(1, y) = 1$
    (b) $-t z_t + x z_x = xtz, \quad z(x, 1) = \sin x$

(c) $z_t + zz_x = 0, \ z(x, 0) = (1 - x)$

(d) $(x - y)z_x + (y - x - z)z_y = z, \ z = 1$ along $x + y = -1$

5.   Obtain and graph the solution for $t > 0$ of the IVP

$$k_t + c(k)k_x = 0 \qquad k(x, 0) = 50H(-x)$$

where $c(k) = 60 + 0.6k$ and $H(z)$ is the Heaviside unit step function.

6.   **(a)**   Obtain the solution of the IVP

$$k_t + c(k)k_x = 0$$

$$k(x, 0) = \begin{cases} 150 & (|x| > 1) \\ 150[1 + (1 - |x|)/5] & (|x| \le 1) \end{cases}$$

where $c(k) = 60 - 0.6k$. [*Note:* There is a difference in sign in the expression for $c(k)$ between this and the previous problem.]

(b)   Graph the distribution of k for values of t less than, equal to, and greater than $\frac{1}{18}$, say $\frac{1}{24}, \frac{1}{18},$ and $\frac{1}{16}$.

7. $xz_x + yz_y = -z \qquad (x > 0, y > 0)$

(a)   Solve the above PDE subject to the initial condition

$$z(x, 1) = x.$$

(b)   Show that there is more than one solution for the IVP with the initial condition: $x_0(\eta) = \eta, y_0(\eta) = \eta,$ and $z_0(\eta) = 1/\eta \ (0 < \eta < \infty)$.

(c)   Give a geometrical interpretation for the result of part (b) (in terms of the characteristics of the PDE).

(d)   You should expect no solution for the IVP if the initial condition is $x_0(\eta) = \eta, \ y_0(\eta) = \eta,$ and $z(\eta) = e^\eta (0 < \eta < \infty)$. Give an explanation (in terms of the characteristics of the PDE).

8.   *The Red-Light Problem.* Suppose the traffic is flowing smoothly with $k(x, t) = 50$ for $t < 0$ and suppose $q(k) = k(60 - 0.3k)$. At $t = 0$, a traffic light located at $x = 0$ turns red.

(a)   Based on your daily experience and/or intuition, sketch a typical graph of $k(x, t_1)$ for some $t_1 > 0$ and $-\infty < x < \infty$.

(b)   The graph of part (a) should have one or more shocks. Calculate the *shock speed* for each shock.

(c)   Graph the (space-average) speed for individual cars, $u = q/k$, as a function of x at the same instant which gives the graph of part (a).

(d) In a t,x plane (with t vertical), graph the position of the shock(s) as a function of time. The graph is called the *shock locus;* it divides up the t,x plane into several regions. Sketch the characteristic base curves in the different regions.

Remark: You should also think about the way to obtain an analytic solution of this problem which involves a *boundary condition* at $x = 0$, namely $k(0, t) = k_j = 200$ for $t > 0$.

9. Shocks form at the first intersection of the characteristic base curves of a first-order PDE. This intersection gives the time t_s and position x_s of shock formation. But it is not always easy to locate such an intersection; an alternative analytical method for determining t_s and x_s would be useful. The following development outlines such a method: The solution of the IVP in the Lighthill-Whitham kinematic theory of traffic flow may be given in the parametric form $x = c(k_0(x_0))t + x_0$ and $k = k_0(x_0)$ $(-\infty < x_0 < \infty)$ where $k_0(x_0)$ is the initial car density distribution. From these we can get $k = k_0(x_0(x, t)) \equiv K(x, t)$ after solving the first equation for x_0 and then use the result in the second.

(a) Show that

$$k_x = \frac{k_0'(x_0)}{1 + \dfrac{dc}{dk_0} k_0'(x_0) t_s} \qquad (\ \ )' \equiv \frac{d(\ \ )}{dx_0}$$

Since k has a vertical slope at t_s and x_s, k_x becomes unbounded there. Show that

$$t_s = \min_{-\infty < x_0 < \infty} \frac{-1}{\dfrac{dc}{dk_0} k_0'(x_0)} \qquad \text{and} \quad x_s = c(k_0(\bar{x}))t_s + \bar{x}$$

where $\bar{x}$ is the value of x_0 for t_s.

(b) Find t_s and x_s for an initial distribution $k_0(x) = 50 + 25 \tanh x$ with $c(k) = 60 - 0.6k$.

(c) Find t_s and x_s for the triangular hump initial density distribution

$$k_0(x) = \begin{cases} 50 & (|x| > 1) \\ 25(3 - |x|) & (|x| \leq 1) \end{cases}$$

again with $c(k) = 60 - 0.6k$.

10. The basic equations of one-dimensional gas dynamics are

$$\frac{\partial u}{\partial t} + u \frac{\partial u}{\partial x} = -\frac{1}{\rho} \frac{\partial p}{\partial x} \tag{1}$$

$$\frac{\partial \rho}{\partial t} + u\frac{\partial \rho}{\partial x} = -\rho\frac{\partial u}{\partial x} \tag{2}$$

$$p = K\rho^{\gamma} \tag{3}$$

where u, ρ, and p are the velocity, density, and pressure of the gas, and K and γ in the equation of state (3) are known constants.

(a) Show that the system of equations (1), (2), and (3) possesses special solutions (representing simple waves) which satisfy the equation

$$\frac{\partial u}{\partial t} + V(u)\frac{\partial u}{\partial x} = 0 \tag{4}$$

where

$$V(u) = \pm c_0 + \tfrac{1}{2}(\gamma + 1)u \tag{5}$$

The density distribution is given through $\rho(c)$ by

$$c = c_0 \pm \tfrac{1}{2}(\gamma - 1)u \tag{6}$$

(b) Show that the general solution of Equation 4 is given by

$$u = F(x - V(u)t) \tag{7}$$

or

$$x = \phi(u) + V(u)t \tag{8}$$

where F and ϕ are arbitrary functions of a single variable.

11. In an infinitely long tube, gas initially at rest (in the region $x < 0$) and at constant density ρ_0 is separated from a vacuum (in the region $x > 0$) with $\rho = 0$ by a diaphragm at $x = 0$. Now the diaphragm is suddenly removed (but the side of the tube is sealed immediately). Use the equations of gas dynamics in Exercise 10 to determine the subsequent motion of the gas by following the steps outlined below.

(a) Does the gas move forward or backward?
(b) Does the *state* of expansion (or decompression, i.e., state of reduced density from ρ_0) propagate forward or backward?
(c) For the given problem, explain why we should use the solution (5) and (6) with the following choice of signs:

$$V(u) = -c_0 + \tfrac{1}{2}(\gamma + 1)u \tag{5'}$$

$$c = c_0 - \tfrac{1}{2}(\gamma - 1)u \tag{6'}$$

(d) From (5'), infer that the wavefront propagating into the initially compressed gas ($\rho = \rho_0$, $u = 0$) is given by the straight line

$$x = -c_0 t \tag{9}$$

(e) From (6') and (5'), infer that the wavefront propagating into the vacuum ($\rho = 0$) is given by

$$x = \frac{2c_0}{\gamma - 1} t \tag{10}$$

(f) From parts (d) and (e), infer that the solution to our problem is probably given by setting $\phi(u) \equiv 0$ in (8) and applying (8) to a proper domain. Confirm it by checking initial conditions and boundary conditions for the domain in question [see part (g)].

(g) Show that the flow field at any instant is given by

$$u = 0 \qquad\qquad (\text{for } x < -c_0 t)$$

$$u = \frac{2}{\gamma + 1}\left(\frac{x}{t} + c_0\right) \qquad \left(\text{for } -c_0 t < x < \frac{2c_0}{\gamma - 1} t\right)$$

$$u \text{ meaningless} \qquad \left(\text{for } \frac{2c_0}{\gamma - 1} < x\right)$$

(h) What is the value of ρ/ρ_0 in each of the above domains?

12. Let $q = (4q_m/k_j^2)(k_j - k)$ where q_m is the maximum flow rate and k_j is the jam density. At time zero, the car density is specified as $k(x, 0) = k_0 + \epsilon k_j f(x)$ with

$$f(x) = \begin{cases} 0 & (|x/D| > 1) \\ (1 - x/D)^2 & (|x/D| < 1) \end{cases}$$

If ϵ is a small number (e.g., $\epsilon = 0.1$), use linear theory to determine the density $k(x, t)$ at a later time. In particular, set $k_0 = 3k_j/4$ and find the wave speed; draw in the x,t plane the characteristics marking the front and back of the disturbance. On the same graph, plot the trajectories of the vehicles (which move with the space-average velocity) whose initial positions are $x = 0$ and $x = 2D$. Locate the trajectory of a pedestrian standing at $x = -D$. Repeat the exercise for $k_0 = \frac{1}{4}k_j$ and $k_0 = \frac{1}{2}k_j$. What is unusual about the last case?

13. Use the method of characteristics to find the exact nonlinear solution of Exercise 12. Discuss the accuracy of the linear solution and its range of validity. Is the nonlinear solution always meaningful?

14. Suppose $k(x, 0) = 0$ and

$$k(0, t) = \begin{cases} k_* t/\tau & (t \leq \tau) \\ k_* & (t \geq \tau) \end{cases}$$

where $k_* = k_j/2$ (k_j = the jam density). Find the solution of

$$k_t + \frac{4q_m}{k_j^2}(k_j - 2k)k_x = 0$$

for $t \geq 0$, $x \geq 0$. If $k_* > k_j/2$, is it still possible to prescribe the density $k(0, t)$?

15. At $t = 0$, a traffic light at $x = 0$ turns from red to green and a platoon of cars, including yours, sitting at $x < 0$ begins to move. For this road, we have $q = k(60 - 0.3k)$ and therefore $c = 60 - 0.6k$ mph. The light will turn red again at $t = T$ (and, for simplicity, stay red thereafter).

    (a) Graph $k(x, t)$ and $u(x, t)$ versus x for a typical t, $0 < t < T$. (*Hint:* $k_j = 200$.)
    (b) Graph $k(x, t)$ and $u(x, t)$ versus x for a typical $t > T$.
    (c) It is not difficult to show that

    $$u(x, t) = \begin{cases} 0 & (x < -60t) \\ 30 + x/2t & (-60t < x < 60t) \\ 60 & (x > 60t) \end{cases}$$

    If your initial position on the road is $X_0 (< 0)$, when will you begin to move? (*Hint:* Suppose you start moving at $t = t_0$, find t_0 in terms of X_0.)
    (d) Find your position $X(t)$ for $t_0 < t < T$.
    (e) If $T = 2$ min ($= \frac{1}{30}$ h) and you were $\frac{1}{2}$ mile back from the light at $t = 0$, will you make the light before it turns red? Explain your answer.

16. A traffic light is red for $0 < t \leq T$ and green at all other times. If $k(x, 0) = k_0$ and $q = 4q_m k(k_j - k)/k_j^2$, solve for $k(x, t)$ and determine the shock locus.

17. Let

$$k(x, 0) = \begin{cases} k_0 & (|x| \geq l) \\ k_0 + (k_1 - k_0)[1 - (x/l)^2] & (|x| \leq l) \end{cases}$$

where $k_1 < k_0$. If $q = 4q_m k(k_j - k)/k_j^2$, find the point (x_s, t_s) at which a shock forms in the traffic flow.

18.   If

$$k(x, 0) = \begin{cases} k_0 & (x \le 0) \\ \dfrac{k_0}{1 + (x/l)^2} & (x > 0) \end{cases}$$

and $q(k)$ is given in Exercise 17, find $k(x, t)$. (An implicit functional relationship is acceptable.) Formulate this problem using dimensionless variables.

19.

$$k(x, 0) = \begin{cases} \dfrac{k_0}{1 + (x/l)^2} & (x \le 0) \\ k_0 & (x > 0) \end{cases}$$

and $q = q(k)$ is as given in Exercise 17, find the point (x_s, t_s) at which a shock forms. Write the differential equation that governs the location of the shock. How would you solve this equation *if* you had to? Draw the characteristics in the x,t plane, indicate the shock locus, and roughly sketch the density $k(x, t)$ at some time *after* the shock forms.

20.   At time $t = 0$, the traffic pattern on a long highway consists of two sections of constant density joined by a shock located to the left of $x = 0$ and moving in the positive direction. If the traffic light at $x = 0$ turns and remains red, describe the resulting motion, assuming

$$q = \frac{4q_m}{k_j^2} k(k_j - k) \qquad k_0(x) = \begin{cases} k_0 & (x < 0) \\ k_1 & (x > 0) \end{cases}$$

with $(k_1 < k_0 < k_j/2)$.

21.   Find the solution of the following problems:

  (a) $x\psi_x - y\psi_y + \psi = 2x$, with $\psi = 0$ on $y = 1$ (curve Γ in Figure 7.13). Draw the characteristics which touch Γ.
  (b) $x^2\psi_x + xy\psi_y = \psi^2$, subject to $\psi = 1$ on $x = y^2$. Draw the characteristic curves which cross $x = y^2$.
  (c) $\psi_x + \psi\psi_y = -\psi$, with $\psi = y$ on $x = 0$.

22.   Let $q = 4q_m k(k_j - k)/k_j^2$ and

$$k(x, 0) = \begin{cases} k_0 & (x \le 0) \\ k_0 + (k_1 - k_0)(x/a) & (0 \le x \le a) \\ k_1 & (x \ge a) \end{cases}$$

with $k_0 < k_1 < \tfrac{1}{2}k_j$. Find the solution $k(x, t)$. In particular determine the point (x_s, t_s) of a shock formation and find the same shock locus thereafter. Sketch the solution at times $t_1 < t_s$ and $t_2 > t_s$.

23.

$$k_t + c_0 k_x = 0 \qquad (c_0 > 0)$$

$$k(x, 0) = 2 + \sin x \qquad (x > 0)$$

$$k(0, t) = 1 + \cos t \qquad (t > 0)$$

(a) Parametrize the two segments $\{t = 0, x > 0\}$ and $\{x = 0, t > 0\}$ as one continuous initial base curve $t = t(\eta)$ and $x = x(\eta)$ for $-\infty < \eta < \infty$ with $\eta > 0$ corresponding to the $t = 0$ segment and $\eta < 0$ to the $x = 0$ segment.

(b) Parametrize the initial data by writing the two auxiliary conditions in terms of η consistent with part (a).

(c) Obtain the solution of the initial-boundary value problem in parametric form, that is, obtain t, x, and k as functions of η and another parameter ξ with $\xi \geq 0$.

(d) Eliminate ξ and η to get $k(t, x)$.

24.

$$k_t + c(k) k_x = 0 \qquad c = 60 - 0.6k$$

$$k(x, 0) = 50 \qquad (x > 0)$$

$$k(0, t) = 25(2 + \sin t) \qquad (t > 0)$$

(a) Parametrize the initial base curve and the initial data as in Exercise 23.

(b) Obtain the solution of the initial-boundary value problem in parametric form consistent with part a.

(c) Sketch the characteristic base curves. Is it possible to have shocks in the region $x > 30t$ (≥ 0)? In the region $(0 \leq) x < 30t$? In $0 \leq t \leq \pi/2$?

(d) Sketch k for all $x \geq 0$ at $t = 0$, $\pi/6$, and $\pi/2$. Do not try to eliminate ξ and η to get $k(t, x)$.

(e) At the entrance $x = 0$, it is more natural to prescribe car flow than car density. Suppose $q(0, t) = 300(2 + \sin t)$ cars/h, determine the two possible $k(0, t)$ (which we need for the solution of the PDE) from the equation of state $q = 60k - 0.3k^2$. Which one is the correct $k(0, t)$ if you know $k(x, 0)$ to be of the order of 10 cars/mile.

25. Suppose the traffic density $k(x, t)$ along a long and crowded single-lane highway is obtained in parametric form:

$$x = (30 - 15 \tanh x_0)t + x_0 \qquad k = 50 + 25 \tanh x_0$$

(a) With the table below for $\tanh x_0$ (or any other means at your disposal), calculate a few values of $k(x, \frac{1}{5})$ and use them to sketch the corresponding density profile for all x. (*Note:* $x \simeq 6$ when $x_0 = 0$, -3, and 3.)

x_0	$\tanh x_0$
0	0
± 0.255	± 0.250
± 0.550	± 0.501
± 0.973	± 0.750
± 2.19	± 0.975
± 3.00	± 0.995
± 4.00	± 1.000

(b)   Insert a shock at an appropriate position x_s for the above profile. Describe your method.

(c)   Calculate the shock speed at that instant if $q = k(60 - 0.3k)$.

Appendix:

Lagrange's Method of Characteristics

The governing PDE, $k_t + c(k)k_x = 0$, for our traffic-flow problem is an equation for an unknown function k of two independent variables t and x. It is a *first-order* PDE since only first-order partial derivatives of the unknown appear in the equation. When $c(k)$ is a constant, the equation is said to be *linear;* no product of the unknown, differentiated or undifferentiated, appears in the equation. The most general first-order linear PDE of two independent variables is of the form

$$P(x, t)k_t + Q(x, t)k_x = R(x, t)k + S(x, t) \qquad (A7.1)$$

where P, Q, R, and S are known functions of t and x.

When $c(k)$ actually depends on k, the equation is said to be a first-order *quasilinear* PDE; it is not linear but contains no product of the derivatives of the unknown, for example, k_x^2, k_t^2, or $k_x k_t$. The most general first-order quasilinear PDE in two independent variables is of the form

$$P(x, t, k)k_t + Q(x, t, k)k_x = R(x, t, k) \qquad (A7.2)$$

In this section, a method of solution for (A7.2) will be described. Needless to say, the method also applies to the linear PDE (A7.1) which is just a special case of (A7.2) with P and Q independent of k and R linear in k.

The solution k of (A7.2) is evidently a function of two independent variables x and t, indicated by $k = K(x, t)$, and may be thought of as a surface in the x, t, k space. For example, the graph of the function

$$k = K(x, t) \equiv (1 + t)\sin x + 1 \qquad (-\infty < x < \infty, t \geq 0) \qquad \text{(A7.3)}$$

is roughly a slanted wavy roof in the t, x, k space. An alternative description of the *solution surface* of (A7.2) is

$$w(x, t, k) = w_0 \qquad \text{(a constant)} \qquad \text{(A7.4)}$$

For example, Equation A7.3 may be put in this form, say

$$w(x, t, k) \equiv k - (1 + t)\sin x - 1 = 0$$

or
$$\qquad \text{(A7.5)}$$

$$w(x, t, k) \equiv k - (1 + t)\sin x = 1$$

If we have the solution surface in the form (A7.4), we can usually solve for k in terms of x and t to get the solution of (A7.2) although sometimes we may not be able to execute the necessary calculations.

Whether in the form $k = K(x, t)$ or $w(x, t, k) = w_0$, the solution of (A7.2) is a function of x and t. Keeping this in mind, we get, with the help of the chain rule,

$$w_x + w_k k_x = 0 \quad \text{and} \quad w_t + w_k k_t = 0$$

or
$$\qquad \text{(A7.6)}$$

$$k_x = -\frac{w_x}{w_k} \quad \text{and} \quad k_t = -\frac{w_t}{w_k}$$

by differentiating (A7.4) partially with respect to x and t. In (A7.6), w_x, w_t, and w_k are partial derivatives of w with x, t, and k all treated as independent variables. If we now use (A7.6) to eliminate k_t and k_x from (A7.2), we get

$$P(x, t, k)w_t + Q(x, t, k)w_x + R(x, t, k)w_k = 0 \qquad \text{(A7.7)}$$

In its equivalent form (A7.7), the original PDE (A7.2) now has a simple geometric interpretation. Its left-hand side can be thought of as the dot product of two vectors $\mathbf{v}$ and ∇w where

$$\mathbf{v} = P\mathbf{i}_t + Q\mathbf{i}_x + R\mathbf{i}_k \qquad \nabla w = w_t\mathbf{i}_t + w_x\mathbf{i}_x + w_k\mathbf{i}_k \qquad \text{(A7.8)}$$

Equation A7.7 itself becomes

$$\mathbf{v} \cdot \nabla w = 0 \qquad \text{(A7.9)}$$

From vector calculus, we know ∇w is a vector normal to the solution surface $w(x, t, k) = w_0$ at the point (x, t, k). Therefore, (A7.9) says that for any point (x, t, k) on that surface, $\mathbf{v}$ is a vector tangent to the surface at that point. Starting with a fixed point (x_1, t_1, k_1) on the surface, there is a curve C_1 on the solution surface which has $\mathbf{v}$ as its tangent vector at any point along the curve, keeping in mind that the vector $\mathbf{v}$ itself depends on the point (x, t, k). We call such a curve on the solution surface a *characteristic curve* of the PDE (A7.2). For another initial point (x_2, t_2, k_2) on the solution surface, another such characteristic curve C_2 can be traced out on the surface. Evidently, the solution surface $w(x, t, k) = w_0$ itself is made up of such characteristic curves. To find the solution surface [which in turn gives us the solution $k = K(x, t)$], all we have to do is to find these characteristic curves and piece them together. In what follows, we describe an analytical method for constructing the characteristic curves of (A7.2) and combining them to get the solution $k = K(x, t)$.

A curve in the x, t, k space may be characterized by the parametric representation

$$x = x(\xi) \qquad t = t(\xi) \qquad k = k(\xi) \qquad (a \le \xi \le b) \qquad \text{(A7.10)}$$

or in the more compact vector notation $\mathbf{r}(\xi) = (x(\xi), t(\xi), k(\xi))$. The corresponding vector $d\mathbf{r}/d\xi$ is tangent to the curve at the point $\mathbf{r}(\xi)$. For the curve to be a characteristic curve of the solution surface [and, of the PDE (A7.2)], we want $d\mathbf{r}/d\xi = \lambda\mathbf{v}$, or

$$\frac{dt}{d\xi} = \lambda P(t, x, k) \qquad \frac{dx}{d\xi} = \lambda Q(t, x, k) \qquad \frac{dk}{d\xi} = \lambda R(t, x, k) \qquad \text{(A7.11)}$$

for an arbitrary multiplicative factor λ. We can use λ for scaling purposes; for example, choose $\lambda^2(P^2 + Q^2 + R^2) = 1$ to make ξ a measure of arc length. We will set $\lambda = 1$ for the subsequent development. Under suitable conditions on the functions P, Q, and R, we can solve (A7.11) to get a solution curve which satisfies the initial condition $\mathbf{r}(a) = \mathbf{r}_0 \equiv (x_0, t_0, k_0)$. We denote this characteristic curve by $\mathbf{r}(\xi; x_0, t_0, k_0)$ to indicate its dependence on the initial point $\mathbf{r}_0$. As we vary the initial point $\mathbf{r}_0$, we get different characteristic curves of the PDE (A7.2).

Now, if we manage to obtain a characteristic curve for every point on a given (noncharacteristic) curve Γ, we would have enough of them to make up a portion of the solution surface (see Figure 7.13). The curve Γ may also be prescribed in the form of a vector function $\mathbf{r}_0(\eta) = (x_0(\eta), t_0(\eta), k_0(\eta))$, $c \leq \eta \leq d$. For a particular value of $\eta = \bar{\eta}$, we solve (A7.11) to get a characteristic curve passing through the point $\mathbf{r}_0(\bar{\eta})$ on the curve Γ. We denote the characteristic curves for different values of η in (c, d) by $\mathbf{r}(\xi; x_0(\eta), t_0(\eta), k_0(\eta))$ or simply $\mathbf{r}(\xi, \eta) = (x(\xi, \eta), t(\xi, \eta), k(\xi, \eta))$ since for all initial points on Γ, it is really the value of η which specifies a particular initial condition and therefore a particular characteristic curve.

At this point, we have the solution of (A7.11) in the form

$$x = x(\xi, \eta) \qquad t = t(\xi, \eta) \qquad k = k(\xi, \eta) \tag{A7.12}$$

for $a \leq \xi \leq b$ and $c \leq \eta \leq d$. As we vary ξ, we have, for a fixed η, a characteristic curve of the solution surface $k = K(x, t)$ or $w(x, t, k) = w_0$ which passes through the point $(x_0(\eta), t_0(\eta), k_0(\eta))$ on Γ. And as we vary η, we get all the characteristic curves passing through all the points on Γ. The totality of these characteristic curves forms a surface which is the solution surface, or at least that portion of it which contains Γ. Indeed, Equations A7.12 themselves constitute a parametric representation of the solution surface.

In many cases, we can solve the first two equations in (A7.12) to get

$$\xi = \xi(x, t) \quad \text{and} \quad \eta = \eta(x, t) \tag{A7.13}$$

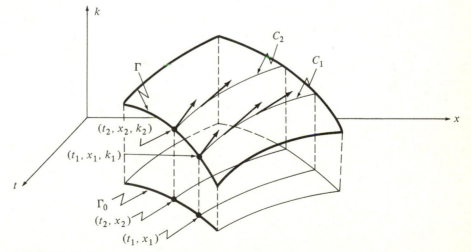

FIGURE 7.13

It is not always possible to do this, but if we manage to accomplish the inversion, we may then use (A7.13) to eliminate ξ and η from the last equation of (A7.12) to get

$$k = k(\xi(x, t), \eta(x, t)) \equiv K(x, t) \qquad (A7.14)$$

which is the more conventional form of the solution.

The above rather abstract description of Lagrange's method of solution for (A7.2) is meant to convey the essence of the method. A few examples and some exercises will show that there is nothing abstruse or difficult about the method itself. The difficulty lies in knowing when the inversion of $t = t(\xi, \eta)$ and $x = x(\xi, \eta)$ is theoretically possible.

8

It's a Bore

Shallow Water Waves

Concepts and solution techniques are developed for the linear wave equation and for a second-order quasi-linear system in one spatial variable.

8.1 Shallow Water Theory

With a little imagination or good computer graphics, we can almost see the propagation of a traffic-density wave (with a traffic hump) through heavy traffic on a long highway as discussed in the last chapter. However, there is nothing more demonstrative of wave propagation than the wave motion created by dropping a stone into still water. In fact, the very concept of wave motion was probably initiated by our experience with wave phenomena in water, whether it be the choppy waves in a relatively calm sea or the steepening waves moving up and breaking over a gently sloping beach. We would be somewhat remiss if something were not said in this book about the dynamics of water waves. More spectacular hydrophenomena include twisting tornados and roaring bores, the latter being a high and steadily advancing abrupt tidal wave in a channel. Just why these and other equally intriguing fluid motions occur in nature deserves a scientific explanation. Their possible significant impact on our daily life (such as weather, air travel, etc.) should be understood and exploited. *Fluid dynamics* is a branch of mechanics which investigates the physical behavior of water and other fluid media. It has been responsible for much of the development of (applied) mathematics over the years. For an introduction to this important source and catalyst for applied mathematics, we discuss in this chapter a class of problems in hydrodynamics known as shallow water waves.

Consider the motion of a body of water of infinite (horizontal) extent in the z direction but of finite depth in the vertical direction of gravity taken to be the y direction. There may or may not be any boundary in the third

(x) direction. We restrict our attention to fluid motion which is uniform in the z direction and for which the undisturbed depth of the water H is small compared to the distance L_0 over which some significant changes in the motion in the x direction, say the velocity u, take place. Typically, L_0 is taken to be the wavelength of the wave motion in the fluid (see Figure 8.1).

Let ρ be the mass density of water, $y = \zeta$ be the moving upper free surface measured from the undisturbed surface at $y = 0$, and $p(x, y, t)$ be the pressure at a point (x, y, z) in the fluid at time t. (It is known from elementary physics that pressure magnitude at any point in a fluid such as water is the same in all directions.) For the fluid phenomena analyzed in this chapter, we may take the velocity field to be independent of the depth coordinate y. Given that the fluid motion of interest here is independent of z, we have $u = u(x, t)$ and $\zeta = \zeta(x, t)$. In that case, force equilibrium in the y direction for any cylindrical column of water of base area A between $y = y_1$ and $y = y_2$ $(>y_1)$ requires a balance between the weight of the column and the vertical pressure differential between the bottom face $y = y_1$ and the top face $y = y_2$:

$$\iint_A [p(x, y_2, t) - p(x, y_1, t)]\, dx\, dz = \iint_A \int_{y_1}^{y_2} (-\rho g)\, dy\, dx\, dz$$

or

$$\iint_A \int_{y_1}^{y_2} \left(\frac{\partial p}{\partial y} + \rho g\right) dy\, dx\, dz = 0 \tag{8.1}$$

As the column may be arbitrarily chosen, the integrand must vanish for all x and t and $-H < y < \zeta$. Otherwise, we would have a contradiction by taking the column to cover a region where the integrand is of one sign only. Now, unlike materials such as rubber, water may be taken to be an *incompressible*

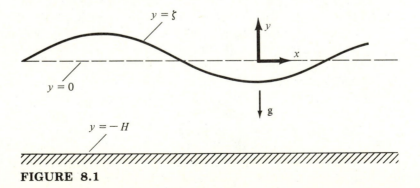

FIGURE 8.1

medium for our problems. However you may force a body of water to deform, the volume of that body remains (nearly) unchanged. Hence the density ρ is assumed to be a constant and we have from (8.1)

$$p = \rho g (\zeta - y) \tag{8.2}$$

if p is measured relative to the ambient (atmospheric) pressure above the free upper surface $y = \zeta$.

Water mass can neither be created nor destroyed. Consider a fixed volume $V \equiv \{z_1 \leq z \leq z_2, x_1 \leq x \leq x_2, -H \leq y \leq \zeta\}$. As water enters and leaves this column, the rate of water accumulation in V is

$$\frac{d}{dt} \int_{z_1}^{z_2} \int_{x_1}^{x_2} \int_{-H}^{\zeta} \rho \, dy \, dx \, dz = \Delta z \frac{d}{dt} \int_{x_1}^{x_2} \rho h \, dx$$

where $\Delta z \equiv z_2 - z_1$ and $h(x, t) \equiv \zeta + H$ is the height of water from the solid bottom of the disturbed free surface at point x and time t. At the same time, the net flux of water entering into V through its walls is

$$\left[\int_{z_1}^{z_2} \int_{-H}^{\zeta} \rho u \, dy \, dz \right]_{x=x_1} - \left[\int_{z_1}^{z_2} \int_{-H}^{\zeta} \rho u \, dy \, dz \right]_{x=x_2} = -\Delta z \left[\rho u h \right]_{x_1}^{x_2}$$

Note that there is no flux through the faces $y = -H$ and $y = \zeta$ while there is no net flux through $z = z_1$ and $z = z_2$. Conservation of water mass applied to this volume requires that the two expressions for water accumulation be identical so that

$$\frac{d}{dt} \int_{x_1}^{x_2} \rho h \, dx + \left[\rho u h \right]_{x_1}^{x_2} = 0 \tag{8.3}$$

Another conservation law needed in the analysis of water waves is Newton's second law of motion formulated for deformable media by Euler. The rate of increase of horizontal momentum (in the x direction) in our control volume V must equal the sum of the net influx of momentum into the volume and the net horizontal force acting on the column. The rate of increase of momentum is given by

$$\frac{d}{dt} \int_{z_1}^{z_2} \int_{x_1}^{x_2} \int_{-H}^{\zeta} \rho u \, dy \, dx \, dz = \Delta z \frac{d}{dt} \int_{x_1}^{x_2} \rho u h \, dx$$

The net influx of momentum through the faces $x = x_1$ and $x = x_2$ is given by

$$\left[\int_{z_1}^{z_2}\int_{-H}^{\zeta}(\rho u)u\,dy\,dz\right]_{x=x_1} - \left[\int_{z_1}^{z_2}\int_{-H}^{\zeta}(\rho u)u\,dy\,dz\right]_{x=x_2} = -\Delta z\left[\rho u^2 h\right]_{x_1}^{x_2}$$

It is customary (and permissible) to ignore the frictional force at the bottom $y = -H$ for our problems. In that case, the only contributions to horizontal forces come from the pressure at $x = x_1$ and $x = x_2$ so that the net horizontal force acting on the body is given by

$$\left[\int_{z_1}^{z_2}\int_{-H}^{\zeta}p\,dy\,dz\right]_{x_2}^{x_1} = -\left[\Delta z\int_{-H}^{\zeta}\rho g(\zeta - y)\,dy\right]_{x_1}^{x_2}$$

$$= -\left[\Delta z\rho g\left(\zeta y - \tfrac{1}{2}y^2\right)\Big|_{-H}^{\zeta}\right]_{x_1}^{x_2} = \left[-\tfrac{1}{2}\,\Delta z\rho g h^2\right]_{x_1}^{x_2}$$

Upon combining the above three effects, Newton's law on the horizontal momentum balance requires

$$\frac{d}{dt}\int_{x_1}^{x_2}\rho uh\,dx + \left[\rho u^2 h + \tfrac{1}{2}\rho g h^2\right]_{x_1}^{x_2} = 0 \tag{8.4}$$

The three conditions (8.2)–(8.4) for the pressure magnitude p, horizontal velocity u, and the free surface height h form the governing equations of what is known as *shallow water theory*. Nothing has been assumed about the smoothness of these three quantities in space. If u and h are continuous and have continuous first partial derivatives [in which case so is p by (8.2)], more familiar governing equations of the theory can be deduced from (8.2) to (8.4). We will carry out this deduction in the next section.

Many important problems in shallow water theory involve discontinuous solutions for u and h, at least at a few points in space. A typical situation is one where a free surface of the fluid ζ has a simple jump discontinuity at $x = x_s$ (Figure 8.2). The values of u and h on the two sides of this discontinuity are not equal and are sometimes indicated by $\{u+, h+\}$ and $\{u-, h-\}$, respectively. Under suitable conditions, the discontinuous free surface profile may move steadily and without alterations, say in the positive x direction. Such a steadily moving free surface profile with a jump discontinuity provides a mathematical idealization of a *bore*. We will return to a discussion of this phenomenon later.

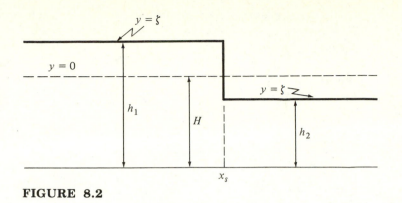

FIGURE 8.2

8.2   Smooth Solutions and Linearization

If u and h are both smooth (i.e., continuous and with continuous first partial derivatives), we may write (8.3) and (8.4) as

$$\int_{x_1}^{x_2} \left[\frac{\partial h}{\partial t} + \frac{\partial}{\partial x}(uh) \right] dx = 0 \tag{8.3'}$$

$$\int_{x_1}^{x_2} \left[\frac{\partial}{\partial t}(uh) + \frac{\partial}{\partial x}(u^2 h + \tfrac{1}{2}gh^2) \right] dx = 0 \tag{8.4'}$$

The interval of integration being arbitrary and the integrands being continuous require the integrands to vanish so that

$$h_t + (uh)_x = 0 \tag{8.5}$$

$$(uh)_t + (u^2 h + \tfrac{1}{2}gh^2)_x = 0 \tag{8.6}$$

We may use (8.5) to eliminate h_t from (8.6). So long as $h \neq 0$, we deduce from (8.6)

$$u_t + uu_x + gh_x = 0 \tag{8.7}$$

For a body of water of infinite extent in the x direction, $-\infty < x < \infty$, the two equations (8.5) and (8.7) completely specify the motion of the water whenever we know u and h at some instant in time, say at $t = 0$,

$$u(x, 0) = u_0(x) \qquad h(x, 0) = h_0(x) \qquad (-\infty < x < \infty) \tag{8.8}$$

If the body of water extends in the x direction only up to a boundary, say $x = x_0$, then some suitable boundary condition(s) will have to be prescribed there as well.

When the deviation from the undisturbed depth is small compared to the undisturbed depth H itself and steep gradients are not present in the wave motion, we have for a flat bottom (so that H is a constant)

$$h = (H + \zeta) = H\left(1 + \frac{\zeta}{H}\right) \simeq H \qquad h_t = \zeta_t \qquad h_x = \zeta_x \qquad (8.9)$$

If, in addition, $|u|$ is also small (on the basis of some dimensionless consideration), products of u and its partials may be neglected. Under these assumptions, we may linearize (8.5) and (8.7) to get

$$\zeta_t + Hu_x = 0 \qquad\qquad\qquad (8.10)$$

$$u_t + g\zeta_x = 0 \qquad\qquad\qquad (8.11)$$

These two equations are the governing differential equations for the small-amplitude theory of shallow water phenomena.

We may use (8.11) to eliminate u_x from (8.10) to get a single equation for ζ. From (8.11), we have $u_{tx} = -g\zeta_{xx}$ and from (8.10) we have $\zeta_{tt} = -Hu_{xt}$. If u has continuous mixed second derivatives, adding these two results gives

$$\zeta_{tt} = c^2\zeta_{xx} \qquad c^2 = Hg \qquad\qquad (8.12)$$

where c has the dimension of L/T and is in fact the wave speed of the problem. The partial differential equation (8.12) for ζ is called the one-dimensional *wave equation*. For other problems, it may be more convenient to work with a single second-order equation for u. Manipulations similar to those which lead to (8.12) allow us to use (8.10) to eliminate ζ_x from (8.11) to get

$$u_{tt} = c^2 u_{xx} \qquad\qquad\qquad (8.13)$$

This equation is of the same form as (8.12) and hence is also a one-dimensional wave equation. The wave equation appears in many different problems involving deformable media. We shall see the wave equation (8.13) again in the next chapter when we discuss vibrating strings and membranes.

The initial conditions (8.8) may be written in terms of ζ alone:

$$\zeta(x, 0) = h_0(x) - H \equiv \zeta_0(x)$$

$$\zeta_t(x, 0) = -Hu_x(x, 0) = -H[u_0(x)]_x \equiv v_0(x)$$
(8.14)

They may also be written in terms of u alone:

$$u(x, 0) = u_0(x)$$

$$u_t(x, 0) = -g\zeta_x(x, 0) = -g[h_0(x)]_x \equiv v_0(x)$$
(8.15)

The PDE (8.12) and the initial conditions (8.14) define an initial-value problem which determines $\zeta(x, t)$. Similarly, the PDE (8.13) and the initial conditions (8.15) also define an IVP for $u(x, t)$. Of course, we do not have to solve both IVPs. With the solution for either ζ or u, we can obtain the other from (8.10) or (8.11).

8.3   D'Alembert's Solution for the One-Dimensional Wave Equation

Our experience with the linear traffic-flow problem in Chapter 7 suggests that for the solution of the one-dimensional wave equation, we change the independent variables by setting $\xi = x - ct$ and $\eta = x + ct$. Observe that

$$\frac{\partial}{\partial x} = \frac{\partial}{\partial \xi} + \frac{\partial}{\partial \eta} \qquad \frac{\partial}{\partial t} = -c\frac{\partial}{\partial \xi} + c\frac{\partial}{\partial \eta}$$

$$\frac{\partial^2}{\partial x^2} = \left(\frac{\partial}{\partial \xi} + \frac{\partial}{\partial \eta}\right)^2 = \frac{\partial^2}{\partial \xi^2} + 2\frac{\partial^2}{\partial \xi\,\partial \eta} + \frac{\partial^2}{\partial \eta^2}$$
(8.16)

$$\frac{\partial^2}{\partial t^2} = \left(-c\frac{\partial}{\partial \xi} + c\frac{\partial}{\partial \eta}\right)^2 = c^2\left(\frac{\partial^2}{\partial \xi^2} - 2\frac{\partial^2}{\partial \xi\,\partial \eta} + \frac{\partial^2}{\partial \eta^2}\right)$$

In terms of ξ and η, the wave equation (8.13) becomes

$$4\frac{\partial^2 u}{\partial \xi\,\partial \eta} = 0$$
(8.17)

which may be integrated to get

$$u = f(\xi) + g(\eta) = f(x - ct) + g(x + ct)$$
(8.18)

For the IVP over the infinite interval $-\infty < x < \infty$, we have from the initial conditions

$$u(x, 0) = f(x) + g(x) = u_0(x) \tag{8.19a}$$

or

$$\dot{f}(x) + \dot{g}(x) = \dot{u}_0(x) \tag{8.19b}$$

and

$$u_t(x, 0) = -c\dot{f}(x) + c\dot{g}(x) = v_0(x) \tag{8.20}$$

where a dot on top indicates differentiation with respect to the argument of the function. We solve (8.19b) and (8.20) for $\dot{f}$ and $\dot{g}$ to get

$$2\dot{g}(\alpha) = \dot{u}_0(\alpha) + \frac{1}{c} v_0(\alpha) \qquad 2\dot{f}(\alpha) = \dot{u}_0(\alpha) - \frac{1}{c} v_0(\alpha) \tag{8.21}$$

or

$$g(\alpha) = \frac{1}{2}\left[u_0(\alpha) + \frac{1}{c} \int^{\alpha} v_0(\tau)\, d\tau \right]$$

$$f(\beta) = \frac{1}{2}\left[u_0(\beta) - \frac{1}{c} \int^{\beta} v_0(\tau)\, d\tau \right] \tag{8.22}$$

It follows that $u(x, t)$ is given by

$$u(x, t) = \tfrac{1}{2}\left[u_0(x - ct) + u_0(x + ct) \right] + \frac{1}{2c} \int_{x-ct}^{x+ct} v_0(\tau)\, d\tau \tag{8.23}$$

As an application of this general solution, consider the special initial conditions

$$u_0(x) = H(x + 1) - H(x - 1) \qquad v_0(x) \equiv 0 \qquad (-\infty < x < \infty) \tag{8.24}$$

where $H(\cdot)$ is the unit step function. For these initial conditions, we have

$$u(x, t) = \tfrac{1}{2}[H(x - ct + 1) + H(x + ct + 1)$$
$$- H(x - ct - 1) - H(x + ct - 1)]$$
$$= \tfrac{1}{2}[H(x - ct + 1) - H(x - ct - 1)]$$
$$+ \tfrac{1}{2}[H(x + ct + 1) - H(x + ct - 1)] \tag{8.25}$$

Observe that the first two terms correspond to a square pulse half the magnitude of $u_0(x)$ propagating to the right with speed c. The last two terms correspond to another square pulse half the magnitude of $u_0(x)$ propagating to the left also with wave speed c. Together, we have the graphs shown in Figure 8.3. More generally, we see from (8.23) that

8A: When $v_0(x) \equiv 0$ and $u_0(x)$ is localized, the solution $u(x, t)$ of the IVP for the one-dimensional wave equation consists of two pulses of the same shape as $u_0(x)$ but half its magnitude propagating in opposite directions with the same wave speed c.

Another example which is of fundamental importance involves the initial conditions

$$u_0(x) \equiv 0 \quad \text{and} \quad v_0(x) = \delta(x) \qquad (-\infty < x < \infty) \tag{8.26}$$

For this case, we have

$$u(x, t) = \frac{1}{2c} \int_{x-ct}^{x+ct} \delta(z)\,dz = \frac{1}{2c}\left[H(x + ct) - H(x - ct)\right] \tag{8.27}$$

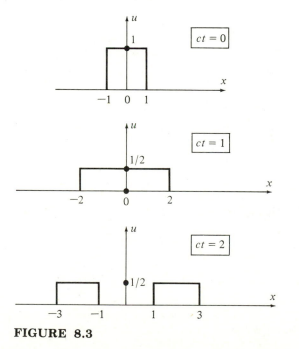

FIGURE 8.3

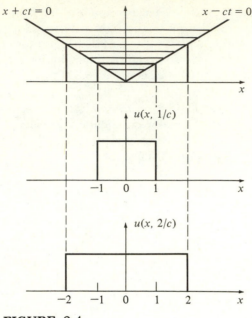

FIGURE 8.4

The solution (8.27) corresponds to a square pulse of magnitude $1/2c$ centered at the origin $x = 0$ and spreading as t increases as shown in Figure 8.4. The shaded triangular region in the x,t plane is called the *domain of influence*. At a given time t_k, the width of the $t = t_k$ line inside the shaded region is the extent in space which experiences the influence of the initial data located at point $x = 0$.

Suppose the impulse is located at $x = x_0$ instead of $x = 0$ and applied at a later time $t = t_0$ instead of $t = 0$. It is not difficult to see that the solution of (8.13) with

$$u(x, t_0) = 0 \qquad u_t(x, t_0) = \delta(x - x_0) \tag{8.28}$$

is (8.27) with x and t replaced by $x - x_0$ and $t - t_0$, respectively:

$$u(x, t) = \frac{1}{2c} \left\{ H[(x - x_0) + c(t - t_0)] - H[(x - x_0) - c(t - t_0)] \right\}$$

$$\tag{8.29}$$

In the next section, we will see how (8.29) can be used to solve the inhomogeneous wave equation.

8.4   The Fundamental Solution and the Inhomogeneous Wave Equation

The IVP, (8.13) and (8.28), is equivalent to the forced problem

$$G_{tt} = c^2 G_{xx} + \delta(x - x_0)\delta(t - t_0) \tag{8.30}$$

$$G\Big|_{t=t_0-} = 0 \qquad G_t\Big|_{t=t_0-} = 0 \tag{8.31}$$

since $\delta(x - x_0)\delta(t - t_0) = 0$ for $t > t_0$ and

$$\begin{aligned}
G_t\Big|_{t=t_0+} &= \left[G_t\right]_{t_0-}^{t_0+} = \int_{t_0-}^{t_0+} G_{tt}\,dt \\[2mm]
&= \int_{t_0-}^{t_0+} [c^2 G_{xx} + \delta(x - x_0)\delta(t - t_0)]\,dt \\[2mm]
&= c^2 \int_{t_0-}^{t_0+} G_{xx}\,dt + \delta(x - x_0) = \delta(x - x_0) \tag{8.32}
\end{aligned}$$

It follows from (8.29) that we also have

$$G(x, t; x_0, t_0) = \frac{1}{2c}\{H[(x - x_0) + c(t - t_0)] - H[(x - x_0) - c(t - t_0)]\}$$

$$\equiv G(x - x_0, t - t_0) \tag{8.33}$$

The solution given by (8.33) is called the *fundamental* solution (or Green's function for the infinite line) of the one-dimensional wave equation.

The function G is *fundamental* because the general IVP for the inhomogeneous wave equation

$$w_{tt} = c^2 w_{xx} + f(x, t) \qquad (-\infty < x < \infty) \tag{8.34}$$

with

$$w(x, 0+) = w_0(x) \qquad w_t(x, 0+) = v_0(x) \qquad (-\infty < x < \infty) \tag{8.35}$$

can be expressed in terms of G and other known quantities. For our water wave problem, the forcing term $f(x, t)$ summarizes forces in the x direction (due to chemical reactions in the fluid, for example) distributed throughout the fluid. For $w_0(x) \equiv 0$ and $v_0(x) \equiv 0$, the solution of (8.34) and (8.35) is given by superposition to be

$$w(x, t) = \int_0^t \int_{-\infty}^{\infty} G(x - x_0, t - t_0)f(x_0, t_0)dx_0 \, dt_0 \equiv w^{(p)}(x, t) \qquad (8.36)$$

A derivation of (8.36) without appealing to the principle of superposition is left as an exercise at the end of this chapter.

For any fixed point of observation (x, t), we have $G \equiv 0$ for $(x - x_0) + c(t - t_0) < 0$ [or $x_0 > x + c(t - t_0)$] and for $(x - x_0) - c(t - t_0) > 0$ [or $x_0 < x - c(t - t_0)$]; in the interval $x - c(t - t_0) < x_0 < x + c(t - t_0)$, we have $G = 1/2c$. Hence, (8.36) reduces to

$$w^{(p)}(x, t) = \frac{1}{2c} \int_0^t \int_{x-c(t-t_0)}^{x+c(t-t_0)} f(x_0, t_0)dx_0 \, dt_0 \qquad (8.37)$$

Thus, $w^{(p)}(x, t)$ depends only on the values of $f(x_0, t_0)$ in the triangular domain in Figure 8.5. This triangular domain, bounded by $x_0 = x - c(t - t_0)$, $x_0 = x + c(t - t_0)$ and $t_0 = 0$, is called the *domain of independence*.

When the initial conditions are also inhomogeneous, the solution for the forced problem is the sum of $w^{(p)}$ and the D'Alembert solution (8.23):

$$w(x, t) = w^{(p)}(x, t) + w^{(c)}(x, t) \qquad (8.38)$$

where

$$w^{(c)}(x, t) \equiv \frac{1}{2c} \int_{x-ct}^{x+ct} v_0(\tau)d\tau + \tfrac{1}{2}[w_0(x + ct) + w_0(x - ct)]$$

$$(8.39)$$

Because of the linearity of the PDE (8.13), it is a straightforward calculation to verify that the wave equation is satisfied:

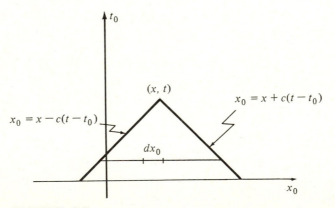

FIGURE 8.5

$$w_{tt} - c^2 w_{xx} = [w_{tt}^{(p)} - c^2 w_{xx}^{(p)}] + [w_{tt}^{(c)} - c^2 w_{xx}^{(c)}]$$

$$= [f(x, t)] + [0] = f(x, t)$$

Also, the initial conditions are satisfied:

$$w(x, 0) = w^{(p)}(x, 0) + w^{(c)}(x, 0) = 0 + \tfrac{1}{2}[w_0(x) + w_0(x)] = w_0(x)$$

$$w_t(x, 0) = w_t^{(p)}(x, 0) + w_t^{(c)}(x, 0) = 0 + \tfrac{1}{2}[v_0(x) + v_0(x)] = v_0(x)$$

In contrast to the buckling phenomenon of Chapter 5, it is significant that

> **8B**   To a given set of initial data and horizontal forcing, there corresponds only one form of propagating small amplitude shallow water waves.

This assertion can be proved by contradiction. Suppose there are two solutions $w^{(1)}$ and $w^{(2)}$. Let $w \equiv w^{(1)} - w^{(2)}$; then w satisfies the homogeneous wave equation (8.13) and homogeneous initial conditions. Without external forcing, we expect w and its first partials to tend to zero as $|x| \to \infty$ as the unforced system cannot have infinite energy (Kevorkian, 1989). Let

$$I(t) \equiv \frac{1}{2} \int_{-\infty}^{\infty} (w_t^2 + c^2 w_x^2) dx \tag{8.40}$$

$$\frac{dI}{dt} = \int_{-\infty}^{\infty} (w_t w_{tt} + c^2 w_x w_{xt}) dx$$

$$= \left[c^2 w_x w_t \right]_{-\infty}^{\infty} + \int_{-\infty}^{\infty} (w_{tt} - c^2 w_{xx}) w_t \, dx = 0 \tag{8.41}$$

Upon integration, we get $I(t) = I_0$ (a constant). From the homogeneous initial conditions for w and w_t, we obtain the additional information that

$$I(0) = \int_{-\infty}^{\infty} \{ [w_t(x, 0)]^2 + c^2 [w_x(x, 0)]^2 \} dx = 0 \tag{8.42}$$

It follows that $I_0 = 0$, that is,

$$I(t) \equiv \frac{1}{2} \int_{-\infty}^{\infty} (w_t^2 + c^2 w_x^2) dx = 0 \tag{8.43}$$

As the integrand is nonnegative, we must have

$$w_t = w_x = 0 \quad \text{or} \quad w(x, t) = w_0 \tag{8.44}$$

But $w(x, 0) = 0$ (by the homogeneous initial condition) requires $w_0 = 0$ so that $w(x, t) = 0$. It follows that we have $w^{(2)} = w^{(1)}$ for all $|x| < \infty$ and $t \geq 0$.

8.5   The Signaling Problem

The small-amplitude motion generated by a vertical wave maker installed at $x = 0$ offers a situation modeled by the wave equation on the half-line. The movement of the wave maker is described by its lateral displacement $x_w(t)$ (see Figure 8.6). For a prescribed $x_w(t)$, we are interested in the motion of the fluid to the right of the wave maker. Suppose the fluid is at rest when the wave maker begins its oscillation, say at $t = 0$. If the displacement of the wave maker is small compared to the undisturbed depth and if its speed of oscillation is sufficiently slow, the amplitude of the motion of the fluid will be small and the PDE (8.13) applies. The fluid is initially at rest so that we have the homogeneous initial conditions

$$u(x, 0) = u_t(x, 0) = 0 \tag{8.45}$$

At the position of the wave maker, we require that the surface of the wave-maker and the fluid in contact with it move at the same velocity:

$$u(x_w(t), t) = \frac{dx_w}{dt} \equiv u_b(t) \tag{8.46a}$$

For a small-amplitude motion wave maker, we may retain only the leading term of the series expansion of $u(x_w, t)$ in x_w:

$$u(x_w, t) = u(0, t) + u_x(0, t)x_w + \tfrac{1}{2}u_{xx}(0, t)x_w^2 + \cdots \simeq u(0, t)$$

so that the appropriate boundary condition at $x = 0$ is

$$u(0, t) = u_b(t) \tag{8.46b}$$

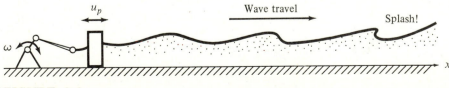

FIGURE 8.6

The linear initial boundary-value problem, (8.13), (8.45) and (8.46b), may be solved by the method of Laplace transforms. Multiplying (8.13) by e^{-st} and integrating from 0 to ∞, we obtain

$$\int_0^\infty c^2 u_{xx}(x, t) e^{-st}\, dt = \int_0^\infty u_{tt}(x, t) e^{-st}\, dt$$

$$= \left[u_t(x, t) e^{-st} \right]_0^\infty + s \int_0^\infty e^{-st} u_t(x, t)\, dt$$

$$= \left[u_t(x, t) e^{-st} + s u(x, t) e^{-st} \right]_0^\infty + s^2 \int_0^\infty e^{-st} u(x, t)\, dt$$

$$(8.47)$$

We denote the Laplace transform of $u(x, t)$ by $U(x, s)$:

$$\int_0^\infty e^{-st} u(x, t)\, dt \equiv U(x, s) \qquad (8.48)$$

In order that the improper integral (8.48) will converge for $\text{Re}(s) > \alpha$, u must have at most an exponential growth with $u \to C_0 e^{\alpha t}$ as $t \to \infty$. In that case, (8.47) may be written, after interchanging integration with respect to t and differentiation with respect to x, as

$$c^2 U_{xx} = s^2 U \qquad [\text{Re}(s) > \alpha] \qquad (8.49)$$

where we have used the homogeneous initial conditions (8.45) and the growth condition at infinity to eliminate the first two terms on the right side of (8.47). The solution of (8.49) is

$$U(x, s) = c_1 e^{sx/c} + c_2 e^{-sx/c} \qquad (8.50)$$

With $\text{Re}(s) > \alpha > 0$, we take $c_1 = 0$ to eliminate exponential growth as $x \to \infty$. The boundary condition at $x = 0$ requires

$$U(0, s) = c_2 = \int_0^\infty e^{-st} u_b(t)\, dt \equiv U_b(s) \qquad (8.51)$$

so that

$$U(x, s) = U_b(s) e^{-sx/c} \qquad (8.52)$$

The inversion formula then gives

$$u(x,\, t) = \frac{1}{2\pi i} \int_{\gamma - i\infty}^{\gamma + i\infty} U_b(s) e^{s(t - x/c)}\, ds \qquad (\gamma > \alpha) \qquad (8.53)$$

which may be evaluated by contour integration. In particular, we know from complex function theory that $u(x,\, t) \equiv 0$ for $t < x/c$.

For the special form of the transformed solution (8.52), we may obtain $u(x,\, t)$ without any knowledge of complex integration. This is accomplished by the observation

$$\int_0^\infty e^{-st} f(t - a) H(t - a)\, dt = \int_{-a}^\infty e^{-s(\tau + a)} f(\tau) H(\tau)\, d\tau$$

$$= e^{-sa} \int_0^\infty e^{-s\tau} f(\tau)\, d\tau \qquad (8.54)$$

where $H(\tau)$ is the Heaviside unit step function. Hence, *the Laplace transform of $f(t - a)H(t - a)$ is e^{-sa} times the Laplace transform of $f(t)H(t)$.* Given the one-to-one correspondence between a function and its Laplace transform, $U_b(s) e^{-sx/c}$ must be the transform of $u_b(t - x/c) H(t - x/c)$ or

$$u(x,\, t) = \begin{cases} u_b\left(t - \dfrac{x}{c}\right) = \dot{x}_w\left(t - \dfrac{x}{c}\right) & (t > x/c) \\[2ex] 0 & (t < x/c) \end{cases} \qquad (8.55)$$

Correspondingly, we get from (8.10)

$$\zeta(x,\, t) = \begin{cases} \dfrac{H}{c}\, \dot{x}_w\left(t - \dfrac{x}{c}\right) & (t > x/c) \\[2ex] 0 & (t < x/c) \end{cases} \qquad (8.56)$$

The parameter H in (8.56) is the undisturbed depth of the water, not to be confused with the Heaviside unit step function $H(\cdot)$. Thus, we have for small-amplitude motion:

8C: A point $x > 0$ inside the fluid experiences at $t = t_k + x/c$ the same velocity as the wavemaker at $t = t_k$.

8.6 A Steadily Advancing Shock Wave

In hydraulics and other engineering areas, we are sometimes interested in the motion of a fluid with a free surface having a steep gradient. For these problems, the linear equation (8.12) (or (8.13)) may not apply while the nonlinear system (8.5) and (8.7) appears intractable. Fortunately, considerable information can be obtained by idealizing a steep gradient in the free surface profile as a finite jump discontinuity. As in chapter 7, such a jump discontinuity is called a *shock wave* or *shock* for brevity when it is in motion and a *hydraulic jump* when it is stationary. An example of a hydraulic jump is the relative position of water on the two sides of a dam at the instant right after the lifting or breaking of the dam—a situation reminiscent of the green light problem in the traffic flow theory. An example of a shock wave is a bore. As we have previously explained in Section 8.2, a bore is a steadily advancing wave with a steep front; it occurs when a rising tide pushes water into a narrowing estuary at the mouth of a river. The height of a bore (at the steep front) is known to reach more than 30 feet in the Qian Tang River in China. A photograph of this bore can be found in Stoker's book on water waves (1957). Another example of a shock wave is an ocean wave steepening as it moves toward a gently sloping beach. In contrast to the dam-breaking problem, we have here an initially smooth wave evolving into a discontinuous profile (and eventually breaking). We limit ourselves here to an analysis of the river bore phenomenon.

Let the jump discontinuity of a possible shock phenomenon be located at $x = x_s$ and $x_1 < x_s < x_2$. For simplicity, we take u and h of the fluid to be uniform on the two sides of the shock with

$$\{u, h\} = \{u_1, h_1\} \qquad (x < x_s)$$
$$\{u, h\} = \{u_2, h_2\} \qquad (x > x_s) \tag{8.57}$$

In that case, we can write (8.3) and (8.4) as

$$\frac{d}{dt}\left(\int_{x_1}^{x_{s-}} \rho h \, dx + \int_{x_{s+}}^{x_2} \rho h \, dx\right) + \left[\rho u h\right]_{x_1}^{x_2} = 0 \tag{8.58}$$

$$\frac{d}{dt}\left(\int_{x_1}^{x_{s-}} \rho u h \, dx + \int_{x_{s+}}^{x_2} \rho u h \, dx\right) + \left[\rho u^2 h + \tfrac{1}{2}\rho g h^2\right]_{x_1}^{x_2} = 0 \tag{8.59}$$

or, with $v_k \equiv u_k - \dot{x}_s$,

$$\frac{d}{dt}\left[\rho h_1(x_s - x_1) + \rho h_2(x_2 - x_s)\right] + \left[\rho u h\right]_{x_1}^{x_2} = \rho h_2 v_2 - \rho h_1 v_1 = 0 \tag{8.60}$$

$$\frac{d}{dt}\left[\rho u_1 h_1(x_s - x_1) + \rho u_2 h_2(x_2 - x_s)\right] + \left[\rho u^2 h + \tfrac{1}{2}\rho g h^2\right]_{x_1}^{x_2}$$

$$= \rho h_2 u_2 v_2 - \rho h_1 u_1 v_1 + \tfrac{1}{2}\rho g(h_2^2 - h_1^2) = 0 \qquad (8.61)$$

We use the first relation, written as

$$u_1 = x_s^{\cdot} + \frac{h_2}{h_1} v_2 = x_s^{\cdot}\left(1 - \frac{h_2}{h_1}\right) + \frac{h_2}{h_1} u_2 \qquad (8.62)$$

to eliminate u_1 from the second to get

$$v_2^2 = (u_2 - x_s^{\cdot})^2 = g h_1 \frac{1 + \sigma}{2\sigma} \qquad \sigma \equiv \frac{h_2}{h_1} \qquad (8.63)$$

Suppose the water level at x_2 is lower (so that $h_2 < h_1$) and both u_1 and u_2 are nonnegative. In that case, we have $\sigma < 1$. For $u_2 = 0$, we have from (8.63) and (8.62)

$$x_s^{\cdot} = \pm\sqrt{gh_2}\sqrt{\frac{1 + \sigma^{-1}}{2\sigma}} \qquad (8.64)$$

$$u_1 = (1 - \sigma)x_s^{\cdot} \qquad (8.65)$$

Since $0 < \sigma < 1$ and $1 + \sigma^{-1} > 2$, we get from (8.64) $|x_s^{\cdot}| > \sqrt{gh_2}$ so that

8D: A bore can only propagate at a steady speed greater than the wave speed of the low-level (still) water. Furthermore, the propagation is in the direction of the still water if $u_1 > 0$.

If we know only h_1 and h_2 (and $u_2 = 0$) but nothing about the sign of u_1, then the shock velocity is determined up to a sign by (8.64) and the relation (8.65) tells us only that the shock velocity must have the same sign as the velocity of water behind it and (given $\sigma < 1$) is greater in magnitude. In principle, u_1 and $x_s^{\cdot}$ may both be negative. However, our intuition suggests (and energy considerations show) that a bore cannot be maintained unless the water level is lower on the front side of the bore than on the back side (Stoker, 1957). In other words, it would not be possible to have a bore if $u_1 < 0$ and $h_1 > h_2$.

We may rewrite the relation between u_1 and $x_s^{\cdot}$ as

$$x_s^{\cdot} - u_1 = \sigma x_s^{\cdot} = \sqrt{\frac{\sigma(1 + \sigma)gh_1}{2}} < \sqrt{gh_1} \qquad (8.66)$$

It follows from this expression that

> **8E:**   The speed of the shock relative to water particles behind the shock is less than the wave propagation speed $\sqrt{gh_1}$ behind the shock.

Hence any small disturbance behind the shock will eventually catch up with it.

Suppose we know u_1 instead of h_2 (and we still know h_1 and $u_2 = 0$). In that case we write the conservation of mass (8.60) as

$$h_2 = h_1 \frac{v_1}{v_2} = h_1 \left(1 - \frac{u_1}{\overset{\cdot}{x}_s} \right) \tag{8.67}$$

and use it to eliminate h_2 from the expression (8.61) to get

$$F(\overset{\cdot}{x}_s) \equiv (\overset{\cdot}{x}_s)^3 - u_1(\overset{\cdot}{x}_s)^2 - gh_1 \overset{\cdot}{x}_s + \tfrac{1}{2}gh_1 u_1 = 0 \tag{8.68}$$

The problem is now more complicated as the shock speed is a root of a cubic equation. For this equation, we have the following observations:

1. $F(\overset{\cdot}{x}_s) < 0$ for $\overset{\cdot}{x}_s \ll 0$ and $F(0) = \tfrac{1}{2}gh_1 u_1 > 0$; so there is a real root along the negative $\overset{\cdot}{x}_s$ axis.

2. $F(u_1) = -\tfrac{1}{2}gh_1 u_1 < 0$; so there is a root in the interval $0 < \overset{\cdot}{x}_s < u_1$.

3. $F(\overset{\cdot}{x}_s) > 0$ for $\overset{\cdot}{x}_s \gg 0$; so there is a third root in the interval $u_1 < \overset{\cdot}{x}_s < \infty$.

Thus, the equation $F(z) = 0$ has three real roots; $z_s < 0 < z_m < u_1 < z_l$, as shown in Figure 8.7. The middle root $z_m < u_1$ is not acceptable for it would require $h_2 = h_1(1 - u_1/z_m) < 0$ which is not physically permissible. The other two roots give

$$h_{2l} = h_1 \left(1 - \frac{u_1}{z_l} \right) < h_1 \qquad (z_l > u_1 > 0)$$

$$h_{2s} = h_1 \left(1 - \frac{u_1}{z_s} \right) > h_1 \qquad (z_s < 0)$$

As $h_2 < h_1$ was stipulated for our problem, we must have $\overset{\cdot}{x}_s = z_l$.

> **8F:**   If h_1, u_1, and $u_2 = 0$ are prescribed and $h_2 < h_1$, then a bore can be maintained with a positive shock speed $\overset{\cdot}{x}_s = z_l$ provided $h_2 = h_1(1 - u_1/z_l)$.

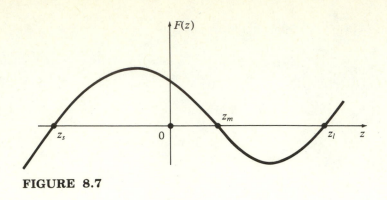

FIGURE 8.7

If we do not stipulate $h_2 < h_1$, then z_l and z_s are both acceptable and an appropriate choice of $x_s^{\bullet}$ depends on the physical situation. An example for which z_s is appropriate will be discussed in the next section.

8.7   Reflection of a Bore from a Rigid Wall

Suppose a vertical rigid wall such as a dam is positioned in front of a bore as indicated in Figure 8.8. The water in front of the shock is at rest. At the wall, positioned at $x = 0$, the velocity of the fluid must vanish so that $u(0, t) = 0$. It follows from (8.7) that $h_x(0, t) = 0$. A solution of this problem in the framework of the small-amplitude theory of Section 8.2 can be obtained

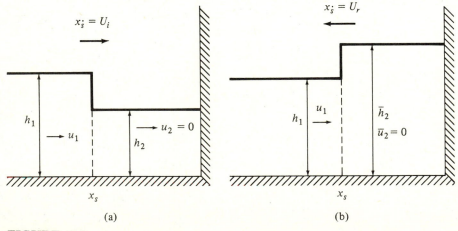

(a)   (b)

FIGURE 8.8

from the initial-boundary value problem consisting of the two PDEs (8.10) and (8.11), the boundary conditions

$$u(0, t) = 0 \qquad \left[\text{or } \zeta_x(0, t) = \frac{1}{g} u_t(0, t) = 0 \right]$$

and the initial conditions

$$u(x, 0) = \begin{cases} u_1 & (x < x_s < 0) \\ 0 & (x_s < x < 1) \end{cases}$$

$$\zeta(x, 0) = \begin{cases} \zeta_1 & (x < x_s < 0) \\ 0 & (x_s < x < 1) \end{cases}$$

where we have positioned the x axis at the still-water level adjacent to the rigid wall with $h_2 = H$ and $h_1 = \zeta + H$ being the depth of water in front of the shock front and behind the shock front, respectively. The above problem can be solved by the Green's function method introduced in Section 8.4 as outlined in the exercises at the end of the chapter. The Green's functions needed here are for the semi-infinite line $-\infty < x < 0$, one with $G = 0$ at $x = 0$ and the other with $G_x = 0$ at $x = 0$. Both are to be obtained from the infinite-line Green's function of Section 8.4 by the method of images (see exercises).

Intuitively (or from the above analysis of the small-amplitude motion), we expect the bore to be reflected from the wall leaving still water ($\bar{u}_2 = 0$) of depth $\bar{h}_2$ behind it. Since the water in contact with the wall should be at rest, this expectation is not unreasonable. We will show presently that the expected motion is compatible with the shock conditions deduced from the two conservation laws. We will also calculate the height $\bar{h}_2$ of the reflected wave.

Similar to the derivation of (8.67), we get from the conservation of mass and $\bar{u}_2 = 0$

$$\bar{h}_2 = \frac{v_1 h_1}{\bar{v}_2} = h_1 \left(1 - \frac{u_1}{\dot{x}_s^*} \right) \tag{8.71}$$

where $\dot{\bar{x}}_s^*$ is the velocity of the shock front of the reflected wave. We use this expression to eliminate $\bar{h}_2$ from the relation (8.61) deduced from the conservation of momentum. The result is again

$$F(\dot{\bar{x}}_s^*) = 0 \tag{8.72}$$

where $F(\cdot)$ is as defined in (8.68). We again reject the middle root but now choose $\bar{x}_s^\bullet = z_s < 0$, as the reflected wave is expected to propagate in the negative x direction. In that case, we have from (8.71)

> **8G:** The level of still water behind the reflected bore is higher than the water level in front of the shock wave, and the shock speed is $\bar{x}_s^\bullet = z_s < 0$.

Given h_1, u_1, and $u_2 = 0$ (and $\bar{u}_2 = 0$), both h_2 and $\bar{h}_2$ are determined in terms of these known parameters. It is possible to construct a plot of $\bar{h}_2/h_2$ versus h_1/h_2 by obtaining these two quantities for different values of u_1. For $(h_1 - h_2)/h_1 \ll 1$ (called *weak shocks*), this plot shows that $\bar{h}_2 - h_2 \simeq 2(h_1 - h_2)$ [see Stoker (1957)] and hence $\bar{h}_2 = 2h_1 - h_2$. It follows that for weak shocks we have

$$\frac{\bar{x}_s^\bullet}{x_s^\bullet} = \frac{h_1 - h_2}{h_1 - \bar{h}_2} \simeq -\frac{h_1 - h_2}{h_1 + h_2} \tag{8.73}$$

so that $|\bar{x}_s^\bullet/x_s^\bullet| \ll 1$. Hence,

> **8H:** In the weak shock limit of $(h_1 - h_2)/h_1 \ll 1$, the height of the reflected bore $\bar{h}_2$ is given approximately by $2h_1 - h_2 = h_1 + (h_1 - h_2)$ (and hence just slightly higher than the incident bore height) and the reflected shock speed is considerably less than the speed of the incident shock.

The situation is of course different if $h_1 - h_2$ is not small compared to h_1. For the *strong shock* limit of $\beta^2 \equiv gh_1/u_1^2 \gg 1$, we may set $x_s^\bullet/u_1 - \frac{1}{2} = z$ and write (8.68) as

$$(z + \tfrac{1}{2})^2 (z - \tfrac{1}{2}) - \beta^2 z = 0 \tag{8.74}$$

For $\beta \gg 1$, we may approximate (8.73) by $z^3 - \beta^2 z \simeq 0$ so that

$$z \simeq \pm\beta = \pm\frac{\sqrt{gh_1}}{u_1} \tag{8.75a}$$

$$\bar{x}_s^\bullet = (\tfrac{1}{2} + z)u_1 \simeq \tfrac{1}{2}u_1 - \sqrt{gh_1} \;(\simeq -\sqrt{gh_1}) \tag{8.75b}$$

$$\frac{\bar{h}_2}{h_1} = 1 - \frac{u_1}{\bar{x}_s^\bullet} \simeq 1 + \frac{u_1}{\sqrt{gh_1}} = 1 + \frac{1}{\beta} \tag{8.75c}$$

Hence,

> **8I:**   In the strong shock limit of $\sqrt{gh_1} \gg u_1$, the reflected bore speed $\bar{x}_s^*$ is only slightly less (in magnitude) than the wave speed of the water behind the incident bore and the reflected bore height is only slightly higher than the height of the incident bore.

8.8   Characteristic Coordinates

The pair of nonlinear partial differential equations (8.5) and (8.7) governing the motion of shallow-water waves can be solved by the method of characteristics introduced in Chapter 7. For the application of this method, it is more convenient to work with the unknown wave speed $c = \sqrt{gh}$ instead of the wave surface height h. In terms of c, Equations 8.5 and 8.7 take the form

$$2(c_t + uc_x) + cu_x = 0 \qquad (8.78a)$$

$$u_t + uu_x + 2cc_x = 0 \qquad (8.78b)$$

We may add these two equations to get

$$(u + 2c)_t + (u + c)(u + 2c)_x = 0 \qquad (8.79)$$

or subtract one from the other to get

$$(u - 2c)_t + (u - c)(u - 2c)_x = 0 \qquad (8.80)$$

Equations 8.79 and 8.80 are still coupled and cannot be solved separately. However, applying the method of characteristics to (8.79) gives

$$\frac{dx}{dt} = u + c \qquad (8.80a)$$

$$\frac{d(u + 2c)}{dt} = 0 \qquad (8.80b)$$

so that $u + 2c$ is constant along the characteristic projections defined by $dx/dt = u + c$. Similarly, we have from (8.80)

$$\frac{dx}{dt} = u - c \qquad (8.81a)$$

$$\frac{d(u - 2c)}{dt} = 0 \qquad\qquad (8.81b)$$

so that $u - 2c$ is constant along the characteristic projections defined by $dx/dt = u - c$.

To see how we can take advantage of these observations to solve the original equations for u and c, consider two auxiliary problems

$$\xi_t + (u + c)\xi_x = 0 \qquad\qquad (8.82a)$$

$$\eta_t + (u - c)\eta_x = 0 \qquad\qquad (8.82b)$$

Evidently, ξ is constant along $dx/dt = u + c$ and η is constant along $dx/dt = u - c$. Moreover, members of the family of ξ = constant curves in the x,t plane intersect those from the other family to form a grid system. The two families of curves are evidently the characteristic projections of the system (8.79) and (8.80). It follows that, in terms of the *characteristic coordinates* ξ and η, $u + 2c$ is only a function of ξ (and not a function of η) and $u - 2c$ is only a function of η. This can be shown to be the case by using the chain rules

$$\frac{\partial}{\partial x} = \xi_x \frac{\partial}{\partial \xi} + \eta_x \frac{\partial}{\partial \eta} \qquad \frac{\partial}{\partial t} = \xi_t \frac{\partial}{\partial \xi} + \eta_t \frac{\partial}{\partial \eta}$$

to write the two equations in (8.78) as

$$u_\xi(\xi_t + u\xi_x) + c_\xi(2c\xi_x) + u_\eta(\eta_t + u\eta_x) + c_\eta(2c\eta_x) = 0$$
$$u_\xi(c\xi_x) + 2c_\xi(\xi_t + u\xi_x) + u_\eta(c\eta_x) + 2c_\eta(\eta_t + u\eta_x) = 0$$

Upon using (8.82) to eliminate ξ_t and η_t, we get

$$c[\xi_x(2c_\xi - u_\xi) + \eta_x(2c_\eta + u_\eta)] = 0 \qquad\qquad (8.83a)$$

$$c[\xi_x(u_\xi - 2c_\xi) + \eta_x(u_\eta + 2c_\eta)] = 0 \qquad\qquad (8.83b)$$

As c, ξ_x, and η_x do not vanish identically, the sum and difference of these last two relations give

$$u_\eta + 2c_\eta = 0 \qquad u_\xi - 2c_\xi = 0$$

so that

$$u + 2c = f(\xi) \tag{8.84a}$$

$$u - 2c = g(\eta) \tag{8.84b}$$

To illustrate how we may exploit (8.84*a, b*) for the solution of u and c, consider the problem of a fluid which is initially at rest to the right of $x = 0$. At $t = 0+$, the wall at $x = 0$ moves to the left with its position given by $x_w = -\beta t^2$ where the subscript w is used to indicate information about the wall. The auxiliary conditions for the partial differential equations (which come from known conditions about the fluid in $x > 0$) take the form

$$u \equiv 0 \quad \text{and} \quad c = c_0 \equiv \sqrt{gH} \qquad (x > 0, t \le 0) \tag{8.85}$$

prior to the wall moving to the left and

$$u = -2\beta t \quad \text{along} \quad x = -\beta t^2 \qquad (t > 0) \tag{8.86}$$

which describes the motion of (fluid particles at) the wall initially at $x = 0$.
For this problem, we can show that

> **8J:** u and c are constant along the characteristic coordinate curves $\xi = $ constant.

Take any two points A and B along an $\xi = \xi_0$ curve in the x,t plane. These two points also lie on two different $\eta = $ constant curves, say $\eta = \eta_A$ and $\eta = \eta_B$, respectively (see Figure 8.9). These curves emanate from two points x'_A and x'_B along the positive x axis to the right of x_A and x_B, respectively (for at $t = 0$ we have $u = 0$ and $c = c_0$ so that $dx/dt = u - c = -c_0 < 0$ along $\eta = $ constant). Now, $u - 2c$ is constant along any $\eta = $ constant curve and hence

$$u_A - 2c_A = [u - 2c]_{t=0, \eta=\eta_A} = -2c_0$$

$$u_B - 2c_B = [u - 2c]_{t=0, \eta=\eta_B} = -2c_0$$

It follows that

$$u_A - 2c_A = u_B - 2c_B \tag{8.87}$$

On the other hand, as points along the $\xi = \xi_0$ curve (where $u + 2c = $ constant), we have also

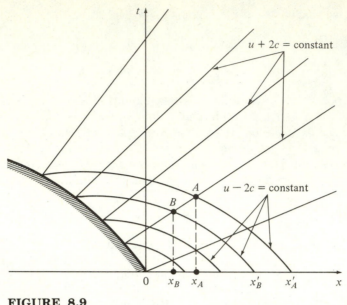

FIGURE 8.9

$$u_A + 2c_A = u_B + 2c_B \tag{8.88}$$

The two equations (8.87) and (8.88) imply

$$u_A = u_B \quad \text{and} \quad c_A = c_B \tag{8.89}$$

as we set out to prove.

Among the consequences of (8.89), we have

8K: The ξ = constant curves in the x,t plane are straight lines (as $dx/dt = u + c$ = constant there).

We take this family of straight characteristic coordinate curves in the form

$$x = x_\xi + (u_\xi + c_\xi)(t - t_\xi) \tag{8.90}$$

where (x_ξ, t_ξ) specifies the point from which a particular member of the straight projections emanates and where u_ξ and c_ξ denote the value of u and c at (x_ξ, t_ξ). Now, the initial data for the partial differential equations are prescribed on the initial curve in the x,t plane consisting of $\{t_\xi = 0, x_\xi > 0\}$ and $\{t_\xi > 0, x_\xi = x_w(t_\xi)\}$. The ξ = constant lines emanating from points along the positive x axis all have the same positive slope as we have $u = 0$

and $c = c_0 > 0$ along $x \geq 0$. With $u = $ constant and $c = $ constant along $\xi = $ constant, we have

$$u \equiv 0 \qquad c = c_0$$

in the region of the x,t plane between the positive x axis and the characteristic projection through the origin $x = c_0 t$. Thus, we conclude (not surprisingly) that

> **8L:** The fluid to the right of $x = c_0 t$ remains at rest.

In the complementary region in the upper half-plane, the velocity u is given by

$$u = \frac{dx_w(t_\xi)}{dt_\xi} = -2\beta t_\xi \tag{8.91}$$

as u is constant along $\xi = $ constant and must therefore assume its value on the initial curve. While c is also constant along $\xi = $ constant, we do not know its value on the initial curve. We do know, however, that $u - 2c = -2c_0$ along $\eta = $ constant. It follows that

$$c = c_0 + \tfrac{1}{2}u = c_0 + \tfrac{1}{2}u_\xi = c_0 - \beta t_\xi \tag{8.92}$$

for $u = u_\xi$ as ξ varies along $\eta = $ constant. The expression (8.92) is consistent with the fact that c is constant along $\xi = $ constant; it also gives the value of c on the portion of the initial curve to the left of the origin.

The information on u_ξ and c_ξ in turn allows us to rewrite the expression (8.90) as

$$x = -\beta t_\xi^2 + (c_0 - 3\beta t_\xi)(t - t_\xi)$$

$$= (c_0 - 3\beta t_\xi)t - t_\xi(c_0 - 2\beta t_\xi) \tag{8.93}$$

We may now solve (8.93) for t_ξ in terms of x and t to get

$$t_\xi = \frac{1}{4\beta}\, c_0 + 3\beta t - \sqrt{(c_0 + 3\beta t)^2 + 8\beta(x - c_0 t)} \tag{8.94}$$

The minus sign is chosen so that the right-hand side reduces to t_ξ when we have $t = t_\xi$ and $x = \beta t_\xi^2$. The expression (8.94) can then be used to eliminate t_ξ from (8.91) and (8.92) to get u and c in terms of x and t alone.

EXERCISES

1. **(a)** With the help of two new independent variables $u = x - c_0 t$ and $v = x + c_0 t$, show that the (spatially) one-dimensional *wave equation* $k_{tt} = c_0^2 k_{xx}$ is equivalent to $k_{uv} = 0$ and the general (twice continuously differentiable) solution for $k(x, t)$ is therefore $k(x, t) = f(x - c_0 t) + g(x + c_0 t)$, where f and g are arbitrary twice continuously differentiable functions.

   (b) Solve the IVP

   $$k_{tt} = c_0^2 k_{xx} \qquad (-\infty < x < \infty, t > 0)$$

   with

   $$k(x, 0) = 2 \sin x$$
   $$k_t(x, 0) = 6x^2 \qquad (-\infty < x < \infty)$$

2. Repeat part (b) of Exercise 1 with the general initial conditions

   $$k(x, 0) = k_0(x) \qquad k_t(x, 0) = j_0(x) \qquad (-\infty < x < \infty)$$

3. With the help of a few graphs, describe the behavior of the solution for $t > 0$ of the initial value problem

   $$k_{tt} = c_0^2 k_{xx} \qquad (-\infty < x < \infty, t > 0)$$

   $$k(x, 0) = k_0(x) \quad \text{and} \quad k_t(x, 0) = 0 \qquad (-\infty < x < \infty)$$

   where $k_0(x)$ is the triangular hump given in Section 7.8.

4. A long straight canal partially filled with an incompressible, inviscid liquid has a cross-sectional area $A_0(x)$ below the mean liquid level which is a slowly changing function of distance x along the canal (see Figure 8.10 for example).

   (a) If at any time t, $u(x, t)$ is the average velocity over the cross section at x, and $A(x, t)$ is the cross-sectional area of the fluid, show that the differential equations which approximately describe the propagation of long waves on the canal, are

   $$\rho \frac{\partial u}{\partial t} + \rho u \frac{\partial u}{\partial x} = -\frac{dp}{dA} \frac{\partial A}{\partial x} \qquad \frac{dp}{dA} = \rho g \frac{dh}{dA} \qquad \frac{\partial A}{\partial t} + \frac{\partial}{\partial x}(uA) = 0$$

   Here dh/dA is the rate of increase of the liquid surface level with respect to cross-sectional area and it is dependent only on the geometry of the canal cross section.

   (b) With the stated assumptions, show that these equations can be linearized to give

$$\rho \frac{\partial u}{\partial t} + \frac{\partial A}{\partial x}\left(\frac{dp}{dA}\right)_{A=A_0} = 0 \qquad \frac{\partial A}{\partial t} + \frac{\partial}{\partial x}(uA_0) = 0$$

(c) If we let $A = \partial \psi / \partial x$, $uA_0 = -\partial \psi / \partial t$, show that ψ obeys the wave equation with variable coefficients:

$$\frac{\partial^2 \psi}{\partial x^2} = \frac{1}{c^2}\frac{\partial^2 \psi}{\partial t^2} \quad \text{where} \quad c^2 = \frac{A_0}{\rho}\left(\frac{dp}{dA}\right)_{A=A_0}$$

Note that c^2 is a function of x.

(d) Consider the special case $A_0 = $ constant. Then c is a constant equal to the long wave propagation speed in the canal. If the canal cross section is triangular, as shown in Figure 8.10 with an angle θ_0 at the bottom, find the wave propagation speed.

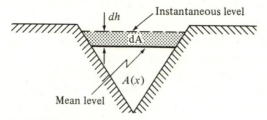

FIGURE 8.10

5. At time zero, a piston that borders one end of a semi-infinite channel begins a small-amplitude oscillation. The locus of the piston is

$$s = \begin{cases} \epsilon \sin \omega t & (t \geq 0) \\ 0 & (t \leq 0) \end{cases}$$

where $|\epsilon| \ll 1$. If the fluid motion is governed approximately by the shallow-water equations, solve to find the surface wave height, $\eta(x, t)$.

6. The initial wave height in a channel of length l and depth h is $\eta(x, 0) = f(x)$; the initial velocity is zero, $u(x, 0) = 0$. Show that the solution of this problem can be obtained from an "infinite channel" problem where the initial wave height is constructed by mirror reflections. Express the solution strictly as a mathematical formula.

7. **(a)** A shallow-water wave of arbitrary shape impinges upon a shelf from the left as shown in Figure 8.11. Part of the wave is transmitted and part is reflected. Show that this can be described by a solution of the form

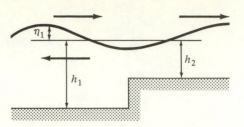

FIGURE 8.11

$$\phi = \begin{cases} F(x - c_1 t) + R(x + c_1 t) & (x < 0) \\ T(x - c_2 t) & (x > 0) \end{cases}$$

where $c_1 = \sqrt{gh_1}$ and $c_2 = \sqrt{gh_2}$, with $h_1 > h_2$.

(b) Reasonable conditions at the shelf are that the wave height be continuous and that mass be conserved across the discontinuity $x = 0$. Formulate these conditions in mathematical terms as relations among the functions F, R, and T.

(c) Express the functions $R(x)$, $T(x)$ in terms of the function $F(x)$ which describes the incoming wave. What conclusions do you draw about the relative amplitudes of the reflected and transmitted waves?

8. Solve (or at least discuss the features of the solution of) the nonlinear shallow-water equations taken in the form

$$\begin{aligned} u_t + uu_x + 2cc_x &= 0 \\ 2(c_t + uc_x) + cu_x &= 0 \end{aligned} \quad \text{with} \quad c^2 = gh$$

for conditions corresponding to the instantaneous collapse of a dam so that at $t = 0$, we have

$$u = 0 \quad \text{and} \quad c = \begin{cases} c_0 & (x \le 0) \\ c_1 & (x > 0) \end{cases}$$

with $c_0 > c_1$ (since we have $h_0 > h_1$).

9. Determine the curves in the x,y plane which can support a weak discontinuity in the most highly differentiated term of the equation $A\psi_{xx} + 2B\psi_{xy} + C\psi_{yy} = 0$, where A, B, and C are constants. For what values of the constants are waves possible? In this case, express the equation in characteristic coordinates.

10. The initial surface elevation of water in an infinite channel is

$$\eta(x, 0) = \begin{cases} \gamma(|x| - 1)^2 & (|x| < 1) \\ 0 & (|x| > 1) \end{cases}$$

Moreover $u(x, 0) = 0$. Let γ be either ± 1 and describe the resultant motion. Find positions of shock onset if they occur. (Are there any?)

11. A piston is moved into a channel at constant speed. Show that the position of shock formation corresponds to the point in the space-time plane where the slope of the wave height becomes infinite, $\eta_x(x, t) = \infty$.

12. Water waves are often made in the laboratory (for the purpose of testing ship models, beach erosion, and the like) by simply oscillating one end wall of a very long tank back and forth in a sinusoidal manner (see Figure 8.6). Suppose that the end wall moves with the small (but *not* infinitesimal) speed $u_p(t) = \epsilon c_0 \cos \omega t$, where ω is a known constant and $c_0 = \sqrt{gh_0}$ is the wave speed corresponding to the mean depth h_0 of the water. Consider the characteristic form

$$\left[\frac{\partial}{\partial t} + (u \pm c)\frac{\partial}{\partial x}\right](u \pm 2c) = 0$$

of the shallow-water equations with $c^2 = gh$, and ignore all conceivable reflections from any far end of the tank.

(a) Explain qualitatively why the resulting waves do not remain nearly sinusoidal, but instead *steepen* at the front, with their increasing distance from the wave maker.

(b) *Estimate* as a function of the relative amplitude ϵ the distance (measured in wavelengths) from the wall at which the formal solution first becomes multivalued, or where the real wave "breaks."

13. Solve for the Green's function as defined in the problem below by the *method of images*:

$$\bar{G}_{tt} = c^2 \bar{G}_{xx} + \delta(x - x_0)\delta(t - t_0) \qquad (0 < x, x_0 < \infty, 0 < t, t_0 < \infty)$$

$$\bar{G}\bigg|_{t < t_0} = 0 \qquad \bar{G}_t\bigg|_{t < t_0} = 0 \qquad \bar{G}(0, t) = 0$$

[*Hint:* Consider the related problem for $-\infty < x < \infty$ with an additional forcing term $-\delta(x + x_0)\delta(t - t_0)$ located at the image point $-x_0$.]

14. Find the Green's function defined by Exercise 12 with the condition $\bar{G}(0, t) = 0$ replaced by $\bar{G}_x(0, t) = 0$.

15. Let $G^*(x, t; y, \tau)$ be the solution of

$$G_{tt}^* = c^2 G_{xx}^* + \delta(x - x_0)\delta(t - t_0)$$

$$(-\infty < x, x_0 < \infty, -\infty < t, t_0 < \infty)$$

$$G^*\Big|_{t > t_0} = 0 \qquad G_t^*\Big|_{t > t_0} = 0$$

Show that $G(x, t; y, \tau) = G^*(y, \tau; x, t)$. {*Hint:* Form $G^*(x, t; y, \tau)$ $\times [G_{tt} = c^2 G_{xx} + \delta(x - x_0)\delta(t - t_0)]$ and integrate by parts. Note that G^* is called the *adjoint* Green's function for the wave equation.}

16. $u_{tt} = c^2 u_{xx} + f(x, t) \qquad u(x, 0) = 0$

$$u_t(x, 0) = 0 \qquad (|x| < \infty, t > 0)$$

(a)   Show that

$$u(x, t) = \int_0^t \int_{-\infty}^{\infty} G^*(x_0, t_0; x, t) f(x_0, t_0) dx_0\, dt_0$$

(b)   Apply the reciprocal relation of Exercise 14 to the result in (a) to get (8.36).

17. The infinite-space Green's function for the wave equation in three spatial dimensions is defined by

$$G_{tt} = c^2 \nabla^2 G + \delta(\mathbf{x} - \mathbf{x}_0)\delta(t - t_0) \qquad G\Big|_{t < t_0} = 0 \qquad G_t\Big|_{t < t_0} = 0$$

where

$$\nabla^2(\ ) \equiv (\ )_{xx} + (\ )_{yy} + (\ )_{zz}$$

$$\delta(\mathbf{x} - \mathbf{x}_0) = \delta(x - x_0)\delta(y - y_0)\delta(z - z_0)$$

(a)   Fourier transform the problem in x, y, and z and solve the resulting problem in ODE.

(b)   Evaluate the inverse transform by lining up the $(\mathbf{x} - \mathbf{x}_0)$ vector with the k_z (the wave-number variable associated with z) axis. Use

$$\frac{1}{2\pi} \int_{-\infty}^{\infty} e^{-i\omega z} d\omega = \delta(z)$$

to get

$$G(x, t; x_0, t_0) = \frac{1}{4\pi c r} \{\delta[r - c(t - t_0)] - \delta[r + c(t - t_0)]\}$$

where

$$r = |\mathbf{x} - \mathbf{x}_0| = \sqrt{(x - x_0)^2 + (y - y_0)^2 + (z - z_0)^2}$$

The Sound of Music

Vibrating Strings and Membranes

The eigenfunction expansion technique is developed for a single second-order linear partial differential equation in one and two spatial variables.

9.1 Why Is a Guitar More Musical than a Drum?

When a wave propagates within a medium of finite extent, it sooner or later reaches the boundary of the medium. Depending on what kind of a medium we have beyond the boundary, the wave incident to the boundary may continue to propagate in the same direction crossing the boundary and passing into the neighboring medium, or it may be reflected back into the first medium traveling in the reverse direction. As we shall see in Section 9.3, both transmission and reflection may occur. Now, the propagating *reflected wave* soon reaches another boundary of the first medium and may be reflected back into the same body again. This process of repeated reflection may continue indefinitely if there is no dissipation and little or no transmission through the body's boundary. The various reflected waves often cancel one another at some points in the body for all time while the net amplitude at other points oscillates periodically in time. This gives rise to the *standing-wave* phenomenon in finite bodies not encountered in wave propagation models of the last two chapters.

To illustrate the possibility of standing waves, take two traveling waves $\sin[k(x - ct)]$ and $\sin[k(x + ct)]$ which are known to be propagating in opposite directions. Their combination in the form of the sum $w(x, t) = \sin[k(x - ct)] + \sin[k(x + ct)]$ may be rearranged with the help of trigonometric identities to read $w(x, t) = 2 \cos kct \sin kx$. The combined effect is therefore a fixed sinusoidal distribution in space with node points at $x_j =$

$2j\pi/k$, $j = 0, \pm 1, \pm 2, \ldots$. The amplitude of $w(x, t)$, given by $2 \cos kct$, is oscillating periodically in time. We have then an example of a standing-wave form (see Figure 9.1) with node points at x_j, $j = 0, \pm 1, \pm 2, \ldots$.

In the models studied in the last two chapters, the wave phenomena of interest are so localized spatially and temporally that the propagating wave does not reach a boundary point for the time interval of interest. For such problems, we can effectively idealize the deformable medium as if it had no boundary, just the opposite of what Newton did by modeling the Sun and the planets as mass points. In other situations, the duration of the wave phenomenon of interest is very long compared to the travel time of the wave over the extent of the carrier medium. In that case, a study of the complicated effects of the boundary including the standing-wave phenomenon is unavoidable. Standing waves constitute an important part of our daily life. For example, the production, transmission, and reception of sound form a multibillion dollar industry and standing waves lie at the very heart of this industry. A short chapter in this book alone obviously does not do justice to such a broad subject. However, it may serve as a stimulus for further study of the vast literature on different aspects of the theory of sound and at the same time introduce us to some of the basic mathematical methods for standing waves.

To provide a focus for our discussion, we ask in this chapter the following simple question: Why is a guitar (or any string instrument) more

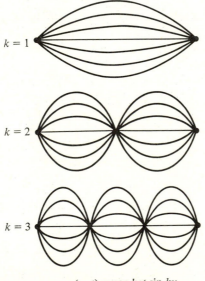

$$w_k(x, t) = \cos kct \sin kx$$

FIGURE 9.1

musical than a drum? The question allows us to discuss standing waves in vibrating strings and membranes. The vibration in these mechanical systems produces periodic disturbances in the surrounding medium (normally air or water). The disturbances propagate in the surrounding carrier medium away from their sources until they reach our ears. It is the vibration induced in our eardrums by these propagating disturbances that produces the sensation of sound. We do not yet fully understand the neurophysiology of hearing and will not discuss this part of the theory of sound here. (It is experimentally observed, however, that the shorter the period of the source oscillation, that is, the higher its frequency, the higher is the pitch of the sound heard.) We also do not wish to discuss the better-known theory of the transmission of sound waves through air or other media. Our concern in this chapter will be only with the kind of (standing) sound waves generated by the vibration of the strings of a guitar or violin and of the membrane of a drumhead. In the small-amplitude range of vibration, these are classical problems in mathematical physics. To the extent that the physical principles involved are relatively simple and well known, problems in vibration are mainly problems in applied mathematics. Through these string and membrane problems, we will have our only exposure in this book to the principal mathematical techniques involved in the linear analysis of mechanical vibration. These techniques turn out to be important for a vast number of other problems in science and engineering as well.

9.2   Small-Amplitude Motion of an Elastic String

Consider an elastic string of uniform mass density (per unit string length) ρ located between $x = 0$ and $x = L$ along the x axis. The string is in static equilibrium under equal and opposite tensile forces at the two ends. At $t = 0$, the string is set in motion in the x,y plane by external disturbances while its two ends are held fixed. Some of these disturbances are momentary in the form of initial conditions; others are continuing in the form of concentrated or distributed forcing along the length of the string for $t > 0$. We are interested in situations where the disturbances induce only a small-amplitude motion in the string. Evidently, we have in the above arrangement an idealization of the essential aspect of string musical instruments such as violins, guitars, etc., whose motion is of technical and general interest.

   At some instant $t > 0$, we look at a small segment of the displaced string which was in the interval $(x, \bar{x} = x + \Delta x)$ before the string was disturbed from its equilibrium position. Suppose $(x + v, u)$ is the displaced position of the point $(x, 0)$ in the x,y plane and θ is the angle formed by the tangent to the segment at $(x + v, u)$ with the x axis (see Figure 9.2). Assuming that

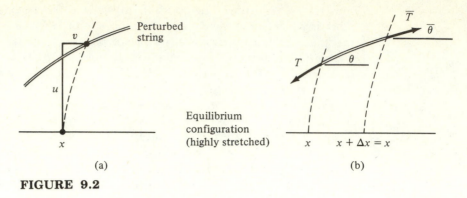

FIGURE 9.2

there is no forcing along the string in the x direction, Newton's second law applied to the horizontal motion of this segment requires

$$\rho \, \Delta x v_{tt}(x, \, t) \simeq \bar{T} \cos \bar{\theta} - T \cos \theta \tag{9.1}$$

where T and $\bar{T}$ are the string tension at the two ends of the segment. Again, we will use the notations $(\quad)_z \equiv \partial(\quad)/\partial z$ and $\bar{f} = f(\bar{x})$ in this chapter. In the vertical direction, the law of motion requires

$$\rho \, \Delta x u_{tt} = \bar{T} \sin \bar{\theta} - T \sin \theta + F(x, \, t) \Delta x \tag{9.2}$$

where $F(x, \, t)$ summarizes the external excitations (forcing, loading) per unit length along the string. In the limit as $\Delta x \to 0$, we have

$$\rho v_{tt} = [T \cos \theta]_x \qquad \rho u_{tt} = [T \sin \theta]_x + F(x, \, t) \tag{9.3}$$

The two equations in (9.3) contain four unknowns u, v, θ, and T. Geometrical considerations require

$$\tan \theta \simeq \frac{\bar{u} - u}{(\bar{x} + \bar{v}) - (x + v)} \tag{9.4}$$

In the limit as $\Delta x \to 0$, we obtain from (9.4) a third relation for the four unknowns:

$$\tan \theta = \frac{u_x}{1 + v_x} \tag{9.5}$$

The remaining relation for u, v, θ, and T comes from the elastic property of the string.

We consider in this chapter only problems in which the horizontal movement of the string is negligible. In that case, we have from the first equation of (9.3) $T \cos \theta =$ constant where the constant is determined to be T_0 by the original tension at the two ends. With $\sin \theta \simeq \tan \theta \simeq u_x$ for small-amplitude (and slope) motion, the second equation of (9.3) then becomes

$$\rho u_{tt} = T_0 u_{xx} + F(x, t) \tag{9.6}$$

or

$$u_{tt} = c^2 u_{xx} + f(x, t) \tag{9.7}$$

where $f = F/\rho$ and $c^2 = T_0/\rho$. The quantity c gives the *wave speed* of the string.

The PDE (9.7) is the (one-dimensional) inhomogeneous wave equation we encountered in studying linearized shallow-water theory in Chapter 8. As in that theory, the equation here is supplemented by two initial conditions; they are conditions on the initial displacement and initial velocity at each point on the string:

$$u(x, 0) = u_0(x) \qquad u_t(x, 0) = v_0(x) \tag{9.8}$$

In addition, we know that the two ends of the string are constrained from movement so that

$$u(0, t) = u(L, t) = 0 \tag{9.9}$$

Thus, (9.7)–(9.9) define a mixed initial-boundary value problem in linear PDE. The solution of this problem $u(x, t)$ describes the transverse motion of the string for all $t > 0$.

9.3 Wave Reflection at a Density Change

Before we embark on an investigation of standing waves in a string of finite length, we consider in this section a traveling-wave problem in the infinitely long string along $-\infty < x < \infty$ without external load. We take the string to have a mass density per unit length ρ_1 for $x < 0$ and a different density ρ_2 for $x > 0$. Suppose a wave train of the form $u_1 = A_1 e^{-i(\omega t - k_1 x)}$ propagates[1]

[1] That the wave train has not gone past the junction $x = 0$ prior to $t = 0$ may be indicated by taking $u_1 = A_1 \sin(\omega t - k_1 x) H(\omega t - k_1 x)$ where $H(\cdot)$ is the Heaviside unit step function. For simplicity, we omit the factor $H(\omega t - k_1 x)$ and work with the complex exponentials instead of sines.

from $-\infty$ toward ∞. For a given frequency ω, the homogeneous wave equation $[f(x, t) \equiv 0]$ for $x < 0$ requires $k_1^2 = \rho_1 \omega^2 / T_0$. For this choice of k_1, the wave train u_1 cannot propagate beyond $x = 0$ for it does not satisfy the homogeneous wave equation with $\rho = \rho_2$ $(\neq \rho_1)$.

The transmission of u_1 through the junction of density change into the region $x > 0$ is possible if it is accompanied by a reflected wave $u_3 = A_3 e^{-i(\omega t + k_1 x)}$. The reflected wave propagates to the left toward $-\infty$ and u_3 satisfies the homogeneous wave equation with $\rho = \rho_1$. The transmitted wave $u_2 = A_2 e^{-i(\omega t - k_2 x)}$ with $k_2^2 = \rho_2 \omega^2 / T_0$ satisfies the homogeneous wave equation with $\rho = \rho_2$ and propagates toward ∞. The amplitudes of u_2 and u_3 are determined by continuity conditions at the junction $x = 0$. The string displacement is continuous so that we require $u_1(0, t) + u_3(0, t) = u_2(0, t)$ or

$$A_1 + A_3 = A_2 \tag{9.10}$$

The tension in the string is also continuous so that

$$\left[T_0 \frac{\partial u_1}{\partial x} + T_0 \frac{\partial u_3}{\partial x} \right]_{x=0} = \left[T_0 \frac{\partial u_2}{\partial x} \right]_{x=0} \tag{9.11}$$

or

$$k_1(A_1 - A_3) = k_2 A_2 \tag{9.12}$$

The two conditions (9.10) and (9.12) may be solved for A_2 and A_3 in terms of A_1, ρ_1, and ρ_2 to get

$$\frac{A_2}{A_1} = \frac{2}{1 + \sqrt{\rho_2 / \rho_1}} \qquad \frac{A_3}{A_1} = \frac{1 - \sqrt{\rho_2 / \rho_1}}{1 + \sqrt{\rho_2 / \rho_1}} \tag{9.13}$$

keeping in mind that $k_n = \rho_n \omega^2 / T_0$, $n = 1, 2$.

Observe that $A_2/A_1 \rightarrow 0$ and $A_3/A_1 \rightarrow -1$ as $\rho_2 \rightarrow \infty$. There is no transmitted wave into an infinitely dense medium and the reflected wave has only a phase shift of π radians with no change in amplitude. At the other extreme, we have $k_2 \rightarrow 0$, $A_3/A_1 \rightarrow 1$ as $\rho_2 \rightarrow 0$ so that there is also no transmitted wave propagating toward ∞. There is no carrier medium for the wave in the region $x > 0$ in this case.

Reflected and transmitted waves of the more general form

$$u_3 = A_3 e^{-i(\omega_3 t + \bar{k}_3 x)} \qquad u_2 = A_2 e^{-i(\omega_2 t + \bar{k}_2 x)} \tag{9.14}$$

with

$$\bar{k}_3^2 = \frac{\rho_1 \omega_3^2}{T_0} \quad \text{and} \quad \bar{k}_2^2 = \frac{\rho_2 \omega_2^2}{T_0}$$

satisfy the relevant wave equation in their respective range of x. The continuity requirement on the displacement at the junction $x = 0$ implies

$$A_1 e^{-i\omega t} + A_3 e^{-i\omega_3 t} = A_2 e^{-i\omega_2 t} \tag{9.15}$$

which cannot be satisfied unless $\omega_2 = \omega_3 = \omega$. It follows that

> **9A:**   Transmitted and reflected waves of the form (9.14) are possible only if their frequencies are identical to the frequency of the incoming wave. Correspondingly, we must also have $\bar{k}_3 = k_1$ and $\bar{k}_2 = k_2$.

9.4   Standing Waves in a Finite String

We now consider the case $f(x, t) \equiv 0$ in (9.7) and seek a solution of the corresponding homogeneous equation in the form $X(x) T(t)$. Upon substituting this solution into (9.7) and after some rearrangements, we get

$$\frac{X''(x)}{X(x)} = \frac{T^{\cdot\cdot}(t)}{c^2 T(t)} \tag{9.16}$$

where $(\ )'$ indicates differentiation with respect to x. The left side of (9.16) is a function of x only and the right side is a function of t only. The two sides can be equal for $0 < x < L$ and $t > 0$ only if they both are equal to a constant. We take this constant to be $-\lambda$ with no loss of generality (as λ is allowed to be a negative number). In that case, we have

$$X'' + \lambda X = 0 \tag{9.17a}$$

$$T^{\cdot\cdot} + \lambda c^2 T = 0 \tag{9.17b}$$

The boundary conditions (9.9) now become $u(0, t) = X(0) T(t) = 0$ and $u(L, t) = X(L) T(t) = 0$ for $t > 0$. They require

$$X(0) = X(L) = 0 \tag{9.18}$$

The ODE (9.17a) and the boundary conditions (9.18) define an eigenvalue problem with λ as the eigenvalue parameter. It is exactly the same problem as that for Euler buckling in Chapter 5. The results obtained there give

$$\lambda = \frac{n^2 \pi^2}{L^2} \equiv \lambda_n \qquad (n = 1, 2, \ldots) \tag{9.19}$$

as the eigenvalues and

$$X(x) = \sin \frac{n\pi x}{L} \equiv X_n(x) \qquad (n = 1, 2, \ldots) \tag{9.20}$$

as the corresponding eigenfunctions up to an amplitude factor.

For each λ_n, the solution of Equation 9.17b is

$$T = A_n \cos \omega_n t + B_n \sin \omega_n t \equiv T_n(t) \qquad \omega_n = \frac{n\pi c}{L} \tag{9.21}$$

where A_n and B_n are arbitrary constants of integration. We have then as a particular solution of the homogeneous wave equation for the finite string

$$u_n \equiv X_n(x) T_n(t) = (A_n \cos \omega_n t + B_n \sin \omega_n t) \sin \frac{n\pi x}{L} \tag{9.22}$$

The expression on the right is a standing wave with a frequency $\omega_n = n\pi c/L = (n\pi/L)\sqrt{T_0/\rho}$.

For a particular physical problem with no body force acting on the string so that $f(x, t) = 0$, we still have to satisfy the initial conditions (9.8) to complete the solution process. These conditions require

$$A_n \sin \frac{n\pi x}{L} = u_0(x) \qquad \omega_n B_n \sin \frac{n\pi x}{L} = v_0(x) \tag{9.23}$$

They cannot be satisfied for general $u_0(x)$ and $v_0(x)$.

Because of the linearity of the problem, we can attempt to satisfy the initial conditions by taking u as a sum of all the particular solutions. We have in that case

$$u(x, t) = \sum_{n=1}^{\infty} (A_n \cos \omega_n t + B_n \sin \omega_n t) \sin \frac{n\pi x}{L} \tag{9.24}$$

with $\omega_n = n\pi c/L = (n\pi/L)\sqrt{T_0/\rho}$. At $t = 0$, we must have

$$u(x, 0) = \sum_{n=1}^{\infty} A_n \sin \frac{n\pi x}{L} = u_0(x) \tag{9.25}$$

$$u_t(x, 0) = \sum_{n=1}^{\infty} B_n \omega_n \sin \frac{n\pi x}{L} = v_0(x) \tag{9.26}$$

Two questions naturally suggest themselves. Can we always choose A_n and B_n, $n = 1, 2, \ldots$, so that these two conditions are satisfied for arbitrary $u_0(x)$ and $v_0(x)$? If we can, how do we compute the unknown coefficients A_n and B_n?

The answer to the second question is relatively easy once we observe the *orthogonality* condition between any two eigenfunctions $\sin(m\pi x/L)$ and $\sin(n\pi x/L)$:

$$\int_0^L \sin \frac{n\pi x}{L} \sin \frac{m\pi x}{L} \, dx = 0 \qquad (n \neq m) \tag{9.27}$$

and the related result

$$\int_0^L \sin^2 \frac{n\pi x}{L} \, dx = \frac{L}{2} \qquad (n = 1, 2, \ldots) \tag{9.28}$$

To get A_m and B_m, we simply multiply Equations 9.25 and 9.26 through by $\sin(m\pi x/L)$ and integrate over $(0, L)$ to get

$$\frac{L}{2} \{A_m, \omega_m B_m\} = \int_0^L \{u_0(x), v_0(x)\} \sin \frac{m\pi x}{L} \, dx \tag{9.29}$$

Given the initial displacement and initial velocity of the string, $u_0(x)$ and $v_0(x)$, it is only a matter of evaluating the integrals in (9.29) for $m = 1, 2, 3, \ldots$ to completely determine the *Fourier series* solution (9.24) for $u(x, t)$.

Whether the series at $t = 0$ is identical to $u_0(x)$ [and its time derivative at $t = 0$ identical to $v_0(x)$] is a more difficult question in the theory of Fourier series and cannot be thoroughly discussed in this book. Under some mild restriction on $u_0(x)$ [and $v_0(x)$], it can be shown that the series $u(x, 0)$ does converge to $u_0(x)$ [and $u_t(x, 0)$ to $v_0(x)$]. A discussion of these results and the related issues of the differentiability of the *Fourier series* representation for $u(x, t)$ can be found in Whittaker and Watson (1952). The more elementary aspects of the subject are treated in Haberman (1977). Here, we simply record that

9B: The solution for the vibrating string problem can be obtained in the form of a Fourier series when the initial displacement and velocity of the string are continuous and piecewise differentiable.

Before leaving the subject of the small-amplitude free (unforced) vibration of a stretched string, we observe a number of consequences of (9.24):

9C: The (unforced) motion of the string is always a combination of standing waves of different frequencies and, correspondingly, different modal shapes.

9D: The motion is periodic in t with a period $P = 2L\sqrt{\rho/T_0}$, that is, $u(x, t + P) = u(x, t)$.

9E: The frequencies of different standing waves are (integer) multiples of the lowest (fundamental) frequency $f_1 = \sqrt{T_0/4\rho L^2}$ with $f_n = nf_1$, $n = 2, 3, \ldots$ [The higher-frequency components (the overtones) are said to be *higher harmonics* of the *fundamental* (component) of the motion.]

When a string is excited, its periodic motion (vibration) is known to generate a disturbance in the surrounding air which acts as a medium for transmitting the disturbance to the human eardrum. The eardrum is in turn excited by the disturbance into motion giving rise to the sensation of *sound*. The inclusion of overtones gives the sound note its character and quality. That these overtones are higher harmonics of the fundamental makes the transmitted disturbance periodic. This periodicity is felt to make the sound note musical "to our sense of hearing."

What is considered musical may be innate and neurological to all human beings. On the other hand, it may be conditioned and, hence, culture bound. Our understanding of this aspect of hearing is too limited to consider the relationship between periodic disturbances and musical sounds to be anything more than a definition of a sound being musical in most Western societies.

Finally, with the trigonometric identities

$$\cos \omega_n t \sin \frac{n\pi x}{L} = \frac{1}{2}\left[\sin\left(\frac{n\pi x}{L} + \omega_n t\right) + \sin\left(\frac{n\pi x}{L} - \omega_n t\right)\right]$$

$$\sin \omega_n t \sin \frac{n\pi x}{L} = \frac{1}{2}\left[\cos\left(\frac{n\pi x}{L} - \omega_n t\right) - \cos\left(\frac{n\pi x}{L} + \omega_n t\right)\right]$$

the solution (9.24) may be written as

$$u(x, t) = \frac{1}{2}\sum_{n=1}^{\infty}\left(A_n\left\{\sin\left[\frac{n\pi}{L}(x + ct)\right] + \sin\left[\frac{n\pi}{L}(x - ct)\right]\right\}\right.$$

$$\left. + B_n\left\{\cos\left[\frac{n\pi}{L}(x - ct)\right] - \cos\left[\frac{n\pi}{L}(x + ct)\right]\right\}\right) \qquad (9.30)$$

The complex structure of the (complementary) solution notwithstanding, $u(x, t)$ for the string problem is of the form $F(x - ct) + G(x + ct)$ as we found in Chapter 8 with

$$F(z) = \frac{1}{2} \sum_{n=1}^{\infty} \left(A_n \sin \frac{n\pi z}{L} + B_n \cos \frac{n\pi z}{L} \right)$$

$$G(z) = \frac{1}{2} \sum_{n=1}^{\infty} \left(A_n \sin \frac{n\pi z}{L} - B_n \cos \frac{n\pi z}{L} \right)$$

We see from (9.30) that

9F: Standing waves are combinations of traveling waves.

9.5 Forced Vibration of a Finite String

If $f(x, t)$ is not zero so that the string is subject to continuing excitation in time, we may again seek a representation of the string motion in the form of a Fourier (sine) series

$$u(x, t) = \sum_{n=1}^{\infty} A_n(t) \sin \frac{n\pi x}{L} \tag{9.31}$$

where the coefficients $\{A_n(t)\}$ are unknown functions of time. The PDE (9.7) now requires

$$\sum_{n=1}^{\infty} (A_n^{\bullet\bullet} + \lambda_n c^2 A_n) \sin \frac{n\pi x}{L} = f(x, t) \tag{9.32}$$

The orthogonality relation (9.27) is now used to obtain from (9.32)

$$A_m^{\bullet\bullet} + \lambda_m c^2 A_m = f_m(t) \qquad (m = 1, 2, \ldots) \tag{9.33}$$

where $\lambda_m c^2 = \omega_m^2 = (m\pi c/L)^2$ and

$$f_m(t) = \frac{2}{L} \int_0^L f(x, t) \sin \frac{m\pi x}{L} \, dx \tag{9.34}$$

Correspondingly, we have from

$$u(x, 0) = \sum_{n=1}^{\infty} A_n(0) \sin \frac{n\pi x}{L} = u_0(x)$$

$$u_t(x, 0) = \sum_{n=1}^{\infty} A_n^{\bullet}(0) \sin \frac{n\pi x}{L} = v_0(x) \tag{9.35}$$

the initial conditions

$$A_m(0) = \mu_m \qquad A_m^{\cdot}(0) = \nu_m \tag{9.36}$$

for the coefficients $A_m(t)$, where

$$\{\mu_m, \nu_m\} = \frac{2}{L} \int_0^L \{u_0(x), v_0(x)\} \sin \frac{m\pi x}{L} dx \tag{9.37}$$

$m = 1, 2, \ldots$. For each m, the ODE (9.33) and the initial conditions (9.36) define an IVP in ODE. Its solution is found by the method of variation of parameters to be

$$A_m(t) = \mu_m \cos \omega_m t + \nu_m \sin \omega_m t + \int_0^t \sin[\omega_m(t - \tau)] f_m(\tau) d\tau \tag{9.38}$$

As an application of the general result, consider the case $f(x, t) = \sin \omega t \, \delta(x - x_0)$, $0 < x_0 < L$, where $\delta(x)$ is the Dirac delta function. The excitation in this case is a periodic point force for which

$$f_m(t) = \frac{2}{L} \sin \omega t \sin \frac{m\pi x_0}{L} \equiv F_m \sin \omega t$$

For simplicity, let the string be initially at rest in its equilibrium position [with $u_0(x) \equiv v_0(x) \equiv 0$]. We have then $\mu_m = \nu_m = 0$ and

$$A_m(t) = \frac{F_m}{\omega_m} \int_0^t \sin[\omega_m(t - \tau)] \sin \omega\tau \, d\tau$$

$$= \frac{F_m/\omega_m}{\omega^2 - \omega_m^2} (\omega \sin \omega_m t - \omega_m \sin \omega t) \tag{9.39}$$

The Fourier series solution (9.31) for this problem is now determined to be

$$u(x, t) = \sum_{m=1}^{\infty} \frac{F_m}{\omega_m} (\omega \sin \omega_m t - \omega_m \sin \omega t) \frac{\sin(m\pi x/L)}{\omega^2 - \omega_m^2} \tag{9.40}$$

This solution is not valid if $\omega = \omega_k$ for some integer k. For that case, we have

$$u(x, t) = \sum_{m \neq k}^{\infty} \frac{F_m}{\omega_m} (\omega_k \sin \omega_m t - \omega_m \sin \omega_k t) \frac{\sin(m\pi x/L)}{\omega_k^2 - \omega_m^2}$$

$$+ \frac{F_k}{2\omega_k^2} [\sin \omega_k t - (\omega_k t) \cos \omega_k t)] \tag{9.41}$$

The last term in (9.41) increases without bound as $t \to \infty$. We have therefore a *resonance* phenomenon similar to the corresponding phenomenon in dynamical systems such as forced spring-mass systems. In fact, the reduction of the wave equation to the system of ODE (9.33) allows us to call vibrating space-time systems such as the vibrating string "dynamical systems with infinite degrees of freedom." Thus,

9G: When forced at one of its natural frequencies, $\omega_m = m\pi c / L$, resonance occurs; otherwise, the string response is again a combination of standing waves.

9.6   Small-Amplitude Vibration of an Elastic Membrane

Consider an elastic membrane of uniform mass density per unit area ρ lying within the closed curve E in the x,y plane. The membrane is in static equilibrium under uniform tension T_0 per unit arc length along the edge of the membrane. At $t = 0$, the membrane is set in motion in the z direction by an external load intensity (per unit undeformed area) $F(x, y, t)$ in the same direction while its edge remains fixed. We are only interested in situations where the amplitude of the motion is small and motions in directions parallel to the x,y plane are negligible.

As in the analysis of the string, negligible horizontal motion and no external load parallel to the x,y plane imply that the (horizontal) tension is uniform throughout the membrane. Let $u(x, y, t)$ be the vertical displacement of the point (x, y) of the membrane, with the deformed membrane making an angle θ with the x axis and an angle ϕ with the y axis (Figure 9.3). The balance of forces in the vertical direction for a rectangular elemental area of the membrane with edge lengths Δx and Δy before deformation requires, in the limit as Δx and Δy tend to zero,

$$\rho \frac{\partial^2 u}{\partial t^2} = \frac{\partial}{\partial x}(T \sin \theta) + \frac{\partial}{\partial y}(T \sin \phi) + F(x, y, t) \qquad (9.42)$$

with $T \sin \xi = T_0 \tan \xi$. For small deformations, we have $\sin \theta \simeq \tan \theta \simeq u_x$ and $\sin \phi \simeq \tan \phi \simeq u_y$ so that (9.42) may be approximated by

$$u_{tt} = c^2(u_{xx} + u_{yy}) + f(x, y, t) = c^2 \nabla^2 u + f(x, y, t) \qquad (9.43)$$

where $f = F/\rho$, $c^2 = T_0/\rho$ and ∇^2 is the two-dimensional Laplace operator in the x,y plane.

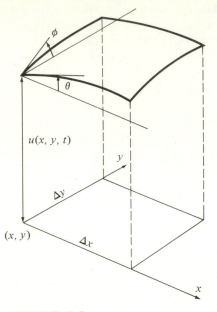

FIGURE 9.3

Equation 9.39 is in the form of an inhomogeneous wave equation in two spatial dimensions. The edge of the membrane, denoted by the curve E, is constrained from motion so that

$$u(x, y, t) = 0 \qquad \text{for } (x, y) \text{ on } E \tag{9.44}$$

At some initial time $t = 0$, we know the displacement and velocity of the membrane so that

$$u(x, y, 0) = u_0(x, y) \qquad u_t(x, y, 0) = v_0(x, y) \tag{9.45}$$

for all points inside E. The PDE (9.43), the boundary condition on E (9.44), and the initial conditions (9.45) define an initial-boundary value problem for $u(x, y, t)$. For a rectangular membrane, this problem can be solved for $u(x, y, t)$ by the method of separation of variables introduced in Section 9.3 leading to a two-dimensional Fourier series representation of the solution [see Haberman (1983)]. This form of the solution also shows that $u(x, y, t)$ is generally composed of (doubly) infinite numbers of two-dimensional standing-wave solutions.

When the edge curve E of the membrane is circular, it is more convenient and fruitful to use cylindrical coordinates with the origin positioned at the center of the membrane. With

$$\nabla^2 u = \nabla \cdot (\nabla u) = \left(\mathbf{i}_r \frac{\partial}{\partial r} + \mathbf{i}_\theta \frac{1}{r} \frac{\partial}{\partial \theta} \right) \cdot \left(\mathbf{i}_r \frac{\partial u}{\partial r} + \mathbf{i}_\theta \frac{1}{r} \frac{\partial u}{\partial \theta} \right)$$

$$= \frac{\partial^2 u}{\partial r^2} + \frac{1}{r} \frac{\partial u}{\partial r} + \frac{1}{r^2} \frac{\partial^2 u}{\partial \theta^2} \tag{9.46}$$

the PDE (9.43) in polar coordinates takes the form

$$u_{tt} = c^2 \left(u_{rr} + \frac{1}{r} u_r + \frac{1}{r^2} u_{\theta\theta} \right) + \bar{f}(r, \theta, t) \tag{9.47}$$

where $\bar{f}(r, \theta, t) \equiv f(r \cos \theta, r \sin \theta, t)$. The boundary condition (9.44) becomes

$$u \mid_{r=a} = 0 \tag{9.48}$$

while the initial conditions (9.45) become

$$u \mid_{t=0} = \bar{u}_0(r, \theta) \qquad u_t \mid_{t=0} = \bar{v}_0(r, \theta) \tag{9.49}$$

where $\bar{u}_0(r, \theta) \equiv u_0(r \cos \theta, r \sin \theta)$, and $\bar{v}_0(r, \theta) \equiv v_0(r \cos \theta, r \sin \theta)$.

9.7   Axisymmetric Modes of a Circular Membrane

Consider a circular elastic membrane of radius a set into unforced axisymmetric motion. In that case, $\partial u / \partial \theta = 0$ and the two-dimensional wave equation in polar coordinates (9.47) with $f(r, \theta, t) \equiv 0$ becomes

$$u_{tt} = c^2 \left(u_{rr} + \frac{1}{r} u_r \right) \tag{9.50}$$

We seek a solution of this equation in the form $u(r, t) = R(r) T(t)$. Upon substitution into (9.50), we obtain

$$\frac{1}{c^2} \frac{T^{\cdot\cdot}}{T} = \frac{R'' + r^{-1} R'}{R} = -\lambda \tag{9.51}$$

By setting $\xi = \sqrt{\lambda} r$, the equation for R may be written in terms of ξ as

$$\frac{d^2 R}{d\xi^2} + \frac{1}{\xi} \frac{dR}{d\xi} + R = 0 \tag{9.52}$$

For auxiliary conditions, we have $u(a, t) = R(a)T(t) = 0$ so that

$$R|_{r=a} = R|_{\xi=\sqrt{\lambda}a} = 0 \qquad (9.53)$$

We also require that $u(r, t)$ be bounded throughout the membrane so that $R(0)$ is finite.

It is not possible to express the solution for R in terms of elementary functions. We may attempt a solution of the ODE (9.52) in a Taylor series in ξ. [Because of the singularity at $\xi = 0$, the Frobenius method would be appropriate for (9.52). But one Taylor series turns out to be possible in this case.] Let

$$R = R_0 + R_1\xi + R_2\xi^2 + \cdots = \sum_{n=0}^{\infty} R_n\xi^n \qquad (9.54)$$

Correspondingly, we have

$$\frac{dR}{d\xi} = R_1 + 2R_2\xi + \cdots = \sum_{n=0}^{\infty} nR_n\xi^{n-1}$$

$$\frac{d^2R}{d\xi^2} = 2R_2 + 3\cdot 2R_3\xi + \cdots = \sum_{n=0}^{\infty} n(n-1)R_n\xi^{n-2} \qquad (9.55)$$

The ODE (9.52) may be written as

$$0\cdot R_0\xi^{-2} + 1\cdot R_1\xi^{-1} + \sum_{n=2}^{\infty} [n(n-1)+n]R_n\xi^{n-2} + \sum_{n=0}^{\infty} R_n\xi^n = 0 \qquad (9.56)$$

A change of the index of summation in the first sum from n to $k+2$ yields

$$R_1\xi^{-1} + \sum_{k=0}^{\infty} [(k+2)^2R_{k+2} + R_k]\xi^k = 0 \qquad (9.57)$$

For (9.57) to hold for all ξ, we must have

$$R_1 = 0 \quad \text{and} \quad R_{k+2} = -\frac{R_k}{(k+2)^2} \qquad (k = 0, 1, 2, 3, \ldots) \qquad (9.58)$$

With $R_1 = 0$, the *recurrence relation* in (9.58) requires $R_{2n+1} = 0$, $n = 0, 1, 2, \ldots$. It follows that R is even in ξ with R_0 arbitrary and all other R_{2n}

expressed in terms of R_0 by way of the recurrence relation. It is a straight-forward calculation to show that

$$R = R_0 \left[1 - \left(\frac{\xi}{2}\right)^2 + \frac{1}{(2!)^2} \left(\frac{\xi}{2}\right)^4 - \frac{1}{(3!)^2} \left(\frac{\xi}{2}\right)^6 + \cdots \right]$$

$$= R_0 \sum_{n=0}^{\infty} \frac{(-1)^n}{(n!)^2} \left(\frac{\xi}{2}\right)^{2n} \equiv R_0 J_0(\xi) \tag{9.59}$$

The series solution in (9.59) occurs so often in applications that it has been named J_0, the zeroth order *Bessel function of the first kind.* The actual series for $J_0(\xi)$ and its many useful properties can be found in Abramowitz and Stegun (1965).

We know that a second-order ODE such as (9.52) should have two complementary solutions. Once we have found one, we can always obtain the second one by the method of reduction of order. Set $R = J_0(\xi)v(\xi)$ and substitute this into the ODE for R to get

$$J_0(\xi) \frac{d^2 v}{d\xi^2} + \left(2 \frac{dJ_0}{d\xi} + \frac{1}{\xi} J_0 \right) \frac{dv}{d\xi} + \left(\frac{d^2 J_0}{d\xi^2} + \frac{1}{\xi} \frac{dJ_0}{d\xi} + J_0 \right) v = 0 \tag{9.60}$$

The last factor in parentheses vanishes because J_0 itself is a complementary solution of (9.52). What remains can be rearranged to read

$$\frac{d^2 v/d\xi^2}{dv/d\xi} + 2 \frac{dJ_0/d\xi}{J_0} + \frac{1}{\xi} = 0 \tag{9.61}$$

which can be integrated once immediately to get

$$\ln \frac{dv}{d\xi} + \ln J_0^2 + \ln \xi = \ln \left(\xi J_0^2 \frac{dv}{d\xi} \right) = C_0$$

or

$$\frac{dv}{d\xi} = \frac{C_1}{\xi J_0^2} \tag{9.62}$$

where $C_1 = e^{C_0}$ is an arbitrary constant of integration. Upon integrating both sides of (9.62) once more, we find

$$v(\xi) = C_2 + C_1 \int^\xi \frac{dz}{z[J_0(z)]^2} = C_2 + C_1 \int^\xi \left[1 + 2\left(\frac{z}{2}\right)^2 + \cdots \right] \frac{dz}{z}$$

$$= C_2 + C_1 \left(\ln \xi + \frac{\xi^2}{4} + \cdots \right) \tag{9.63}$$

The general solution for R is therefore

$$R = J_0(\xi)\left[C_2 + C_1\left(\ln \xi + \frac{\xi^2}{4} + \cdots\right)\right] \tag{9.64}$$

which contains two arbitrary constants of integration C_1 and C_2 as we would expect.

We now note that the solution (9.64) is unbounded at $r = 0 (\xi = 0)$ unless $C_1 = 0$. As we expect the displacement $u(r, t)$ to be generally bounded throughout the membrane, we must eliminate the singularity by setting $C_1 = 0$ leaving us with

$$R = C_2 J_0(\xi) = c_2 J_0(\sqrt{\lambda}r) \tag{9.65}$$

Now, the boundary condition (9.53) requires $C_2 J_0(\sqrt{\lambda}a) = 0$. We do not want a trivial solution for R; so we must choose λ so that $J_0(\sqrt{\lambda}a) = 0$. In other words, we must have

$$\lambda = \frac{z_m^2}{a^2} \equiv \lambda_m \qquad (m = 1, 2, 3, \ldots) \tag{9.66}$$

where z_m is the mth positive root of $J_0(z)$. These roots are arranged in order of increasing magnitude. [Note that $J_0(z)$ is even in z but the negative roots do not add any new eigenvalues.] To the extent that the ODE (9.52) is effectively a spring-mass-dashpot system with monotone decreasing damping, the solution must reflect the underdamped oscillatory nature of the system. Hence it must have an infinite number of positive simple roots. The actual graph and roots of $J_0(z)$ can be found in Abramowitz and Stegun (1965). To three significant figures after the decimal, the first five roots are

m	1	2	3	4	5
z_m	2.404…	5.520…	8.654…	11.792…	14.931…

For each eigenvalue $\lambda_m = z_m^2/a^2$, we have as the corresponding eigenfunction

$$R_m(r) \equiv J_0\left(z_m \frac{r}{a}\right) \tag{9.67}$$

up to an amplitude factor (see Figure 9.4). From the first part of (9.51), we get $T^{\cdot\cdot} + \lambda c^2 T = 0$ so that for $\lambda = \lambda_m$

$$T = A_m \cos \omega_m t + B_m \sin \omega_m t \equiv T_m(t) \qquad (9.68)$$

where $\omega_m^2 = \lambda_m c^2 = T_0 z_m^2 / \rho a^2$. Thus, for a particular eigenvalue λ_n the corresponding solution of the two-dimensional wave equation with axisymmetry (9.50) is $u_n(r, t) = R_n(r) T_n(t)$. As in the string case, a sum of the special solutions corresponding to different eigenvalues

$$u(r, t) = \sum_{n=1}^{\infty} (A_n \cos \omega_n t + B_n \sin \omega_n t) J_0\left(\frac{r}{a} z_n\right) \qquad (9.69)$$

is also a bounded solution of the PDE; it satisfies the boundary condition (9.44) as well.

The fundamental frequency is $f_1 = (z_1/2\pi a)\sqrt{T_0/\rho}$, and the higher frequencies are given in terms of f_1 by $f_m = (z_m/z_1) f_1$. Unlike the string case,

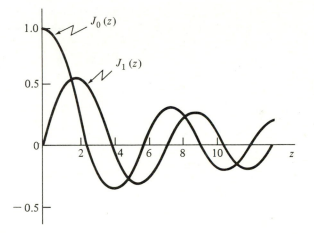

(a)

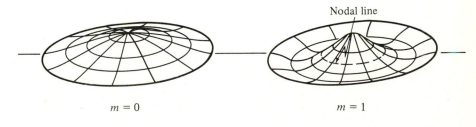

(b)

FIGURE 9.4
(a) Bessel functions of the first kind. (b) Mode shapes.

the higher frequencies are not integer multiples of f_1 as z_m/z_1 are nowhere near an integer value for the five roots listed, for example, $f_2/f_1 = 2.296\ldots$, $f_3/f_1 = 3.599\ldots$, etc. The overtones in the axisymmetric vibration of the circular membrane are therefore not harmonics of the fundamental.

Now, the fundamental component oscillates periodically in time with a period $P_1 = (2\pi a/z_1)\sqrt{\rho/T_0}$. The overtones are also periodic individually with periods $P_m = (2\pi a/z_m)\sqrt{\rho/T_0}$. Since $P_1/P_m = z_m/z_1$ is not an integer, none of the overtones of the circular membrane has a period P_1. In other words, the expression (9.69) is not periodic in time, as (9.24) is for the string case.

> **9H:** There is no periodic combination of the fundamental and overtones in the unforced vibration of circular membranes.

Periodic events are regular and seem to be attractive to human sensation. The periodic motion of a violin string is therefore more appealing (or musical) to our hearing sensation than the nonperiodic motion of the drum head. The only periodic motion of a drum is a pure tone. Pure tones lack the quality, character, and texture of notes with overtones.

A similar analysis can be performed for unsymmetric vibration. We find from such an analysis that the fundamental frequency for the axisymmetric modes is in fact the lowest. Therefore, we are justified to consider the axisymmetric fundamental frequency being responsible for the existence or nonexistence of harmonics. Incidentally, the relevant eigenvalues are difficult to get in all cases concerning circular membranes. The Rayleigh quotient technique introduced in Chapter 5 should be even more useful here.

9.8   The Initial-Value Problem and the Rayleigh Quotient

The solution (9.69) satisfies the PDE (9.50) and the boundary condition $u(a, t) = 0$ as well as the boundedness condition throughout the membrane, particularly at $r = 0$. To solve the initial-value problem [with $\bar{f}(r, t) \equiv 0$], it must also satisfy the initial conditions (9.49) with $\bar{u}_0$ and $\bar{v}_0$ independent of θ (to be consistent with the stipulation of axisymmetry). Therefore, we must choose the unknown coefficients A_n and B_n, $n = 1, 2, 3, \ldots$, in (9.69) so that

$$\sum_{n=1}^{\infty} A_n J_0\left(z_n \frac{r}{a}\right) = \bar{u}_0(r) \qquad \sum_{n=1}^{\infty} \omega_n B_n J_0\left(z_n \frac{r}{a}\right) = \bar{v}_0(r) \qquad (9.70)$$

On the basis of our experience with the string case, it would be desirable to have some kind of orthogonality condition among the eigenfunctions (normal modes). However, unlike the string case, we cannot establish this condition from our knowledge of antiderivatives of elementary functions alone.

To establish the relevant orthogonality, we multiply the ODE for R_n, taken now in the form $R_n'' + r^{-1}R_n' + \lambda_n R_n = 0$, by rR_m and integrate over $(0, a)$ to get

$$0 = \int_0^a (rR_mR_n'' + R_mR_n' + \lambda_n rR_mR_n)\,dr$$

$$= \left[rR_mR_n' - rR_m'R_n \right]_0^a + \int_0^a [(rR_m')' + \lambda_n rR_m]R_n\,dr$$

$$= \int_0^a (-\lambda_m + \lambda_n)rR_mR_n\,dr \tag{9.71}$$

where we have used (i) $R_k(a) = 0$ to eliminate the terms outside the integral sign, and (ii) the ODE for R_m to simplify the integrand. As long as $\lambda_m \neq \lambda_n$ (or $m \neq n$), we have

$$\int_0^a R_m(r)R_n(r)r\,dr = \int_0^a J_0\!\left(z_m\frac{r}{a}\right)J_0\!\left(z_n\frac{r}{a}\right)r\,dr = 0 \tag{9.72}$$

The orthogonality condition (9.72) may now be applied to (9.70)

$$\sum_{n=1}^{\infty} A_n \int_0^a J_0\!\left(z_m\frac{r}{a}\right)J_0\!\left(z_n\frac{r}{a}\right)r\,dr = \int_0^a J_0\!\left(z_m\frac{r}{a}\right)\bar{u}_0(r)r\,dr$$

$$\sum_{n=1}^{\infty} \omega_n B_n \int_0^a J_0\!\left(z_m\frac{r}{a}\right)J_0\!\left(z_m\frac{r}{a}\right)r\,dr = \int_0^a J_0\!\left(z_m\frac{r}{a}\right)\bar{v}_0(r)r\,dr \tag{9.73}$$

The condition (9.72) now simplifies the left sides of (9.73) leaving us with

$$\alpha_m A_m = \int_0^a J_0\!\left(z_m\frac{r}{a}\right)\bar{u}_0(r)r\,dr \qquad \alpha_m\omega_m B_m = \int_0^a J_0\!\left(z_m\frac{r}{a}\right)\bar{v}_0(r)r\,dr \tag{9.74}$$

where

$$\alpha_m \equiv \int_0^a \left[J_0\!\left(z_m\frac{r}{a}\right) \right]^2 r\,dr \tag{9.75}$$

The convergence of the series (9.69) and the convergence of the series in (9.70) to the prescribed data will be assumed in this book. The conditions we should impose on $\bar{u}_0(r)$ and $\bar{v}_0(r)$ for convergence can be found in Whittaker and Watson (1952).

To find the eigenvalues for circular membranes, we need to find the roots of the function $J_0(z)$. Often, an approximate solution by a Rayleigh quotient (similar to that discussed in Chapter 5) is useful. For the present problem, we multiply the ODE for R by $rR(r)$ and integrate over $(0, a)$. The result is

$$
\begin{aligned}
0 &= \int_0^a \left(R'' + \frac{1}{r} R' + \lambda R \right) Rr \, dr \\
&= \left[rR'R \right]_0^a + \int_0^a [\lambda R^2 - (R')^2] r \, dr \\
&= \int_0^a [\lambda R^2 - (R')^2] r \, dr
\end{aligned}
\tag{9.76}
$$

or

$$
\lambda = \frac{\int_0^a (R')^2 r \, dr}{\int_0^a R^2 r \, dr}
\tag{9.77}
$$

It follows from (9.77) that

9I: The fundamental frequency is the minimum value of the relevant Rayleigh quotient.

By taking $R = [1 - (r/a)^2]$ [which satisfies the boundary condition $R(a) = 0$], we get, from (9.77),

$$
\lambda = \frac{1}{a^2/6} = \frac{6}{a^2}
\tag{9.78}
$$

which differs from the lowest eigenvalue $\lambda_1 = (2.404\ldots)^2/a^2$ by about 4 percent.

EXERCISES

1. Solve the initial-boundary value problem (IBVP) defined by (9.7) with $f(x, t) \equiv 0$, (9.9) and (9.8) with $u_0(x) = \sin(2\pi x/L)$ and $v_0(x) = \cos(3\pi x/L)$.

2. Solve the IBVP:

$$u_{tt} = c^2 u_{xx} \qquad (0 < x < L, \, t > 0)$$

$$u(0, t) = 0 \qquad u_x(L, t) = 0 \qquad (t > 0)$$

$$u(x, 0) = u_0(x) \qquad u_t(x, 0) = v_0(x) \qquad (0 < x < L)$$

3. Solve IBVP (9.7)–(9.9) with $f(x, t) = x(L - x)$ and $u_0(x) \equiv v_0(x) \equiv 0$.

4. Obtain the Green's function $G(x, t; x_0, t_0)$ for the one-dimensional wave equation with $G \equiv 0$ at $x = 0$ and at $x = L$:

$$G_{tt} = c^2 G_{xx} + \delta(x - x_0)\delta(t - t_0)$$

$$G|_{x=0} = G|_{x=L} = 0 \qquad G|_{t<t_0} = G_t|_{t<t_0} = 0$$

5. Derive an expression for the solution $u(x, t)$ of the IBVP (9.7)–(9.9) with $u_0 \equiv v_0 \equiv 0$ in terms of the Green's function of Exercise 4.

6. Repeat Exercise 4 with $G = 0$ at $x = L$ replaced by $G_x = 0$ at $x = L$.

7. For $\beta^2 < 4\pi^2 c^2/L^2$, obtain the solution of

$$u_{tt} = c^2 u_{xx} - \beta u_t \qquad (0 < x < L, \, t > 0)$$

$$u(0, t) = u(L, t) = 0$$

$$u(x, 0) = u_0(x) \qquad u_t(x, 0) = v_0(x)$$

8. Use the method of eigenfunction expansion to obtain the solution of the vibrating rectangular membrane problem:

$$u_{tt} = c^2(u_{xx} + u_{yy}) \qquad (0 < x < a, \, 0 < y < b, \, t > 0)$$

$$u(0, y, t) = u(a, y, t) = u(x, 0, t) = u(x, b, t) = 0$$

$$u(x, y, 0) = u_0(x, y) \qquad u_t(x, y, 0) = v_0(x, y)$$

9. Obtain by eigenfunction expansion the Green's function for a vibrating rectangular membrane with fixed edges.

10. **(a)** Appeal to the superposition principle and write down the solution of the forced vibration problem for a rectangular membrane with fixed edges in terms of the Green's function of Exercise 9. You may take the membrane to be at rest initially.

(b) Derive an expression for the same solution in terms of the adjoint Green's function G^* and show that $G(x, t; x_0, t_0) = G^*(x_0, t_0; x, t)$.

11. **(a)** Apply the method of eigenfunction expansion directly to the forced vibration of a rectangular membrane with fixed edges and initially at rest.

    (b) Repeat part (a) for the case of axisymmetric forced vibration of a circular membrane with a fixed edge and initially at rest.

12. For a vibrating circular membrane *without* axisymmetry, use the fact that u must be single-valued and bounded at every point of the membrane to obtain the solution of the IBVP with $f(r, \theta, t) \equiv 0$ in terms of Bessel functions of the first kind $J_m(z)$, $m = 0, 1, 2, \ldots$, which is the solution (unique up to an amplitude factor) of the ODE

$$\frac{d^2 J_m}{dz^2} + \frac{1}{z}\frac{dJ_m}{dz} - \frac{m^2}{z^2} J_m + J_m = 0$$

    which is bounded everywhere in $|z| < \infty$. To be concrete, treat only the case of a membrane fixed at its edge $r = a$.

13. Solve the IBVP for a pie-shaped membrane with fixed edges and no interior forcing. The initial conditions are

$$u(r, \theta, 0) = u_0(r, \theta) \qquad u_t(r, \theta, 0) = v_0(r, \theta)$$

14. Obtain the Rayleigh quotient for a fixed edge membrane in sinusoidal oscillation, that is, $u(x, y, t) = v(x, y)e^{i\omega t}$ so that v is a solution of

$$\nabla^2 v + \lambda v = 0 \qquad \lambda = \frac{\omega^2}{c^2}$$

15. **(a)** Show that $G_t(x, t; x_0, t_0) = -G_{t_0}(x, t; x_0, t_0)$.
    (b) Show that the solution of the IBVP with $f(x, t) \equiv 0$, $u_0(x) = g(x)$, and $v_0(x) \equiv 0$ is the time derivative of the solution of the IBVP with $f(x, t) = 0$, $u_0(x) \equiv 0$, and $v_0(x) = g(x)$.

PART **IV**

Diffusion

If you pick up a metal rod at room temperature and heat it at the far end, the near end of the rod in your hand will become hotter and hotter. Similar to the ripple generated by the pebble, something (heat in this case) seems to be propagating from one end of the rod to the other. However, the spreading of heat in a rod is qualitatively different from the "propagation" of the wave generated by dropping the pebble into a pond. It is not because there is no ripple associated with diffusion of heat in the rod (and note that we do not see any ripple associated with propagation of electromagnetic waves either); rather, we observe (by careful measurement if necessary) that some heat reaches your hand the instant the rod is heated at the far end while it takes a finite time for the ripple in the pond to travel a finite distance. Also, when we stop heating the rod (but continue to hold it in our hand), its temperature eventually redistributes itself and reaches a new steady-state temperature distribution. In contrast, the water at a particular location in the pond returns to its rest position after the ripple passes through that location. The phenomenon of heat conduction is in fact governed by a different mechanism, known as *diffusion*. Diffusion plays an important role in many other phenomena of technical interest, notably the behavior of randomly moving particles (such as smoke particles) and motion of objects extremely sensitive to small changes of initial conditions and system parameter values (e.g., the kinetic theory of gas).

In Chapter 10, the problem of one-dimensional heat conduction is introduced indirectly by way of the traffic-flow problem. The simple model of Chapter 7 for this problem is modified by incorporating the effect of sensitivity to traffic gradient. For a particular class of state equations, the relevant second-order nonlinear PDE can be transformed into the linear one-dimensional equation of heat conduction. The heat equation is then solved by the method of Fourier transforms and again by the use of a similarity variable obtained with the help of a dimensional analysis. The solution allows us to eliminate the overhang of the corresponding multivalued solution which appears in the simple Lighthill-Whitham theory of traffic flows of Chapter 7, leaving us with an acceptable concept of the shock phenomenon.

A different and rather novel occurrence of the diffusion mechanism is discussed in Chapter 11. It is concerned with the effect of diffusivity of fish populations on a fishing moratorium within the 200-mile fishing limit. This effect is studied through a search for possible equilibrium populations and their stability and bifurcation. The perturbation method for eigenvalue problems in differential equations is introduced in conjunction with linear stability analyses.

10

$\overline{\textit{A Hot Rod in Traffic}}$

Sensitivity to a Sharp Traffic Density Gradient

A nonlinear PDE of second order is introduced and transformed into the linear equation of heat conduction. Methods of solution for the latter including Fourier transforms and similarity variables are developed.

10.1 Burger's Equation and the Cole-Hopf Transformation

In Chapter 7, we saw how the Lighthill-Whitham continuum theory of traffic flow may give rise to a multivalued description of the traffic density and flow rate. This is physically unacceptable as there is only one possible traffic density at a given position and time. To remove this deficiency, we modify the simple theory of Chapter 7 by incorporating one additional feature which becomes important when there is a large change in traffic density over a short distance. The equation of state used in the simple theory, $q = Q(k)$, effectively assumes that a driver's speed is only a function of the traffic density at the particular location of the driver. The local dependence allows cars to go at high speed until they get to the edge of the traffic hump; only then do they adjust their speed abruptly to account for the change in traffic density. This leads to a sharper and sharper density gradient at the edge of the traffic hump entrance, eventually a shock profile and its overhang. In real life, drivers do look ahead for change in traffic density; they generally slow down well before the traffic hump once they sense (or see) the heavy traffic at a stretch of road some distance in front. The effect of this "looking ahead" driving habit can be included in a continuum theory by a modified equation of state of the form

$$q = Q(k) - \frac{\nu}{2} k_x \qquad\qquad (10.1)$$

where ν is a constant, usually "small" since drivers are normally sensitive only to a substantial density gradient. Corresponding to this new equation of state, we have

$$q_x = \frac{dQ}{dk} k_x - \frac{\nu}{2} k_{xx} \equiv c(k) k_x - \frac{\nu}{2} k_{xx} \qquad\qquad (10.2)$$

and the conservation law $k_t + q_x = 0$ becomes

$$k_t + c(k) k_x - \frac{\nu}{2} k_{xx} = 0 \qquad\qquad (10.3)$$

This is a second-order PDE and is linear only if $c =$ constant.

For a quadratic $Q(k)$, c and k are related linearly, say $c = \alpha + \beta k$. If we use c as the primary unknown for that case, Equation (10.3) becomes the "cleaner" *Burger's equation:*

$$c_t + cc_x = \frac{\nu}{2} c_{xx} \qquad\qquad (10.4)$$

In particular, this equation applies to the example $Q = 3k(200 - k)/10$ studied in Chapter 7 for which $c = 3(100 - k)/5$. (For a more general Q, it can be argued that Equation 10.4 for c, gives a good first approximation for the exact equation.) If the nonlinear term cc_x were absent in (10.4), then c would be governed by the equation of one-dimensional heat conduction $c_t = \nu c_{xx}/2$ (which can be solved by classical techniques). As heat tends to diffuse over a period of time, we expect the solution of the related Burger's equation also to exhibit a "diffusive" behavior.

It turns out that even with the nonlinear term cc_x, Burger's equation can be transformed into the equation for heat conduction by a change of variable

$$c = -\nu \frac{\psi_x}{\psi} \qquad\qquad (10.5)$$

This so-called Cole-Hopf transformation began to play a significant role in nonlinear partial differential equations when it was rediscovered in 1950–

1951 for problems in gas dynamics. Upon substituting this expression into the PDE (10.4) for c, we get, after some calculation,

$$\psi_t = \frac{\nu}{2}\psi_{xx} \tag{10.6}$$

Equation 10.6 arises naturally in heat conduction problems in one spatial dimension (Haberman, 1983). These problems are easier to visualize and the results are easier to interpret than those for the transformed traffic flow problem. The next few sections will be devoted to this classical problem in diffusion. Meanwhile, we should pause to relish the rather astonishing accomplishment made possible by the Cole-Hopf transformation. It establishes that

10A: The gradient (or looking-ahead) theory of traffic flow is effectively equivalent to the diffusion of heat in a hot rod!

In case it has escaped your notice, we point out here that the heat equation (10.6) is a linear equation while Burger's equation is nonlinear. Strictly speaking, the reduction is accomplished only for a quadratic $Q(k)$ in the equation of state (10.1). But applied mathematicians rejoice every time a difficult nonlinear problem is reduced to a linear one. A linear problem is as good as solved.

10.2 One-Dimensional Heat Conduction in a Straight Rod

Consider a straight rod of *length L, mass density* (per unit volume) ρ, and uniform *cross-sectional area A*. The cylindrical surface of the rod is insulated from heat flow. The rod is sufficiently slender so that the temperature distribution is uniform over a cross section at any point along the *central axis* of the rod. Let the central axis be positioned along the x axis extending from $x = 0$ to $x = L$. To study the *temperature distribution* $u(x, t)$ in the rod, we look at a segment of the rod between $x = a$ (>0) and $x = b$ ($<L$) at time t. Let c_0 be the *specific heat* of the rod material defined to be the amount of heat energy needed to raise the temperature of a unit mass of material by one degree Kelvin (1 K). The specific heat is a material property and can be measured experimentally. The rate of change in the *heat* (thermal) *energy* $H(x, t)$ in this segment is given by

$$\frac{dH}{dt} = \frac{d}{dt}\int_a^b c_0 u(x, t)\rho A \, dx \tag{10.7}$$

There is another way of calculating the same rate of change of heat energy. Let $\phi(x, t)$ be the amount of heat per unit time flowing from left to right across a unit area of the cross section at location x and at time t. Let $F(x, t)$ be the rate of heat generated by a unit mass of the rod within the insulated cylindrical surface by chemical, electrical, or other processes. In that case, we have (see Figure 10.1)

$$\frac{dH}{dt} = A[\phi(a, t) - \phi(b, t)] + \int_a^b F(x, t)\rho A\, dx \qquad (10.8)$$

The two expressions (10.7) and (10.8) for dH/dt must be identical. The rate of increase in heat in the segment (calculated by way of the temperature of the segment) must be equal to the rate of heat gained (calculated by the net heat flow across the two cross sections of the segment plus the rate of heat generated inside the segment). We have therefore

$$\frac{d}{dt} \int_a^b c_0\rho u(x, t)\, dx = \phi(a, t) - \phi(b, t) + \int_a^b \rho F(x, t)\, dx \qquad (10.9)$$

where c_0 and ρ may vary with x and t. Suppose both u and ϕ are both smooth functions and ρ is independent of t. In that case, we write (10.9) as

$$\int_a^b [c_0\rho u_t(x, t) - \rho F(x, t) + \phi_x(x, t)]\, dx = 0 \qquad (10.10)$$

If the integrand is continuous in x and the segment (a, b) is arbitrary, we conclude (as in Chapters 5, 8, and 9) that

$$c_0\rho u_t = -\phi_x + \rho F \qquad (10.11)$$

Equation 10.11 actually holds for all x in $(0, L)$ and all $t > 0$. It may be considered a local conservation law for the heat energy in the rod.

Equation 10.11 is one equation for two unknowns: u and ϕ. We know from our daily experience that there is a relation between the temperature gradient at a point in a body and the heat flow rate (per unit area) across

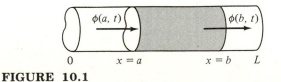

FIGURE 10.1

that position—the sharper the gradient the higher is the flow away from the higher-temperature location. A description of this relation is the *Fourier law:*

$$\phi = -K_0 u_x \tag{10.12}$$

where the *coefficient of thermal conductivity* K_0 is a measure of the ability of a material to diffuse or conduct heat. K_0 has to be determined experimentally and it varies from material to material. For some materials, it may even vary with the temperature of the material. Upon substitution of (10.12) into (10.11), we get a single second-order PDE for $u(x, t)$:

$$c_0 \rho u_t = (K_0 u_x)_x + \rho F \tag{10.13}$$

We limit ourselves in this chapter to the case where K_0 is a known quantity. If, in addition, the rod is homogeneous in its ability to conduct heat, K_0 is a constant and (10.13) simplifies to read

$$c_0 \rho u_t = K_0 u_{xx} + \rho F \tag{10.14}$$

In either form, (10.13) or (10.14), the PDE is supplemented by initial and boundary conditions. We normally know the temperature distribution of the rod at some reference time $t = 0$ so that

$$u(x, 0) = u_o(x) \qquad (0 \le x \le L) \tag{10.15}$$

Since the PDE is first order in t, this is all we can prescribe (and in fact all we know) at the initial time. The end $x = 0$ of the rod may be kept at some ambient temperature of its surroundings (which may change with time) so that

$$u(0, t) = U_0(t) \qquad (t > 0) \tag{10.16}$$

Other physically meaningful end conditions are also possible. For example, the end may be insulated from heat loss (or gain) so that, by Fourier's law, we have

$$u_x(0, t) = 0 \qquad (t > 0) \tag{10.17}$$

The situation for the other end $x = L$ is similar.

In contrast to the presence of shocks and overhangs in the solution for traffic density obtained in Chapter 7, the heat conduction problem does not have a multivalued solution. In fact, it has a unique solution. To see this, suppose it has two different solutions $u_1(x, t)$ and $u_2(x, t)$ (which may agree for some ranges of x and t). Let $u = u_2 - u_1$; then u satisfies the homogeneous PDE (10.13) with $F \equiv 0$. To be definite, suppose both ends are maintained at a fixed temperature, $U_0(t)$ at $x = 0$ and $U_L(t)$ at $x = L$. In that case, we have also

$$u(0, t) = u(L, t) = 0 \tag{10.18}$$

and

$$u(x, 0) = 0 \tag{10.19}$$

Multiply the homogeneous version of the PDE (10.14) by u and integrate over $(0, L)$ to get

$$\int_0^L c_0\rho u u_t \, dx = \int_0^L [K_0 u_x]_x u \, dx = \left[K_0 u_x u \right]_0^L - \int_0^L K_0(u_x)^2 \, dx \tag{10.20}$$

The first term on the right vanishes because of (10.18). (Incidentally, it also vanishes if the heat flow is prescribed at one or both ends instead.) The integrand of the left integral may be written as $c_0\rho(u^2)_t/2$ so that (10.20) becomes (for c_0 and ρ independent of t)

$$\frac{1}{2} \frac{d}{dt} \int_0^L c_0\rho u^2 \, dx = -\int_0^L K_0(u_x)^2 \, dx \equiv -P(t) \tag{10.21}$$

where $P(t)$ is nonnegative. Integrate both sides over $(0, T)$ and we get for $T > 0$

$$\left[\frac{1}{2} \int_0^L c_0\rho u^2 \, dx \right]_{t=T} = -\int_0^T P(t) \, dt \tag{10.22}$$

as the integral inside the brackets on the left vanishes at $t = 0$ because of (10.19). The left side of (10.22) is nonnegative as $u^2 \geq 0$ while the right side is nonpositive because $P(t) \geq 0$. The only way (10.22) can be true is

$$\int_0^L c_0\rho u^2(x, T) \, dx \equiv 0 \tag{10.23}$$

With $c_0\rho > 0$ and $u^2 \geq 0$, (10.23) holds only if $u(x, T) \equiv 0$. As T is arbitrary, we have $u(x, t) = 0$ or $u_1(x, t) = u_2(x, t)$ for $0 \leq x \leq L$ and $t \geq 0$. In other words, the two supposedly different solutions must in fact be the same. Given the initial temperature distribution of the rod,

10B: The temperature at any point of the rod is uniquely determined for all future time.

10.3   Infinite Domain Problem by Fourier Transforms

For constant values of c_0, ρ, and K_0, Equation 10.14 may be written as

$$u_t = \frac{\nu}{2}\, u_{xx} + f(x, t) \tag{10.24}$$

where $\nu/2 = K_0/c_0\rho$ and $f = F/c_0$. As a model for the temperature distribution in a rod, (10.24) is supplemented by an initial condition (10.15) at $t = 0$ and two boundary conditions such as (10.16) or (10.17) at the two ends of the rod. This initial-boundary value problem may be solved by the method of separation of variables of Chapter 9 and will be left as exercises at the end of this chapter.

The homogeneous form of (10.24) [with $f(x, t) \equiv 0$] is identical to the transformed Burger's equation (10.6). The linear PDE for ψ (or u) is considerably simpler to solve than the nonlinear Burger's equation for $c(x, t)$. By the method of separation of variables, we seek a product form solution $u(x, t) = X(x)T(t)$ and deduce from the PDE for $u(x, t)$

$$X'' + \lambda X = 0 \qquad T^{\boldsymbol{\cdot}} + \tfrac{1}{2}\nu\lambda T = 0 \tag{10.25}$$

The solution for $X(x)$ is therefore

$$X = A \cos \mu x + B \sin \mu x \qquad \mu^2 = \lambda \tag{10.26}$$

Now, unlike the heat conduction in a finite rod, there are no boundary conditions for the determination of the unknown eigenvalue parameter λ. The road is assumed to extend indefinitely in both directions so that $-\infty < x < \infty$. There are no special requirements on x at $\pm\infty$ except for boundedness, and the solution (10.26) is bounded for all x. We have effectively a continuous spectrum of eigenvalues $-\infty < \mu < \infty$ (or $\lambda > 0$) for which a solution of the form (10.26) is possible. Note that $\lambda < 0$ is unacceptable as μ would be complex and X would be unbounded in general.

From the second equation in (10.25), we get $T = Ce^{-\nu\mu^2 t/2}$ so that the corresponding solution for u is

$$u_\lambda(x, t; \lambda) = e^{-\nu\mu^2 t/2}[\bar{A}(\mu) \cos \mu x + \bar{B}(\mu) \sin \mu x] \tag{10.27}$$

where $\lambda = \mu^2$ and we have allowed $\bar{A}$ and $\bar{B}$ to change with μ. To satisfy a general initial condition, we sum up the solutions for all real μ:

$$u(x, t) = \int_{-\infty}^{\infty} [\bar{A}(\mu) \cos \mu x + \bar{B}(\mu) \sin \mu x]e^{-\nu\mu^2 t/2}\, d\mu$$

With $A(\mu) = \bar{A}(\mu) + \bar{A}(-\mu)$ and $B(\mu) = \bar{B}(\mu) - \bar{B}(-\mu)$, we may rewrite the above solution as

$$u(x, t) = \int_0^\infty e^{-\nu\mu^2 t/2}[A(\mu)\cos\mu x + B(\mu)\sin\mu x]\,d\mu \qquad (10.28)$$

To satisfy the initial condition, say $u(x, 0) = u_0(x)$ for $-\infty < x < \infty$, we must have

$$\int_0^\infty [A(\mu)\cos\mu x + B(\mu)\sin\mu x]\,d\mu = u_0(x) \qquad (10.29)$$

where $u_0(x)$ may be obtained from (10.5) for the known initial wave speed $c(x, 0)$. The problem now is to find $A(\mu)$ and $B(\mu)$ in terms of $u_0(x)$. The solution is given by the inversion formulas

$$A(\mu) = \frac{1}{\pi}\int_{-\infty}^\infty u_0(\xi)\cos\mu\xi\,d\xi \quad \text{and} \quad B(\mu) = \frac{1}{\pi}\int_{-\infty}^\infty u_0(\xi)\sin\mu\xi\,d\xi$$

$$(10.30)$$

in the theory of Fourier transforms (Hildebrand, 1976; Haberman, 1983). The solution of our IVP for ψ, given by (10.28), may now be expressed in terms of $u_0(x)$:

$$u(x, t) = \frac{1}{\pi}\int_0^\infty e^{-\nu\mu^2 t/2}\left\{\int_{-\infty}^\infty u_0(\xi)\cos[\mu(x-\xi)]\,d\xi\right\}d\mu$$

$$= \frac{1}{2\pi}\int_{-\infty}^\infty e^{-\nu\mu^2 t/2}\left\{\int_{-\infty}^\infty u_0(\xi)\cos[\mu(x-\xi)]\,d\xi\right\}d\mu \qquad (10.31)$$

Upon reversing the order of integration and making use of the identity (Hildebrand, 1976),

$$\int_{-\infty}^\infty e^{-\mu^2}\cos(2a\mu)\,d\mu = \sqrt{\pi}e^{-a^2} \qquad (10.32)$$

we get

$$u(x, t) = \frac{1}{\sqrt{2\pi\nu t}}\int_{-\infty}^\infty u_0(\xi)e^{-(x-\xi)^2/2\nu t}\,d\xi \qquad (10.33)$$

It is instructive to write (10.33) as

$$u(x,\, t) \equiv \int_{-\infty}^{\infty} G(x,\, t;\, \xi,\, 0)\, u_0(\xi)\, d\xi \qquad (10.34)$$

with the *fundamental solution* for the (one-dimensional) heat equation $G(x,\, t;\, \xi,\, 0)$ given by

$$G(x,\, t;\, \xi,\, 0) = \frac{\exp[-(x - \xi)^2/2\nu t]}{(2\pi\nu t)^{1/2}} \qquad (10.35)$$

This fundamental solution, also called the *Green's function* of the (one-dimensional) heat equation *for an infinite domain,* satisfies the one-dimensional heat equation for u and tends to the Dirac delta function $\delta(x - \xi)$ as t tends to 0 (Figure 10.2). In terms of heat flow, we have the interpretation that

10C: $G(x,\, t;\, \xi,\, 0)$ is the response of a rod of infinite length to a unit amount of heat initially concentrated at $x = \xi$. As time goes on, the energy from the localized hot spot is diffused throughout the rod, eventually resulting in a uniform temperature distribution.

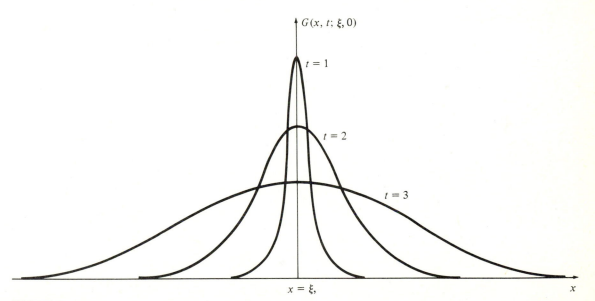

FIGURE 10.2

Also,

10D: The solution for u in the form of (10.34) can be interpreted as the sum of responses to initial hot spots of different strength at different points along the rod.

10.4 Fundamental Solution and Dimensional Analysis

The method described in section 10.3 for the solution of the one-dimensional heat equation requires a knowledge of Fourier transforms and other mathematical results in complex function theory. A more elementary method for obtaining the same solution is by dimensional analysis which we encountered previously in Sections 1.4–1.7. In those sections, we applied dimensional arguments to situations where we did not know the scientific laws governing the phenomenon being investigated. With more information about the phenomenon, we may expect dimensional analysis to yield a complete solution of the problem. In this section, we will show how the fundamental solution for the one-dimensional heat conduction problem may be obtained from the governing PDE by dimensional analysis without reference to Fourier transforms. As the fundamental solution is the response of an infinitely long rod to a unit of heat energy applied at a point $x = 0$ and time $t = 0$, we may take $\rho F(x, t) = F_0 \delta(x) \delta(t)$ with $F_0 = 1/A$ so that

$$\int_{-\infty}^{\infty} \int_{-\infty}^{\infty} \rho F(x, t) A \, dx \, dt = F_0 A = 1$$

Before the instant of application, the rod is at a uniform ambient temperature which we take to be $u = 0$ for $t < 0$. Since the total heat energy in the rod must remain finite, we expect $u \to 0$ as $|x| \to \infty$ [as confirmed by the solution given in (10.35)].

There are five fundamental dimensional units (fdu's) in the above heat conduction problem:

$$M = \text{mass} \qquad L = \text{length} \qquad T = \text{time}$$

$$C = \text{heat (in calories)} \qquad D = \text{temperature (in degrees)}$$

Correspondingly, the dimensions of the variables and parameters in the problem are

$$[\rho] = ML^{-3} \qquad [c_0] = CD^{-1}M^{-1} \qquad [K_0] = CD^{-1}T^{-1}L^{-1}$$
$$[F_0] = CL^{-2} \qquad [u] = D \qquad [x] = L \qquad [t] = T \tag{10.36}$$

Evidently, the solution u may depend on all six remaining quantities:

$$u = f(x, t, \rho, c_0, K_0, F_0) \tag{10.37}$$

By the second part of the Π theorem, we can have only one dimensionless combination of the six quantities on the right. We take it in the form

$$\pi_1 = \frac{x}{\sqrt{2\nu t}} \tag{10.38}$$

where $\nu/2 \equiv K_0/\rho c_0$ is the thermal diffusivity with $[\nu] = L^2 T^{-1}$. Since

$$\pi_0 = \frac{\sqrt{2\nu t}}{F_0/\rho c_0} u \tag{10.39}$$

is also dimensionless, we have from the first part of the Π theorem that

$$\pi_0 = \hat{F}(\pi_1) \quad \text{or} \quad u = \frac{F_0}{\rho c_0} \frac{\hat{F}(x/\sqrt{2\nu t})}{\sqrt{2\nu t}} \tag{10.40}$$

Thus, the application of dimensional analysis reduces the solution for u to a form involving only an unknown function $\hat{F}(\cdot)$ of one variable which is a specific combination of the two independent variables x and t (and the parameter ν). For the fundamental solution, the actual form of $\hat{F}$ can now be determined by the direct substitution of (10.40) into the homogeneous heat equation.

For $t > 0$, the fundamental solution satisfies the homogeneous heat equation [Equation 10.24 with $f(x, t) \equiv 0$]. The expression (10.40) reduces the PDE to an ordinary differential equation (ODE) for $\hat{F}(\pi_1)$:

$$\hat{F}'' + 2\pi_1 \hat{F}' + 2\hat{F} = 0 \qquad (-\infty < \pi_1 < \infty) \tag{10.41}$$

where $(\ )' \equiv d(\ )/d\pi_1$. Rewritten as $\hat{F}'' + 2(\pi_1 \hat{F})' = 0$, the ODE (10.41) may be integrated once immediately to get

$$\hat{F}' + 2\pi_1 \hat{F} = c_1 \tag{10.42}$$

The first-order ODE (10.42) has as its general solution

$$\hat{F}(\pi_1) = c_2 e^{-\pi_1^2} + c_1 e^{-\pi_1^2} g(\pi_1) \quad \text{with} \quad g(\pi_1) = \int^{\pi_1} e^{\xi^2}\, d\xi \qquad (10.43)$$

where c_1 and c_2 are two constants of integration to be determined presently. By L'Hôpital's rule, we have $e^{-\pi_1^2} g(\pi_1) \to 0$ as $|\pi_1| \to \infty$. However, we expect

$$\int_{-\infty}^{\infty} \hat{F}(\pi_1)\, d\pi_1 = c_2 \int_{-\infty}^{\infty} e^{-\pi_1^2}\, d\pi_1 + c_1 \int_{-\infty}^{\infty} e^{-\pi_1^2} g(\pi_1)\, d\pi_1 \qquad (10.44)$$

to be bounded also as only one unit of heat was injected into the rod. By repeated integration by parts, it is not difficult to show

$$e^{-\pi_1^2} g(\pi_1) = e^{-\pi_1^2}\left[\frac{e^{\pi_1^2}}{2\pi_1} + O\!\left(\frac{e^{\pi_1^2}}{\pi_1^3}\right)\right] = \frac{1}{2\pi_1} + O\!\left(\frac{1}{\pi_1^3}\right) \qquad (10.45)$$

as $|\pi_1| \to \infty$. Hence, we have for any large but fixed z:

$$\int_{-\infty}^{\infty} e^{-\pi_1^2} g(\pi_1)\, d\pi_1 = \int_{-z}^{z} e^{-\pi_1^2} g(\pi_1)\, d\pi_1 + 2 \int_{z}^{\infty} e^{-\pi_1^2} g(\pi_1)\, d\pi_1$$

$$\geq 2 \int_{z}^{\infty} \left[\frac{1}{2\pi_1} + O\!\left(\frac{1}{\pi_1^3}\right)\right] d\pi_1 \qquad (10.46)$$

It follows that (10.44) would not be bounded unless $c_1 = 0$. For the fundamental solution, we take $c_1 = 0$ leaving us with

$$\hat{F}(\pi_1) = c_2 e^{-\pi_1^2} \quad \text{or} \quad u = \frac{F_0}{\rho c_0} \frac{c_2 e^{-x^2/2\nu t}}{\sqrt{2\nu t}} \qquad (10.47)$$

To determine c_2, we integrate the PDE for u to get

$$\int_{-\infty}^{t} \int_{-\infty}^{\infty} (\rho c_0 u_t - K_0 u_{xx})\, dx\, dt = \int_{-\infty}^{t} \int_{-\infty}^{\infty} [F_0 \delta(x)\delta(t)]\, dx\, dt$$

or

$$\int_{-\infty}^{\infty} \left[\rho c_0 u\right]_{-\infty}^{t}\, dx - K_0 \int_{-\infty}^{t} \left[u_x\right]_{-\infty}^{\infty}\, dt = F_0 H(t) \qquad (10.48)$$

With $u = 0$ for $t < 0$ and $u_x \to 0$ as $|x| \to \infty$, (10.48) simplifies to read

$$\int_{-\infty}^{\infty} \rho c_0 u(x, t)\, dx = F_0 H(t) \tag{10.49}$$

We now insert the expression for u from (10.47) into (10.49) to get for $t > 0$

$$c_2 \int_{-\infty}^{\infty} e^{-z^2}\, dz = 1 \quad \text{or} \quad c_2 = \frac{1}{\sqrt{\pi}} \tag{10.50}$$

This gives the fundamental solution

$$u(x, t) = \frac{F_0}{\rho c_0 \sqrt{\pi}} \frac{e^{-x^2/2\nu\tau}}{\sqrt{2\nu t}} H(t) \equiv G(x, t; 0, 0) \tag{10.51}$$

where π is the number 3.14159. ... If the unit strength heat energy should be applied at $t = \tau$ and $x = \xi$ instead of $t = x = 0$, it is not difficult to see that the corresponding fundamental solution would be

$$G(x, t; \xi, \tau) = \frac{F_0}{\rho c_0} \frac{e^{-(x-\xi)^2/2\nu(t-\tau)}}{\sqrt{2\pi\nu(t-\tau)}} H(t - \tau) \tag{10.52}$$

instead, consistent with what we obtained earlier by Fourier transforms.

10.5   Shock Speed

With $\psi(x, t)$ already known from (10.33), we can get $c(x, t)$ and $k(x, t)$ from their defining equations

$$c = -\frac{\nu\psi_x}{\psi} \quad \text{and} \quad c = \frac{dQ}{dk} \quad \left[= \frac{3(100 - k)}{5} \right]$$

The expressions for c and k are ratios of two integrals and can be evaluated by numerical or asymptotic methods. They have been analyzed by the method of steepest descent (Whitham, 1974) to study the evolution of the wave speed and car density distribution for small ν. The main conclusions from such an analysis are

1. Shocks form when the diffusive mechanism is ineffective.
2. Steepening of the density wave triggers diffusion, which prevents the shock waves from tipping over.
3. Shock waves move along the road with a *shock speed* $U_s = (c_r + c_l)/2$ where c_r and c_l are the wave speeds slightly to the right and left of the shock position x_s, respectively.

Multivalued solutions which appeared in the Lighthill-Whitham original theory are now avoided by a delicate balance between diffusion and further steepening of the sharp change in car density over a small distance. The details of the shock structure, that is, the profile of c or k within the shock region, can also be obtained. But from a practical point of view, such details are not particularly useful or reliable. Our mathematical model (with or without the diffusion term) is only an idealization (and therefore an approximation) of reality. Also, it is difficult to measure the diffusion coefficient ν accurately.

Once we know the qualitative effect of diffusion on the solution behavior after shock formation it is not difficult to calculate the shock speed on the basis of the simple theory of Chapter 7. In other words, we really do not need the exact solution for $\psi(x, t)$. For a fixed small interval (x_1, x_2) we have from the definition of k and q,

$$\frac{d}{dt} \int_{x_1}^{x_2} k(x, t)\,dx = q_1 - q_2 \qquad q_m = q(x_m, t) \qquad (10.53)$$

If the shock happens to be inside this interval so that $x_1 < x_s < x_2$ where $x_s(t)$ is the position of the shock, then we have

$$\frac{d}{dt} \int_{x_1}^{x_2} k(t, x)\,dx = \frac{d}{dt} \left[\int_{x_1}^{x_s(t)-} + \int_{x_s(t)+}^{x_2} \right] k(x, t)\,dx$$

$$= \left[\int_{x_1}^{x_s-} + \int_{x_s+}^{x_2} \right] k_t(x, t)\,dx$$

$$+ (x_s-)\dot{}\,k(x_s-, t) - (x_s+)\dot{}\,k(x_s+, t)$$

$$= \left[\int_{x_1}^{x_s-} + \int_{x_s+}^{x_2} \right] k_t(x, t)\,dx - \dot{x_s}[k] \qquad (10.54)$$

where $[k] \equiv k_r - k_l$ is the jump in k across the shock. As $\Delta x \equiv x_2 - x_1$ tends to zero, we have

$$q_1 - q_2 \rightarrow -[q] \equiv -(q_r - q_l) \tag{10.55}$$

and

$$\left[\int_{x_1}^{x_s-} + \int_{x_s+}^{x_2}\right] k_t(x, t)\, dx = \left[\int_{x_1}^{x_s-} + \int_{x_s+}^{x_2}\right] q_x(x, t)\, dx \rightarrow 0 \tag{10.56}$$

since q has no jump in (x_1, x_s-) and (x_s+, x_2). Together, we have, from (10.53)–(10.56) as $\Delta x \rightarrow 0$,

$$x_s^{\bullet} = \frac{[q]}{[k]} = \frac{q_r - q_l}{k_r - k_l} \tag{10.57}$$

For a quadratic $Q(k)$, say $Q = \alpha k + \beta k^2$, it is not difficult to see that

$$x_s^{\bullet} = \cdots = \alpha + \beta(k_r + k_l)$$

$$= \tfrac{1}{2}[(\alpha + 2\beta k_r) + (\alpha + 2\beta k_l)] = \tfrac{1}{2}(c_r + c_l) \tag{10.58}$$

which is the same as that obtained from an exact solution of Burger's equation for a small diffusion coefficient ν.

> **10E:** Shock waves in traffic flow move along the road at a speed which is the average of the wave speed in front and in back of the shock.

Note that $x_s^{\bullet}$ is generally a function of t so that we get traveling-wave solutions only in special cases such as the red light problem.

10.6   Traveling-Wave Solution and Shock Structure

The exact solution of the IVP for Burger's equation contains all the information about the evolution of $c(x, t)$, and therefore $k(x, t)$, starting with $t = 0$. For example, it gives the behavior of c and k before and after shock formation in the case of an initial traffic hump. Less complete but adequate information about the traffic flow can be obtained with much less effort than that expended in the last few sections. In particular, if we are only interested in knowing whether a shock wave can be maintained and propagated with a constant shock speed U_s, we can find that out also without obtaining the rather messy exact solution for $\psi(x, t)$. The maintenance and propagation of a shock corresponds to the existence of a traveling-wave solution $c(x, t) = S(y)$, with $y \equiv (x - Vt)/\nu \equiv \xi/\nu$, for some $S(\cdot)$ and some constant V. Upon substituting this into Burger's equation, we get

$$(-V + S)S' = \tfrac{1}{2}S'' \tag{10.59}$$

where $(\ )' \equiv d(\ )/dy$. The above equation can be integrated once right away to get

$$-2VS + S^2 = S' + A_0 \tag{10.60}$$

where A_0 is a constant of integration. To determine the unknown constants A_0 and V, we make use of the fact that S is c_r and c_l on the two sides away from the shock, say, $\xi = x - Vt = \pm\bar{\xi}$ or $y = \pm\bar{\xi}/\nu$. Now, the speed gradient $c_x = S_x = S'/\nu$ outside the shock is small compared to its value inside the shock where a sharp change takes place over a small distance. Therefore, $S' = \nu c_x$ is small, of order ν at most, allowing $c_x = O(1)$ outside the shock. As a good approximation for sufficiently small ν, we may write the conditions on c and c_x away from the shock (and therefore for large y) as the following limiting behaviors:

$$\text{(i)} \quad y \to \infty: \qquad S \to c_r \quad \text{and} \quad S' \to 0 \tag{10.61}$$

$$\text{(ii)} \quad y \to -\infty: \qquad S \to c_l \quad \text{and} \quad S' \to 0 \tag{10.62}$$

giving us

$$\text{(i)} \quad A_0 + 2Vc_r = c_r^2 \qquad \text{(ii)} \quad A_0 + 2Vc_l = c_l^2 \tag{10.63}$$

or

$$V = \frac{c_l + c_r}{2} \qquad A_0 = c_r c_l \tag{10.64}$$

Therefore, the wave speed of the traveling (shock) wave is the same as the shock speed U_s obtained in Section 10.5 from the exact solution of the improved theory and from the simpler Lighthill-Whitham theory.

Having found A_0 and V, we can now solve the first-order ODE for S (since it is separable) to get

$$c(x, t) = S = \frac{c_l + c_r \exp\left[-\tfrac{1}{2}(c_r - c_l)y\right]}{1 + \exp\left[-\tfrac{1}{2}(c_r - c_l)y\right]} \qquad y = \frac{x - Vt}{\nu} \tag{10.65}$$

where we have used the condition $c = (c_r + c_l)/2 = U_s = V$ at the location of the shock $y = 0$. It is not difficult to see that $S \to c_l$ to the left of the shock and $S \to c_r$ to the right.

The above expression for a traveling-wave solution of Burger's equation gives the detailed distribution of c across the shock. But as we mentioned before, this information about the shock structure is mainly of academic interest. The important point here is that we have shown that

> **10F:**   Burger's equation admits traveling-wave solutions, hence it is possible to have shock waves propagating at constant speed in vehicular traffic.

In order for the shock speed U_s to be constant, the sum $c_r + c_l$ (which is generally a function of time) must be constant. This is the case for an initial step profile and for the red light problem in the exercises at the end of this chapter.

10.7   The Equal-Area Rule

We mentioned in Sections 10.5 and 10.6 that the exact solution of Burger's equation is too complicated to be useful (notably to those without any experience in the techniques in asymptotic expansions). In Section 10.5, we showed how the main result contained in the exact solution, namely the approximate shock speed expression (10.58), can be deduced directly from the conservation law. We showed further in Section 10.6 that this same expression for the shock speed and the approximate shock profile can also be deduced from Burger's equation by seeking a solution $c = f(x - Vt)$ whenever the shock takes the form of a traveling-wave solution. These results provide us with important qualitative information about our continuum theory of traffic flow such as the possibility of maintaining and propagating shock waves. They generally do not give us the practical information we tried to obtain from the modified theory. (Exceptions include simple cases such as the red light problem with a uniform initial density profile.) For example, what should we take as the values for c_r and c_l to compute U_s? We were deliberately vague about that point in Section 10.5. In the traffic hump problem in Chapter 7, the solution first becomes multivalued at $t = \frac{1}{15}$ (Figure 10.3a). It seems easy to pick $c_l = c \ (k = 50) = 30$; but what should we take for c_r which may vary from 45 to 30? The situation is worse for $t > \frac{1}{15}$ when it is no longer obvious what c_l should be (Figure 10.3-b). Also, we want to know the position of the shock as a function of time as it propagates along

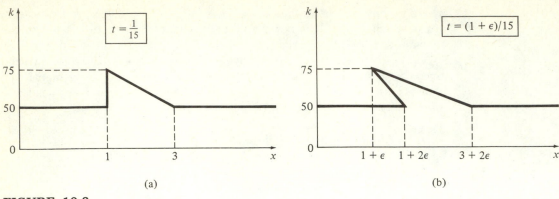

FIGURE 10.3

the x axis. But that position depends on the correct value of the shock speed (which may vary with time), and we have not gotten that yet.

In this section, we introduce the *equal-area rule* of Whitham for the determination of the shock speed and shock position without any reference to the complicated exact solution of the more elaborate gradient theory formulated in Section 10.1. This ingenious rule is based again on the conservation law. Its application tells us how to replace a multivalued density profile by a shock profile. The conservation of cars requires that the total number of cars over the road be the same for both profiles. Now, the total number of cars is given by the integral of $k(x, t)$ over the stretch of the road under consideration. This conservation requirement is equivalent to stipulating the same area under the two profiles. We have then the following geometric rule to handle shocks:

1. Use the simple Lighthill-Whitham theory to obtain the parametric representation of k allowing for the possibility of a multivalued density profile.

2. At each time t after the appearance of the multivalued profile, simply insert a vertical line at a position $x_s(t)$ such that the shaded areas A_1 and A_2 on the two sides of the vertical line in Figure 10.4 are equal.

3. Discard those shaded "overhanging" pieces leaving us an approximate shock profile. (More accurately, use the upper overhang to fill in the lower void.)

4. Take k_r and k_l to be the value of k slightly to the right and left of the vertical line, respectively.

5. Use $\dot{x}_s = [q]/[k]$ to calculate shock speed.

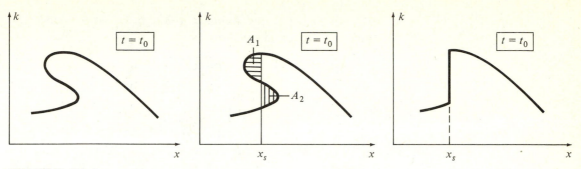

FIGURE 10.4

More analytical details of this recipe can be found in Whitham (1974). It suffices to note here that

> **10G:** The equal-area rule gives an accurate approximate shock position and shock speed in the density wave of traffic with no reference to the numerical value of the diffusion coefficient or the solution of Burger's equation.

EXERCISES

1. Deduce (10.6) from (10.4) with the Hopf-Cole transformation (10.5).

2. Use the Fourier transform method to solve the BVP in the upper half-plane: $u_{xx} + u_{yy} = 0$, $u(x, 0) = f(x)$, and $u(x, \infty) = 0$. (*Hint:* Use the real-valued version of the Fourier transform. Assume that both u and u_x tend to zero as $|x| \to \infty$.)

3. If a model of traffic flow is $c_t + cc_x = c_{xt}$, are there any acceptable traveling-wave solutions of the form $c = F(x - Ut)$ where c approaches constant values as $|x - Ut| \to \infty$? If so, find F and the constant U.

4. Find the spherically symmetric solution of the *three*-dimensional heat equation that represents the response to a unit source. In this case, the equation is

$$\frac{a}{r^2} \frac{\partial}{\partial r} \left(r^2 \frac{\partial T}{\partial r} \right) - \frac{\partial T}{\partial t} = 0$$

where a is a constant. Show that a similarity solution of the form

$$T = \frac{1}{t^\beta} F\left(\frac{r}{t^\alpha} \right)$$

exists for some choice of the constants α and β. Use it to reduce the PDE to an ODE for F and determine F.

5.   The initial temperature of an infinitely long iron bar is

$$T(x, 0) = \begin{cases} 0°C & (x < 0) \\ 100°C & (x > 0) \end{cases}$$

   (a)   The thermal diffusivity of iron is $a = 0.15 \text{ cm}^2/\text{s}$ in $\partial T / \partial t = a(\partial^2 T / \partial x^2)$. Solve for the temperature as a function of position and time and graph $T(x, t)$ at $t = 0, 1, 10, 100$ s. Find the loci of the positions where the temperatures are $T = 25°$, $T = 50°$, and $T = 75°$.

   (b)   Suppose the bar is made of firebrick for which $a = 0.007 \text{ cm}^2/\text{s}$. Compare the temperatures in the two cases at times $t = 0, 1, 10, 100$ s. [*Hint:* The error function is all that is required:

$$\text{erf}(x) = \frac{2}{\sqrt{\pi}} \int_0^x e^{-z^2} \, dz$$

   A table of values for $\text{erf}(x)$ can be found in Abramowitz and Stegun (1965).]

6.   **(a)**   Let $f(x)$ be continuous and piecewise continuously differentiable in $(0, \infty)$. The quantity

$$F_c(\lambda) = \int_0^\infty f(x) \cos \lambda x \, dx \qquad (\lambda \geq 0)$$

   is called the *Fourier cosine transform* of $f(x)$ provided that the integral exists. Find the Fourier cosine transform of $\exp(-ax)$ for all $a > 0$.

   (b)   It is clear from part (a) that there is a unique $F_c(\lambda)$ for a given $f(x)$. Conversely, it can be shown that $F_c(\lambda)$ is the Fourier cosine transform of a unique f given by the *inversion formula*

$$f(x) = \frac{2}{\pi} \int_0^\infty F_c(\lambda) \cos \lambda x \, d\lambda \qquad (x > 0)$$

   [*Note:* If f has a finite jump discontinuity at x_0, then the inversion formula gives $f(x)$ except at the point x_0.] Unfortunately, it is usually difficult to use the *inversion formula* to find $f(x)$ from its Fourier cosine transform $F_c(\lambda)$. Sometimes we can take advantage of the unique correspondence between $F_c(\lambda)$ and $f(x)$ to recover $f(x)$ from $F_c(\lambda)$. Given $F_c(\lambda) = 1/(1 + \lambda^2)$, find $f(x)$ without the inversion formula.

7. **(a)** Find the Fourier sine transform, $F_s(\lambda)$, of $f(x) = e^{-ax}$, $a > 0$:

$$F_s(\lambda) = \int_0^\infty f(x) \sin \lambda x \, dx$$

(b) Find the Fourier sine transform, $Y_s(\lambda)$, of the solution of the BVP

$$y'' - 4y = e^{-x} \qquad y(0) = y(x \to \infty) = 0$$

(c) Find $y(x)$ from $Y_s(\lambda)$ with

$$y(x) = \frac{2}{\pi} \int_0^\infty Y_s(\lambda) \sin \lambda x \, d\lambda$$

8. **(a)** $y'' - 4y = e^{-x}$ $(x > 0)$, with $y'(0) = 0$ and $y(\infty) = 0$. Let $Y_c(\lambda)$ be the (unknown) Fourier cosine transform of the unknown $y(x)$. Then by the inversion formula, we have

$$y(x) = \frac{2}{\pi} \int_0^\infty Y_c(\lambda) \cos \lambda x \, d\lambda \qquad (x > 0)$$

Show that the initial condition $y'(0) = 0$ is satisfied and that the ODE is also satisfied if

$$Y_c(\lambda) = -\frac{1}{1 + \lambda^2} \frac{1}{4 + \lambda^2}$$

(b) Find $y(x)$ from $Y_c(\lambda)$. What is the limit of $y(x)$ as $x \to \infty$?

9. **(a)** Let $f(x)$ be continuous and piecewise continuously differentiable in $(-\infty, \infty)$. The complex-valued function

$$F(\lambda) = \int_{-\infty}^\infty e^{i\lambda x} f(x) \, dx \qquad (-\infty < \lambda < \infty)$$

is called the *Fourier transform* (FT) of $f(x)$ provided that the integral exists. Find the Fourier transform of the function $\exp(-a|x|)$, $a > 0$.

(b) If $f(x)$ is even, that is, $f(-x) = f(x)$, show that $F(\lambda) = 2F_c(\lambda)$, $\lambda > 0$.

(c) If $f(x)$ is odd, that is, $f(-x) = -f(x)$, show that $F(\lambda) = 2iF_s(\lambda)$.

(d) It can be shown that $F(\lambda)$ is the FT of a unique $f(x)$ given by

$$f(x) = \frac{1}{2\pi} \int_{-\infty}^\infty F(\lambda) e^{-i\lambda x} \, d\lambda \qquad (-\infty < x < \infty)$$

(The situation with a discontinuity in f is the same as in Exercise 6.) Without using the inversion formula, find the inverse FT of $F(\lambda) = 1/(1 + \lambda^2)$.

10. A bar of iron 25 cm long is initially at 50°C. The lateral surface is kept insulated and the temperature at one end is raised to 100°C (the other is kept at 50°C). Write the equation governing the flow of heat in dimensionless form. What is the ultimate temperature distribution (at $t = \infty$)? Without solving the time-dependent problem, estimate the essential time period of the transient evolution to the steady state.

11. Find the temperature at the center of a square region, one of whose edges is kept at 100°C while the others are maintained at 0°C.

12. A cold fluid flows past a hot sphere of radius R_0. The equation governing the fluid temperature θ is

$$a \frac{\partial^2 \theta}{\partial x^2} + \frac{\partial^2 \theta}{\partial y^2} - V \frac{\partial \theta}{\partial x} = \frac{\partial \theta}{\partial t}$$

where V is the convective velocity of the fluid.
   (a) Write this equation in dimensionless form. How is the characteristic time chosen? What single dimensionless number arises? When is convection more important than diffusion? When is it less? How is this indicated by the dimensionless number?
   (b) If $V = 0$, what is the only way of choosing a time scale that characterizes the diffusion process? What additional information would you obtain by solving the problem in detail?

13. Use the equal-area rule to find the shock position and shock speed for an arbitrary $t > 0$ for

$$k_t + c(k)k_x = 0 \qquad k(x, 0) = 50H(-x) = \begin{cases} 50 & (x < 0) \\ 0 & (x > 0) \end{cases}$$

with $c(k) = 60 + 0.6k$.

14. Solve $y'' - 4y = e^{-|x|}$, with $y \to 0$ as $|x| \to \infty$, by Fourier transforms. Use the complex exponential form of the transform pairs:

$$F(\lambda) = \int_{-\infty}^{\infty} e^{i\lambda x} f(x)\, dx \qquad f(x) = \frac{1}{2\pi} \int_{-\infty}^{\infty} e^{-i\lambda x} F(\lambda)\, d\lambda$$

15. A nonlinear PDE that arises in many areas of application including water waves is the Korteweg-deVries equation:

$$u_t + (\alpha + \epsilon u)u_x + \beta u_{xxx} = 0$$

where α, ϵ, and β are constants. For a traveling-wave solution, we take $u(x, t) = f(x - Vt)$ where V is some constant.

(a) Show that f is a solution of the ODE $(\alpha - V)f + \frac{1}{2}\epsilon f^2 + \beta f'' = C_1$, where C_1 is a constant of integration.

(b) Assume that f, f', and $f'' \to 0$ as $\xi \equiv x - vt \to \pm\infty$. Show that the ODE in part (a) can be integrated to give the solitary wave solution

$$u = f(\xi) = A \operatorname{sech}^2\left(\sqrt{\frac{\epsilon A}{12\beta}}\, \xi\right)$$

where $V = \alpha + \epsilon A/3$.

11

Fishing Is Strictly Prohibited

The 200-Mile Fishing Limit

Equilibrium solutions and their stability are deduced for a nonlinear PDE of the reaction-diffusion type by methods developed in the chapter. Among the new mathematical features is the solution of an eigenvalue problem by the perturbation method.

11.1 The Depletion of East Coast Fisheries

The fish population off the east coast of North America has been severely depleted by overfishing in the last few decades. This is particularly true for the more valuable fish stocks such as cod, haddock, and red fish. Various estimates suggest that it would take a 20-year program of intensive experimental management, including a fishing moratorium, to rebuild these fisheries to the carrying capacity of the region (Larkin, 1975). The west coast of the Americas is also not free from the depletion problems. There were the collapses of the sardine fishery near San Francisco and the anchovies fishery in Peru. Many similar situations around the world have given considerable impetus to the insistence of a 200-mile fishing limit by various countries with a significant fishing industry. Among the many advantages, these extended territorial rights give the host countries a more favorable environment in which to regulate the nearby fishing grounds. Many fishing grounds extend well beyond the conventionally recognized 12-mile limit. It is simply not possible to declare a fishing moratorium without exclusive jurisdiction over the entire fish habitat.

Protective management, such as a complete closure, poses many new questions in fishery research. In the past, the growth dynamics of fish populations have been analyzed by treating an entire fish population as a single entity; the spatial distribution of the population over a body of water plays no role in these models. The situation is analogous to treating a planet as a point mass in the study of its orbit around the sun. Until recently, fishery management has been concerned almost exclusively with dynamical system models: their evolution, stability, and control. (We shall see some of these in the next few sections.) The continental shelf goes out a long way off the east coast, especially around Nova Scotia and Newfoundland, and populations of fish often meander beyond artificial political boundaries such as the 200-mile offshore limit in these areas. There is nothing to stop foreign fishing fleets from lining up along the 200-mile limit and catching any fish that come their way. Thus, the spatial distribution of fish populations is important in this context. It is not at all certain that we can rebuild our depleted fishing grounds, nevermind to the carrying capacity of the region, by a fishing moratorium inside the political boundary. Before adopting any kind of management strategy, we should undertake a more adequate analysis of the situation, especially one which includes the spatial aspect of the problem. The time required to restock the fishing ground by the natural growth process within the political boundary cannot be properly estimated if we do not consider the effect of the fishing activities at the boundary and beyond.

In Sections 11.4–11.7, we will describe and analyze a mathematical model for fishery dynamics which allows for both temporal and spatial variations in the fish population. Although still not an adequate model for actual application, it does serve to indicate the types of mathematical problems to be expected in this area of scientific endeavor. It will show also how the different mathematical techniques used in planetary motion, Euler buckling, etc., are needed again in a completely unrelated area of mathematical science. More realistic models are discussed in Section 11.8. To lead up to the space-time models, we begin with a brief discussion of some conventional dynamical system models in the next two sections and introduce the issues and analyses peculiar to fishery problems. We will see in particular why the forces of economics alone are not enough to prevent overfishing and why central planning (or government regulations) becomes necessary. This sets the stage for a study of the effect of the 200-mile fishing limit by our diffusion-reaction model.

In our space-time model, fish are allowed a one-dimensional (seaward) movement which is random in nature. This random movement may be modeled by a diffusion term in the growth dynamics of the fish population. The relation between diffusion and random walk is discussed in Lin and Segel (1974).

11.2 Compensatory Growth of Fish Populations

In conventional analyses of fishery problems, a single-specie fish population with no significant effect of age structure is characterized by its size $P(t)$

(measured in units of fish biomass) which may vary with time. We take both the birth and mortality rates to depend on P and write the growth dynamics of the fish population in the form

$$P^{\cdot} = F(P, t) \tag{11.1}$$

In population theory, it is customary to write $F \equiv rP$ where r is the net *growth rate* per unit of population biomass. At the low-population range, r is nearly a constant in most cases so that the population grows (or decays) exponentially. But for larger population sizes, r itself may depend on P. In the extreme case, P is so large that it taxes the support system (such as available space, food, air, water, etc.) and r would decrease with P, eventually taking on negative values. While we continue to discuss the growth rate density $r(P, t)$ occasionally, we will work mainly with (11.1) as the growth dynamics of the fish population.

A well-known growth function is the logistic growth rate

$$F_L(P, t) \equiv r_0\left(1 - \frac{P}{P_c}\right)P \qquad (r_0 > 0) \tag{11.2}$$

for which $r = F/P = r_0(1 - P/P_c)$ is a decreasing function of P for all $P > 0$. Generally, a growth function F with the corresponding r decreasing with P for all $P > 0$ is called a *compensation* (type) *growth model*. The graph of a compensation growth function is shown in Figure 11.1. It is always unimodal (single hump) and concave with only one crossing of the positive P axis at the *carrying capacity* $P_c \ (> 0)$, in addition to the crossing at $P = 0$. Given the initial population size $P(0) = P_0$, (11.1) determines the evolution of P for all time thereafter.

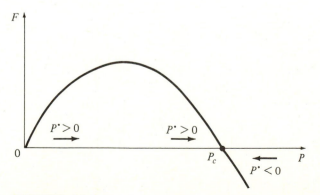

FIGURE 11.1

The dynamical system (11.1) is an autonomous system if $F(P, t)$ does not depend on t explicitly. The logistic growth model (11.2) is autonomous if r_0 and P_c are independent of t. For the purpose of bringing out the basic problems in fishery, it is sufficient to focus our attention on autonomous systems. As we know from Chapter 4, a root $\bar{P}$ of $F(P) = 0$ is an equilibrium state of (11.1) as $\bar{P}^{\cdot} = F(\bar{P}) = 0$. For the logistic growth or any other compensation model, there are two equilibrium populations $\bar{P}_1 = 0$ and $\bar{P}_2 = P_c$. If the population size is larger than P_c at time t, then P tends to P_c as time increases because $P^{\cdot} < 0$ in the range $P > P_c$ (see Figure 11.1). For $0 < P < P_c$ at time t, we have $P^{\cdot}(t) > 0$ so that P increases at all later time tending to P_c. Evidently, $\bar{P}_1 = 0$ is an unstable equilibrium population and $\bar{P}_2 = P_c$ is an asymptotically stable state.

Fish populations are not left to grow naturally; instead, many are harvested more or less continually in time so that their actual growth rate is the natural growth rate minus the harvest rate h:

$$P^{\cdot} = F(P) - h \qquad (11.3)$$

The harvest rate is nonnegative and may vary with time and the fish population size. A constant harvest rate $h_0 > 0$ modifies the equilibrium states and the stability of an unharvested population in a relatively simple way, as described in Figure 11.2. Below a critical level of harvest rate h_c, we generally have two new equilibrium population sizes P'_1 and P'_2, $0 < P'_1 <$

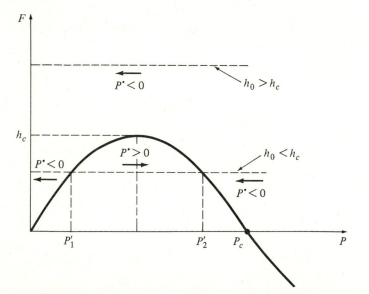

FIGURE 11.2

$P'_2 < P_c$. For logistic growth with a constant r_0, P'_k are the two roots $(P_c/2)(1 \pm \sqrt{1 - 4h_0/r_0 P_c})$, of the quadratic equation $F_L(P) = h_0$. The population would be fished to extinction if it was initially below the smaller unstable equilibrium population P'_1. Otherwise, it would tend to (and remain at) P'_2. For $h_0 > h_c = r_0 P_c/4$, we have $P^{\cdot} < 0$ for all P and any initial P_0 would be fished to extinction. (What happens if $h_0 = h_c$?)

Until recently, fishing was not a regulated industry, at least not for the purpose of management. The harvest rate is more or less dictated by the forces of economics which adjust the effort E expended by the fishing industry on the particular fish population. Fishing effort measures the commitment of equipment and labor to a fishery. The harvest rate is generally an increasing function of E. It is also an increasing function of the fish population P for a fixed E—the more fish available, the easier it is to catch them. Typically, we have $h = \beta PE$ so that (11.3) for logistic growth becomes

$$P^{\cdot} = (r - \beta E)P = P\left[r_0\left(1 - \frac{P}{P_c}\right) - \beta E \right] \tag{11.4}$$

The two equilibrium states are $\bar{P}_1 = 0$ and

$$\bar{P}_2 = P_c\left(1 - \frac{\beta E}{r_0}\right) \tag{11.5}$$

so that the nontrivial steady-state harvest, called the *sustained yield,* from the fishing effort is

$$h(\bar{P}_2, E) = \beta E \bar{P}_2 = \beta E P_c\left(1 - \frac{\beta E}{r_0}\right) \equiv \bar{h}_2 \tag{11.6}$$

The sustained-yield curve $\bar{h}_2(E)$ for logistic growth is shown in Figure 11.3a. The same curve for a general compensation growth model is shown in Figure 11.3b. The effort level $E^* (= r_0/\beta$ for the logistic case) is the upper bound for a nonzero (sustained) yield. There is no equilibrium fish population for $E > E^*$ and hence, no sustained yield.

The graphs of $\bar{h}_2$ in Figure 11.3 suggest that the optimal level of effort would be E_0 determined by $d\bar{h}_2/dE = 0$. For the logistic case, we have $E_0 = E^*/2$ and $\bar{h}_2(E_0) = r_0 P_c/4$. However, if the primary factor influencing the level of effort is profit, then we expect the actual fishing effort expended by the industry would be at a different level E_∞ instead. To see this, suppose the constant price per unit fish is p and cost per unit effort is c. Then the total cost (per unit time) is cE while the profit from the harvest (per unit time) is ph. In a sustained-yield situation, the break-even point would be

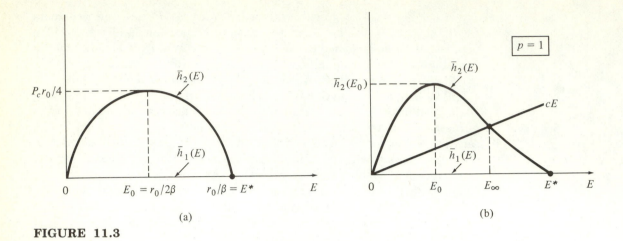

FIGURE 11.3

the intersection of the curve $p\bar{h}_2(E)$ and the straight line cE. With no loss of generality, we can take $p = 1$ (measuring cost in terms of price) and the intersection will be at effort level E_∞, as indicated in Figure 11.3b. For the logistic growth model, we have

$$E_\infty = \frac{r_0}{\beta}\left(1 - \frac{c}{\beta P_c}\right) \qquad p\bar{h}_2(E_\infty) = \frac{cr_0}{\beta}\left(1 - \frac{c}{\beta P_c}\right) \qquad (11.7)$$

If c is too high, say greater than dF/dP at $P = 0$, then $E_\infty = 0$ and no level of positive effort would yield a sustained profit. In this case, the fishery will not be exploited at all; this particular fish species cannot support any commercial fishery. For lower c so that $E_\infty > 0$, there would be profit if fishing is done at a level below E_∞ and this would include E_0 if $E_\infty > E_0$, that is, if the cost per unit effort is sufficiently small. In principle, the fishery may operate at any level E, $0 < E < E_\infty$ [although it should do so at E_0 (for $E_0 < E_\infty$) given the larger profit]. Unfortunately, this is not the case; instead,

11A: Fishing will be done at level E_∞ in the long run!

If the current effort expended were $E < E_\infty$, then there would still be profit for additional effort. Another fishing company or fishing fleet would enter into the fishery and thereby increase the total effort rate. The argument repeats until we get to E_∞. On the other hand, no level of effort greater than E_∞ can be sustained; some fisherman (if not all) would lose money and withdraw from the fishery reducing the level of effort.

Thus, the exploitation of a fish population opened to entrepreneurs is not *economically efficient*. A reduction from E_∞ to E_0 (or any lower level if $E_0 > E_\infty$) would have the double effect of increasing revenues and decreasing fishing cost. On the other hand, the expected inefficiency and overexploitation do not seem to endanger the fish population to the point of a rapid or sudden sharp decrease and possible extinction (see Exercise 5). While there may be some difficulty in divesting excess equipment in cases where a retrenchment is needed toward a sustained yield, there will be some bankruptcies long before the last fish is taken out of the sea. What then is the explanation for the collapse of the sardines and anchovies? We will take up this question in the next section.

Before leaving the economically more efficient effort level E_0, it should be mentioned that a drop from E_∞ to E_0 does not necessarily improve the overall economic efficiency of the fishery. There would be a drop in revenue for a period after the drop in effort level; it would take time for the fish population to grow to the larger size to compensate for the lower level of effort (and more). The net result may not be the maximum total revenue possible. The proper determination of the correct optimal policy is by a dynamic optimization process, discussed in Part V of this book through a study of exhaustible resource economics. The actual application of optimal control theory to the present problem can be found in Clark (1976).

11.3   Depensation Growth Models

The results of the last section did not provide an explanation for the crash of many fisheries in recent years. It turns out that the inadequacy does not lie in the mathematical analysis or the economic forces assumed to be driving our bioeconomic phenomenon. Rather, it lies in the growth process assumed for the fish population. In compensation models, the growth rate density r decreases with P. There is considerable evidence, however, that at a very low population level, r actually increases with P, at least for awhile. (Note that F could be negative for $P > 0$ even if $dr/dP > 0$.) Mathematical models for such growth rate behavior are called *depensation* (growth) models. The term depensation is used by fishery biologists to contrast the strictly compensatory characteristics of the more traditional models. They may have the noncritical form indicated in Figure 11.4a or the *critical* form of Figure 11.4b. In the latter-type growth process, both the growth rate $F(P)$ and the growth rate density $r(P)$ actually become negative for a range of P near the origin, $0 < P < P_i$. A growth function of the critical depensation type is

$$F(P) = -r_D(t)\left(1 - \frac{P}{P_i}\right)\left(1 - \frac{P}{P_c}\right)P \tag{11.8}$$

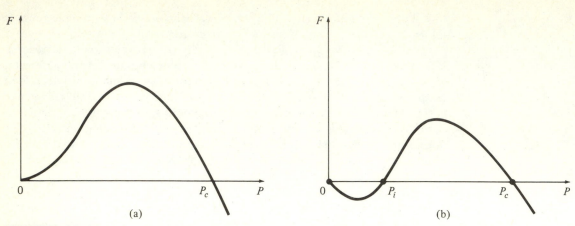

FIGURE 11.4

with $r_D > 0$ and $0 < P_i < P_c$. For both types of depensation models, the growth rate density $r(P, t)$ eventually becomes decreasing for sufficiently large P.

A *critical* depensation model predicts actual extinction of the fish population if it falls below a certain threshold value P_i [see (11.8) for an example]. What natural phenomena could lead to this growth behavior? One possible way of increasing growth density rate with increasing population would be for the mortality density rate to decrease as the population level increases. Natural mechanisms which bring about such a behavior include a predation that becomes less effective at higher levels of the prey population. This mechanism is known to be effective for salmon as well as clupeids (such as sardines, anchovies, and herrings) which form large and closely packed schools (Clark, 1976) leading to both noncritical and critical depensation growth rates. In the development below, we will focus our attention on critical depensation models of the type shown in Figure 11.4b. Corresponding analyses for more general critical depensation models and for noncritical depensation models will be left as exercises.

For the critical depensation model in Figure 11.4b, there are three equilibrium populations 0, $\bar{P}_1$, and $\bar{P}_2$ for each level of harvesting effort E with

$$\bar{P}_k^{\bullet} = F(\bar{P}_k) - \beta E \bar{P}_k = 0 \tag{11.9}$$

We see from Figure 11.5a that $\bar{P}_1(E)$ is unstable while 0 and $\bar{P}_2(E)$ are asymptotically stable. Note that the two nontrivial equilibrium harvest rates (also called *sustained yield*) are given by $\bar{h}_k(E) \equiv h(\bar{P}_k(E),E) = \beta E \bar{P}_k(E) = F(\bar{P}_k(E))$ with $\bar{h}_2(E) > \bar{h}_1(E)$. If the initial fish population P_0 is greater than $\bar{P}_1(E)$ for the particular effort level chosen, then $P(t)$ tends to $\bar{P}_2$; but if $P_0 < \bar{P}_1(E)$, then the population will become extinct eventually (with E held fixed throughout).

As E increases from 0, the graphs of the two sustained yields are shown in Figure 11.5b, the dashed curve being $\bar{h}_1(E)$ and the solid curve being $\bar{h}_2(E)$. These two curves eventually meet at $E = E_{\max}$, as shown in the figure; $E_{\max}$ gives the βEP line (in Fig. 11.5a) with steepest slope which is still in contact with the graph of $F(P)$ (and tangent to it at one point). For larger values of E, the only equilibrium state is extinction as $P^{\cdot} < 0$ for $E > E_{\max}$. The lower (dashed) sustained-yield curve in Figure 11.5b is unstable.

The actual level of effort expended by the fishermen is determined by the intersection of cE with the sustained-yield (versus effort) curve. At high unit effort cost, cE intersects $\bar{h}(E)$ only once at the origin; there is no profit at any level of effort. For smaller c, the line cE intersects the solid portion of the sustained-yield curve in Figure 11.5b with the fishery breaking even for the level of E at the point of intersection, denoted by E_∞. There is a profit for all levels of effort below E_∞. If effort rises beyond the level $E_{\max}$ (possibly induced by a large initial fish population P_0), the fish population may be reduced drastically to some level below P_i, the minimum viable population. Once it falls below P_i, the species is destined to extinction even if we stop fishing altogether.

Before the fish population reaches P_i, however, we may of course reduce fishing effort to such a low level E_l that the remaining fish population at that point is greater than $\bar{P}_1(E_l)$. At that low level of effort, the population would head toward an equilibrium state $\bar{P}_2$ and, hence, back to the solid yield curve. However, with profit still to be made for any $E > E_l$, it is unlikely that any party would be persuaded to reduce its effort level. Hence,

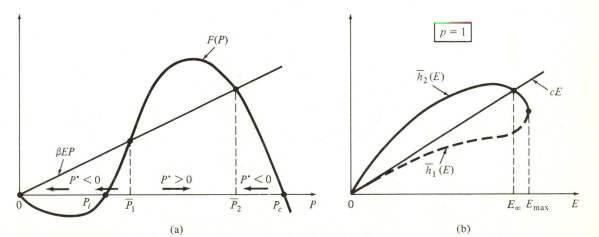

FIGURE 11.5

11B: A catastrophic collapse of the fishery or an outright extinction of the species is a distinct possibility for fish populations with a depensation growth behavior.

With the threat of extinction for fish species governed by critical depensation growth curve, some kind of central planning and government regulation is necessary to induce a proper management and the survival of fisheries. For highly depleted species, a fishing moratorium may be the only way to replenish the stock to a reasonable level. But will a fishing moratorium be effective given that it is only for a region 200 miles from shore?

11.4 One-Dimensional Reaction-Diffusion Model

In the remainder of this chapter, we will take a first step toward answering the question concerning the effectiveness of the 200-mile fishing limit for the implementation of a fishing moratorium or other regulations. Specifically, we will study the simplest model which allows for a spatial distribution of the fish population proposed by Ludwig (1976). For the possibility of replenishing depleted stocks, the existence and local stability of nontrivial equilibrium states for this model will be addressed (Wan, 1978). We will work with a dimensionless population variable by setting $u = P/P_c$. In particular, the logistic model of the last section may then be written as

$$u^{\cdot} = \Gamma f(u) \qquad f(u) = u(1 - u) \qquad (11.10)$$

where $\Gamma = r_0$. We will also take other types of growth dynamics of interest in the form of (11.10) with different $f(u)$. For example, for the depensation model (11.8), we have $u(1 - u)(1 - u/u_i)$ allowing Γ ($= -r_D$) to be negative. In general, Γ is chosen so that $f'(0) = 1$. To include the effect of fish movement, we consider the simplest situation where the only effective movement is in a direction normal to a long straight coastline with x being the distance from shore in this direction. For protecting a fishery from foreign fishing fleets, the movement of fish along the coast is of little or no interest. We take the seaward movement to be random so that it effectively spreads the fish population along the x axis. This random movement gives rise to a diffusion effect [see Lin and Segel (1974)] which modifies the ODE models of the last section into the following reaction-diffusion equation:

$$u_t = \nu^2 u_{xx} + \Gamma f(u) \qquad (0 < x < l, \, t > 0) \qquad (11.11)$$

where $x = l$ is the outer boundary of the regulated fishing zone. The rate of diffusion of fish in the x direction is assumed to be proportional to the gradient of the fish density with a constant of proportionality ν^2. This constant has a role similar to the thermal diffusivity $\nu/2$ in the one-dimensional heat conduction problems of Chapter 10 and is called the *diffusion coefficient*. With $\nu^2 = 0$, (11.11) is the usual ODE characterizing the population growth in the last two sections. For simplicity, we will restrict our discussion to the class of $f(u)$ analytic in a finite neighborhood of $u = 0$, that is, $f(u)$ has a Taylor series expansion about $u = 0$. This class includes most growth rate functions encountered in the fishery literature (Clark, 1976).

Just as the thermal diffusivity varies with the material of the bar, the diffusion coefficient ν^2 takes on different values for different fish habitats and must be measured for a particular habitat. In principle, an estimate of the value of ν^2 may be obtained from empirical data based on the physical interpretation of the diffusion term $\nu^2 u_{xx}$. As the rate of diffusion of fish is taken to be proportional to the gradient of the fish density, the rate of diffusion toward a location x from the population further offshore is approximately proportional to $[u(x + \Delta) - u(x)]/\Delta$. Similarly, the rate of diffusion from x toward shore is proportional to $[u(x) - u(x - \Delta)]/\Delta$. Thus, the net rate of increase of the fish density at x is given by $[u(x + \Delta) - 2u(x) + u(x - \Delta)]/\Delta^2$ which tends to u_{xx} in the limit as $\Delta \to 0$. Hence, measurements of the fish density at selective locations can be made to approximately determine ν^2 as long as complicating factors such as schooling are absent (Clark, 1976).

We will first investigate the worst possible scenario (Ludwig, 1976) where fish are harvested as soon as they go outside the political boundary $x = l$ so that

$$u(l, t) = 0 \qquad\qquad (11.12)$$

Much of our discussion in this chapter is concerned with this situation, although more realistic models will also be formulated (Wan, 1978, 1983). Since there is no flux of fish onshore, we have

$$u_x(0, t) = 0 \qquad\qquad (11.13)$$

Given the distribution of fish population in the region $0 \le x \le l$ at the start of the fishing moratorium, say

$$u(x, 0) = u_0(x) \qquad (0 \le x \le l) \qquad\qquad (11.14)$$

Equations 11.11–11.14 determine the fish population $u(x, t)$ in the region of no fishing for some time thereafter. In general, the solution of this problem can only be obtained by numerical methods. Fortunately, the fishery managers are mainly interested in whether or not the fish population evolves in time toward some equilibrium state [probably below the carrying capacity in view of (11.12)]. In other words, are there time-independent solutions of the boundary-value problem (11.11)–(11.13) which are asymptotically stable?

With $f(0) = 0$, it is clear that $u(x) \equiv 0$ is an equilibrium state of the system. As this corresponds to extinction, we would like to know its stability, at least local stability. Starting with an initial population close to the equilibrium state $u(x) \equiv 0$, does the population remain close to the equilibrium state thereafter or does it evolve away from it as time increases?

Beside $u \equiv 0$, are there other equilibrium states? Note that $u \equiv 1$ ($P = P_c$) is no longer an equilibrium state as it does not satisfy the boundary condition $u(l, t) = 0$. However, there may be other equilibrium states which are not uniform in space, that is, equilibrium states which are nonconstant functions of x. As we shall see, the existence and (local) stability of *spatially nonuniform* equilibrium state(s) depend very much on the magnitude of the diffusion coefficient ν. Global stability theorems for time-independent solutions of the PDE (11.11) have been obtained (Aronson and Weinberger, 1975) for a slightly different set of boundary conditions with the help of a maximum principle. These theorems can be extended to cover our problem. In this chapter, we will be less ambitious and ask only about local stability.

Local stability analyses are usually easier and suggest the appropriate global stability results which may not be apparent. The relevant mathematical problem for a local stability analysis in our case is the same as the one-dimensional heat conduction in a finite bar and can be solved by an elementary method. In fact, the problem of determining nontrivial equilibrium densities itself is mathematically the same as Euler's elastica of Chapter 5 and can be handled in exactly the same way. The methods for these two classes of classical problems also apply, either directly or with some modifications, to more realistic space-time models of the 200-mile-limit problem. It is not an exaggeration to say that, while many emerging problems in mathematical modeling may be new in appearance, their mathematical structures are often the same or similar to problems already treated successfully in the past. The general class of 200-mile fishing limit problems is only one of the many examples. The practitioners of applied mathematics should be forever ready to take a page from history.

11.5   Local Stability of the Trivial State

To analyze the stability of the equilibrium state $u(x, t) \equiv 0$, we consider the evolution of a fish population which deviates slightly from this trivial state at $t = 0$. Let this deviation be $u_0(x)$ with $|u_0(x)| \ll 1$. For such a $u_0(x)$, we expect to have $|u(x, t)| \ll 1$, at least for a short while. In that case, we can neglect terms of order u compared to one and linearize the PDE (11.11). In other words, we expect $u(x, t) \simeq v(x, t)$ where v is the solution of the linearized version of (11.11)–(11.14):

$$v_t = \nu^2 v_{xx} + \Gamma v \qquad (0 < x < l, \, t > 0) \tag{11.15}$$

$$v_x(0, t) = v(l, t) = 0 \qquad (t > 0) \tag{11.16}$$

$$v(x, 0) = u_0(x) \qquad (0 \le x \le l) \tag{11.17}$$

since $f(0) = 0$ and $f'(0) = 1$.

Now, the PDE (11.15) can be written as

$$\psi_t = \nu^2 \psi_{xx} \tag{11.18}$$

with $\psi \equiv e^{-\Gamma t} v(x, t)$. By the method of separation of variables (of Chapter 9), we find the general solution of (11.18) which satisfies the boundary conditions (11.16) to be

$$\psi(x, t) = \sum_{k=0}^{\infty} A_k e^{-\nu^2 \lambda_k^2 t} \phi_k(x) = A_0 e^{-\nu^2 \lambda_0^2 t} \cos \frac{\pi x}{2l} + \cdots \tag{11.19}$$

where

$$\lambda_k^2 = \left[\frac{(2k + 1)\pi}{2l} \right]^2 \qquad \phi_k(x) = \cos \frac{(2k + 1)\pi x}{2l} \tag{11.20}$$

The constants A_k, $k = 0, 1, 2, \ldots$, are determined from the initial condition (11.17) with the help of the orthogonality condition among the eigenfunctions. The function ψ decays with time at a rate dictated by the smallest λ_k, namely λ_0. With $\psi = e^{-\Gamma t} v(x, t)$, we have

$$v(x, t) = e^{\Gamma t} \psi = \sum_{k=0}^{\infty} A_k e^{(1 - \gamma_k/\Gamma)\Gamma t} \phi_k(x)$$

$$= A_0 e^{(1 - \gamma_0/\Gamma)\Gamma t} \cos \frac{\pi x}{2l} + \cdots \tag{11.21}$$

where $\gamma_n = \lambda_n^2 \nu^2 = [(2n + 1)\pi\nu/2l]^2$ with $\gamma_0 = (\pi\nu/2l)^2$.

The stability of the trivial equilibrium state evidently depends on the sign of Γ. If $\Gamma < 0$ (as in depensation growth), we have $v \to 0$ as $t \to \infty$. In this case, the trivial equilibrium is asymptotically stable. On the other hand, if $\Gamma > 0$ (as in the logistic growth case), the steady-state behavior of $v(x, t)$ is dominated by the leading term of the series (11.21). If $\Gamma(l/\nu)^2$ is smaller than $(\pi/2)^2$, then again $v(x, t)$ tends to zero as $t \to \infty$. The fish population will head toward extinction if it ever gets "too small." If $\Gamma(l/\nu)^2 > (\pi/2)^2$, the trivial equilibrium state is unstable and the fish stock, if left undisturbed within the region of regulated fishing, will grow with time beyond the range of applicability of a linearized analysis.

The above conclusions merely tell us in more precise terms what we should have expected all along:

> **11C:**   The fish population will grow in time if the reproduction rate of the fish population is large compared to the rate at which fish leave the regulated fishing zone.

The value of our analysis is that it tells us, in the case of $\Gamma > 0$, how large Γ (or how small ν^2) has to be for the fish stock to increase in spite of attrition due to fish movement across the political boundary.

> **11D:**   For a given fish population so that both Γ and ν are fixed constants, our results for the case $\Gamma > 0$ suggest how far we have to extend the offshore fishing moratorium, that is, the location of the boundary $x = l$, in order for such a moratorium to be effective.

These results and those to be obtained in the subsequent development underscore the importance of having a numerical value for the diffusion coefficient. Without at least an order of magnitude estimate of ν^2, the theoretical result on an effective moratorium is useless. An estimate of ν^2 can be obtained by suitable recordings of the fish movement in a carefully planned set of field observations as indicated in Section 11.1. Such a field project tends to be expensive but must be done for the design of a meaningful management policy.

11.6   Nontrivial Equilibrium States

Consider now the situation where $\Gamma(l/\nu)^2 > (\pi/2)^2$ so that the trivial state is unstable or where the fish population is above the critical level so that $\Gamma f(u) > 0$. In either case, the initially depleted fish stock will grow with the help of a period of no fishing. But this growth cannot be indefinite since there is a limit to the size of the fish stock the region can carry. It is not difficult to see that, unlike the case of no spatial variation in the fish distribution, nontrivial solutions of $f(u) = 0$ are not equilibrium fish densities for our problems since they do not satisfy the boundary condition $u(l, t) = 0$. Therefore, we expect that there must be one or more equilibrium densities which are not uniformly distributed over the interval $(0, l)$.

An *equilibrium* (dimensionless) *fish population* $U(x)$ is a time-independent solution of the PDE (11.11) which satisfies the boundary conditions (11.12) and (11.13). In other words, $U(x)$ is a solution of the BVP:

$$\nu^2 U_{xx} + \gamma^2 f(U) = 0 \qquad U_x(0) = U(l) = 0 \tag{11.22}$$

where we have set $\Gamma \equiv \gamma^2 > 0$. This is nearly the same mathematical problem encountered in the Elastica of Section 5.3. [Physically, the elastic column is now clamped at one end so that $\theta(s = l) = 0$ and free of bending moment (hinged) at the other end so that $d\theta/ds = 0$ at $s = 0$.] A first integral of the nonlinear ODE (11.22) gives

$$U_x \equiv \frac{dU}{dx} = -\frac{\gamma}{\nu}[F(U_0) - F(U)]^{1/2} \tag{11.23}$$

where $dF/dU \equiv 2f(U)$ and where $U_0 \equiv U(0)$ is an unknown constant. The negative square root was chosen since we must have $U(l) = 0$. Incidentally, the condition $U_x = 0$ when $U = U_0$ has been used to fix the constant of integration in (11.23).

The first-order ODE (11.23) is separable and can be integrated to give

$$x = \frac{\nu}{\gamma} \int_U^{U_0} \frac{dY}{\sqrt{F(U_0) - F(Y)}} \tag{11.24}$$

where we have made use of $U(0) = U_0$ to fix the constant of integration. The relation (11.24) may be inverted to give a nontrivial equilibrium density distribution $U(x)$. To determine the yet unknown constant U_0, we use the fact that $U(l) = 0$ to get from (11.24)

$$\alpha \equiv \frac{\gamma l}{\nu} = \int_0^{U_0} \frac{dY}{\sqrt{F(U_0) - F(Y)}} \tag{11.25}$$

The relation (11.25) may be inverted to give U_0 as a function of $\alpha \equiv \gamma l/\nu$.

For the logistic growth model with $f(U) = U(1 - U/u_c)$ [where u_c was taken to be 1 earlier in (11.10)], Equations 11.24 and 11.25 take the form

$$x = \frac{\nu}{\gamma} \int_{U}^{U_0} \frac{dY}{\sqrt{(U_0^2 - Y^2) - \frac{2}{3u_c}(U_0^3 - Y^3)}}$$

$$= \frac{\nu}{\gamma} \int_{w}^{1} \frac{dy}{\sqrt{(1 - y^2) - \frac{2a}{3}(1 - y^3)}} \qquad (11.26)$$

and

$$\alpha \equiv \frac{\gamma l}{\nu} = \int_{0}^{1} \frac{dy}{\sqrt{(1 - y^2) - \frac{2a}{3}(1 - y^3)}} \qquad (11.27)$$

respectively, with $a \equiv U_0/u_c$, $y = Y/U_0$, and $w \equiv U/U_0$. Note that α, a, y, and w are dimensionless quantities. The integral in (11.27) is a monotone increasing function of the parameter a in the range $0 \le a < 1$. The value of α is $\pi/2$ at $a = 0$ and tends to infinity as a tends to unity (Figure 11.6). Hence:

11E: There is exactly one nontrivial equilibrium fish population for the logistic growth model [given by (11.26) and (11.27)] for all $a > \pi/2$. The only equilibrium state for $a < \pi/2$ is the trivial state $U(x) \equiv 0$. The bifurcation from the trivial state occurs at $a = \pi/2$, precisely the value above which the trivial state was found earlier to be unstable.

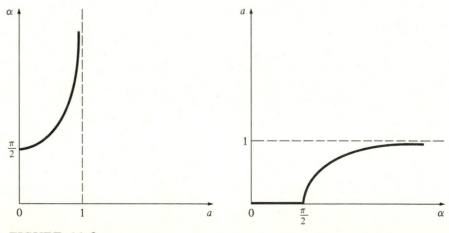

FIGURE 11.6

For α slightly larger than $\pi/2$, say $\alpha = \pi(1 + \epsilon)/2$ with $0 < \epsilon \ll 1$, we can obtain an approximate expression for a and for the nontrivial equilibrium population in terms of elementary functions. Since a $(\alpha = \frac{1}{2}\pi) = 0$, we expand a in a Taylor series in powers of ϵ for the case $\alpha = \pi(1 + \epsilon)/2$:

$$a = a_1\epsilon + a_2\epsilon^2 + \cdots \tag{11.28}$$

and the integral (11.27) can then be written as

$$\frac{\pi}{2}(1 + \epsilon) = \int_0^1 \left[1 + \frac{a_1\epsilon}{3} \frac{1 - y^3}{1 - y^2} + O(\epsilon^2) \right] \frac{dy}{\sqrt{1 - y^2}}$$

$$= \frac{\pi}{2} + \tfrac{2}{3}a_1\epsilon + O(\epsilon^2) \tag{11.29}$$

which can be solved to give

$$a_1 = \tfrac{3}{4}\pi \tag{11.30}$$

While the remaining coefficients a_k, $k = 2, 3, \ldots$, in (11.28) can also be obtained from terms in (11.29) involving higher powers of ϵ, it suffices for our purpose to know

$$a = \tfrac{3}{4}\pi\epsilon + O(\epsilon^2) \tag{11.31}$$

and, correspondingly, from (11.26)

$$U(x) = U_0 w = \tfrac{3}{4}\pi u_c \epsilon \cos \frac{\pi x}{2l}[1 + O(\epsilon)] \qquad \epsilon = \frac{2\alpha}{\pi} - 1 \tag{11.32}$$

The expression (11.32) gives an adequate first approximation for the nontrivial equilibrium population when $\alpha \gtrsim \pi/2$. This expression will be useful in a local stability analysis of the nontrivial equilibrium state *near bifurcation* in the next section. We note in passing that it is often easier to obtain higher-order terms in the parametric series for $U(x; \epsilon)$ by directly seeking a perturbation solution of the BVP (11.22) in the form

$$U(x; \epsilon) = \tilde{U}_0(x) + \epsilon\tilde{U}_1(x) + \epsilon^2\tilde{U}_2(x) + \cdots \tag{11.33}$$

along with the series expansion (11.28) for a and $\alpha \equiv \gamma l/v = \pi(1 + \epsilon)/2$.

11.7 Local Stability near Bifurcation

In order for a fishing moratorium to be effective (in restocking the fishing ground), we need the nontrivial equilibrium fish population to be stable. To see whether this is so, we consider an initial fish population $u_0(x) \equiv U(x) + v_0(x)$ with $|v_0(x)| \ll U(x)$ where $U(x)$ is the nontrivial equilibrium fish population discussed in Section 11.6. We are interested in the time evolution of the fish population $u(x, t) \equiv U(x) + v(x, t)$ starting from a small perturbation from the equilibrium state $U(x)$. Since $|v_0(x)| \ll U(x)$, we expect to have $|v(x, t)| \ll U(x)$ at least for awhile. We can therefore linearize the PDE (11.11) in $v(x, t)$ to get

$$v_t = \nu^2 v_{xx} + \gamma^2 q(x) v \qquad q(x) \equiv \left. \frac{df}{du} \right|_{u=U} \qquad (11.34)$$

where we have set $\gamma^2 = \Gamma > 0$ with no loss in generality as we know from Sections 11.2 and 11.3 that $f(u) > 0$ for u away from the origin. The associated boundary and initial conditions for v follow directly from (11.12)–(11.14) and the definition of $v(x, t)$ and $U(x)$:

$$v_x(0, t) = v(l, t) = 0 \qquad (t > 0) \qquad (11.35)$$

$$v(x, 0) = v_0(x) \qquad (0 \le x \le l) \qquad (11.36)$$

The linear initial-boundary value problem for $v(x, t)$ may be solved by the method of separation of variables. We get

$$v(x, t) = \sum_{k=0}^{\infty} V_k \phi_k(x) e^{(\alpha^2 - \bar{\lambda}_k^2)\nu^2 t/l^2} \qquad \alpha^2 = \left(\frac{\gamma l}{\nu} \right)^2 \qquad (11.37)$$

where $\{\bar{\lambda}_k^2\}$ and $\{\phi_k(x)\}$ are the eigenvalues (ordered in increasing magnitude) and eigenfunctions, respectively, of the eigenvalue problem:

$$l^2 \phi'' + [\alpha^2(q - 1) + \bar{\lambda}^2]\phi = 0 \qquad \phi'(0) = \phi(l) = 0 \qquad (11.38)$$

with $(\;)' = d(\;)/dx$. The constants V_k, $k = 0, 1, 2, \ldots$, in (11.37) are determined by the initial condition (11.36). The local stability of the nontrivial equilibrium state $U(x)$ evidently depends on the sign of $(\alpha^2 - \bar{\lambda}_0^2)$. This sign can be found for most growth models of interest by an elementary method when the system is *near bifurcation* (Wan, 1978), that is, when α is only slightly greater than its critical value for a nontrivial equilibrium solution.

To illustrate, take again the logistic growth case where $f(u) = u(1 - u/u_c)$ so that $q - 1 = -2U/u_c$. Near bifurcation, that is, when $\alpha = \pi(1 + \epsilon)/2 \equiv \lambda_0(1 + \epsilon)$ for $0 < \epsilon \ll 1$, we have from (11.32) $U/u_c = \frac{3}{4}\pi \times \cos(\pi x/2l)[1 + O(\epsilon)]$ so that (11.38) becomes

$$l^2\phi'' + \left\{\bar{\lambda}^2 - \tfrac{3}{2}\pi\epsilon\lambda_0^2 \cos \frac{\pi x}{2l}[1 + O(\epsilon)]\right\}\phi = 0 \tag{11.39}$$

$$\phi'(0) = \phi(l) = 0 \tag{11.40}$$

where $\lambda_0^2 = (\pi/2)^2$ is the lowest eigenvalue for the limiting case with $\epsilon = 0$. For local stability, it suffices to seek a perturbation solution of the first eigenvalue $\bar{\lambda}_0^2$ and the corresponding eigenfunction $\phi_0(x)$. Let

$$\bar{\lambda}_0^2 = \Lambda_0^2(1 + \epsilon\mu_1 + \epsilon^2\mu_2 + \cdots)$$

$$\phi_0(x) = \Phi_0(x) + \epsilon\Phi_1(x) + \epsilon^2\Phi_2(x) + \cdots \tag{11.41}$$

The unknown coefficients Λ_0^2, μ_k, and $\Phi_j(x)$ are to be determined by substituting (11.41) into (11.39) and (11.40). Upon collecting terms of the same powers of ϵ, we get

$$(l^2\Phi_0'' + \Lambda_0^2\Phi_0) + \epsilon\left[l^2\Phi_1'' + \Lambda_0^2\Phi_1 + \left(\Lambda_0^2\mu_1 - \frac{3}{2}\pi\lambda_0^2 \cos \frac{\pi x}{2l}\right)\Phi_0\right] + O(\epsilon^2) = 0 \tag{11.42}$$

$$\Phi_0'(0) + \epsilon\Phi_1'(0) + O(\epsilon^2) = 0 \tag{11.43a}$$

$$\Phi_0(l) + \epsilon\Phi_1(l) + O(\epsilon^2) = 0 \tag{11.43b}$$

These equations are to be satisfied identically in ϵ. For $\epsilon = 0$, they require

$$l^2\Phi_0'' + \Lambda_0^2\Phi_0 = 0 \qquad \Phi_0'(0) = \Phi_0(l) = 0 \tag{11.44}$$

for the leading term solution Λ_0^2 and $\Phi_0(x)$. The solution of the eigenvalue problem (11.44) is straightforward giving

$$\Lambda_0^2 = \left(\frac{\pi}{2}\right)^2 \quad \text{and} \quad \Phi_0(x) = \cos \frac{\pi x}{2l} \tag{11.45}$$

as the lowest eigenvalue and the corresponding eigenfunction, respectively [keeping in mind that the series (11.41) are for the lowest eigenvalue of (11.38) and the associated eigenfunction]. Note that $\bar{\lambda}_0^2 \simeq \Lambda_0^2 = \lambda_0^2$.

From terms multiplied by ϵ, we get

$$l^2\Phi_1'' + \Lambda_0^2\Phi_1 = \Phi_0(x)\left(\lambda_0^2\tfrac{3}{2}\pi\cos\frac{\pi x}{2l} - \mu_1\Lambda_0^2\right)$$

$$\Phi_1'(0) = \Phi_1(l) = 0 \tag{11.46}$$

for the first-order correction terms μ_1 and $\Phi_1(x)$. With Φ_0 and Λ_0 given by (11.45), the general solution of the ODE for Φ_1 which satisfies the condition $\Phi_1'(0) = 0$ is

$$\Phi_1(x) = c_0\cos\frac{\pi x}{2l} + \frac{\pi}{4}\left(3 - \cos\frac{\pi x}{2l}\right) - \frac{\mu_1}{2}\left(\cos\frac{\pi x}{2l} + \frac{\pi x}{2l}\sin\frac{\pi x}{2l}\right) \tag{11.47}$$

The remaining condition in (11.46), $\Phi_1(l) = 0$, requires

$$\pi - \frac{\mu_1}{2}\frac{\pi}{2} = 0 \quad\text{or}\quad \mu_1 = 4 \tag{11.48}$$

(The undetermined constant c_0 is available for a suitable normalization of the approximate eigenfunction.) Therefore, we have

$$\bar{\lambda}_0^2 = \left(\frac{\pi}{2}\right)^2[1 + 4\epsilon + O(\epsilon^2)] \tag{11.49}$$

and

$$\alpha^2 - \bar{\lambda}_0^2 = -\frac{\pi^2}{2}\epsilon[1 + O(\epsilon)] \tag{11.50}$$

ensuring that $v(x, t)$ of (11.37) decays exponentially in time.

The result in (11.50) therefore implies that

11F: Near bifurcation, the nontrivial equilibrium population $U(x)$ for the logistic growth case is locally (asymptotically) stable.

This result appears very upbeat. However, it must be kept in mind that the existence of a nontrivial equilibrium state is possible only if $\alpha \equiv l\sqrt{\Gamma}/v > \pi/2$ or $l > v\pi/2\sqrt{\Gamma}$ [see (11.32)].

11G: If the diffusion rate is much higher than the natural growth rate of the fish population so that $v\pi/2\sqrt{\Gamma} > 200$, it would take more than a 200-mile fishing limit to rejuvenate a depleted stock.

That there exists a stable nontrivial equilibrium population (for a much more distant political boundary) is of little use to fishery management.

The local stability of nontrivial equilibrium states away from bifur-cation, $\alpha \gg \pi/2$, is more difficult to analyze. The difficulty lies mainly in not having an explicit expression for $U(x)$ and not being able to take advantage of the small parameter ϵ in $\alpha = \pi(1 + \epsilon)/2$ by applying the perturbation method to the eigenvalue problem in the stability analysis. On the other hand, we do know that the nontrivial equilibrium state is limited by $0 < U(x) < 1$ for $\alpha > \pi/2$. We can therefore obtain lower and upper bounds on the eigenvalues by the Sturm comparison theorem even without an explicit expression for $U(x)$ (see exercises at the end of this chapter).

11.8   Finite Fishing Effort Rate

Up to now, we have effectively assumed the fishing effort rate at the political boundary $x = l$ (and beyond) to be unlimited so that $u(x, t) \equiv 0$ for $x \geq l$. We comment briefly in this section on the more realistic case of a finite rate of fishing effort E beyond $x = l$.

Suppose the dimensionless harvest rate $h_0(u, E, x)$ in the region $x > l$ is also proportional to Eu. The dynamics of the fish population outside the political boundary $x = l$ is evidently governed by

$$u_t = v^2 u_{xx} + \Gamma_0 f_0(u) - h_0(u, E, x) \qquad (l < x < L) \qquad (11.51)$$

where we have allowed for a different natural growth rate function for the fish population beyond $x = l$. Beyond the line $x = L$, the fish population would find a hostile environment so that there is effectively no flux of fish crossing this outer boundary:

$$u_x(L, t) = 0 \qquad (11.52)$$

An example of this outer boundary is the edge of the continental shelf. If no such boundary of hostile region exists, then L extends to infinity while the total fish population remains finite. At $x = l$, the condition $u(l, t) = 0$ is now replaced by continuity conditions on u and u_x. Given the initial fish popu-lation density, we have an initial-boundary value problem for $u(x, t)$ in the larger domain $0 < x < L$.

The analysis of the above more realistic problem is evidently more complex. However, the methods for obtaining the nontrivial equilibrium state and for performing local stability analysis can still be used for this new problem. Two special cases of the new problem deserve further discussion:

Concentrated Fishing Effort at $x = l$

For the special case of $\Gamma_0 f_0 = \Gamma f$ and concentrated fishing effort at $x = l$, resulting in a total harvest rate $H_0(u(l), E_0)$, we have a simpler problem consisting of the PDE (11.11) for $0 < x < L$, the boundary conditions (11.13) and (11.52), the continuity condition $u(l-, t) = u(l+, t)$, the jump condition

$$\nu^2 u_x(l-, t) + H_0(u(l), E_0) = \nu^2 u_x(l+, t) \qquad (11.53)$$

and a prescribed initial fish population distribution (11.14) for $0 < x < L$. With some modifications, methods used for (11.11)–(11.14) earlier are again applicable to this problem.

When the fish population is severely harvested at the boundary $x = l$, we often have $|u_x(l+, t)| \ll |u_x(l-, t)|$. In that case, we may neglect the $u_x(l+, t)$ term in (11.53) to obtain an initial-boundary value problem in the region $0 \le x \le l$ consisting of (11.11), (11.13), (11.14), and (11.53) with the right-hand side replaced by zero. For a constant fishing effort rate E_0 and with H_0 proportional to $u(l)$ so that $H_0 = c_l E_0 u$, the simplified boundary condition at $x = l$ becomes

$$\nu^2 u_x(l, t) + c_l E_0 u(l, t) = 0 \qquad (11.54)$$

The IBVP defined by (11.11), (11.13), (11.14), and (11.54) is akin to the IBVP (11.11)–(11.14) already analyzed and analogous to the elastica with both ends elastically supported (i.e., attached to a spring). By the methods used in Section 11.5, it is not difficult to show that the trivial equilibrium state $u \equiv 0$ is asymptotically stable for $\Gamma < \nu^2 \alpha_0^2 / l^2$, where α_0 is the smallest positive root of

$$\cot z = \mu z \quad \text{with} \quad \mu \equiv \frac{\nu^2}{E_0 c_l} \qquad (11.55)$$

and is unstable otherwise. For $0 < \mu < \infty$, we have $\alpha_0 < \pi/2$ and the difference decreases with increasing E_0. As expected, the threshold for extinction is reduced by a decrease of fishing effort rate. Furthermore, an asymptotically stable nontrivial equilibrium state corresponding to (11.24) exists for $\Gamma \equiv \gamma^2 > \nu^2 \alpha_0^2 / l^2$ and a bifurcation from the trivial state occurs at $\gamma l / \nu = \alpha_0$. As $E_0 \to 0$ [and $\mu \to \infty$ by (11.55)], we have $\alpha_0 \to 0$, so that we recover the expected results for the natural growth of an unharvested fish population.

Fixed Mortality Rate Beyond the Political Boundary

Another model (Clark, 1976) deals with the situation where the fish population in the region $l < x < L$ is so heavily harvested that it declines at a fixed rate $s^2 > 0$. In that case, we have

$$u_t = v^2 u_{xx} - s^2 u \qquad (l < x < L) \tag{11.56}$$

With the absorbing boundary condition (11.52), the nontrivial equilibrium dimensionless fish population U in the outer region $l < x < L$ is

$$U = \begin{cases} c \cosh\left[\dfrac{s}{v}(L - x)\right] & (L \text{ finite}) \\[2ex] ce^{-sx/v} & (L \text{ infinite}) \end{cases} \tag{11.57}$$

where the unknown constant c is determined as a part of the solution of the BVP in the inner region, consisting of the ODE in (11.22), the no-flux condition (11.13), and the continuity conditions on U and U_x at $x = l$. The solution $U(x), 0 < x < L$, for this problem and its behavior near bifurcation will be left as an exercise [see also Wan (1978, 1983) for the case of an absorbing boundary condition $U(L) = 0$]. The local stability of the equilibrium states can also be analyzed as in Sections 11.5 and 11.7.

Throughout this chapter, we have been concerned exclusively with the fishing moratorium for restocking the fishing ground. Instead of a total ban on fishing inside the region $0 < x < l$, the case of a regulated fishery within the political boundary is also of interest. Whether the regulation is by a catch quota or by a maximum allowed fishing effort (such as the number of vessel-days), the fish population in $0 < x < l$ is harvested at rate h (measured in fractions of P_c) so that a term $-h$ is to be added to (11.11). Other than that, the structure of all the models considered in this chapter remains unchanged and the new models themselves can be similarly analyzed.

EXERCISES

1. Obtain the explicit solution of

$$P^\bullet = r_0 P\left(1 - \frac{P}{P_c}\right) - \beta E P \qquad P(0) = P_0$$

where r_0, β, P_c, and E are positive constants. Deduce the limiting behavior of $P(t)$ as $t \to \infty$. Give the behavior of the harvest rate $h(t)$ as a function of time.

2.
$$P^{\cdot} = r_0 P \ln \frac{P_c}{P}$$

Find the maximum sustainable yield and the corresponding fish population for the above Gompertz law of population growth. Sketch the yield-effort curve and determine E^*.

3.
$$P^{\cdot} = \begin{cases} r_0 P & (P < P_c) \\ -\infty & (P > P_c) \end{cases}$$

Find the yield-effort relation for this growth model.

4.
$$P^{\cdot} = a - bP \qquad (a, b > 0)$$

Find the equation of the yield-effort curve and discuss its features.

5. **(a)** Suppose fishing is initiated at the effort level E^* and P^* is the equilibrium fish population for E^*. Discuss the long-run situation of the fishery for a compensatory-type growth model with an initial fish population P_0 less than P^*.

(b) How does the situation change if fishing is initiated at a level not equal to E^*?

(c) Repeat part (a) for $P_0 > P^*$.

6. In the problem of fish dynamics with diffusion, we were interested in steady-state solution(s) $\bar{P}(x)$ for the reaction-diffusion equation
$$P_t = \nu^2 P_{xx} + \gamma P(1 - P)$$
so that
$$\nu^2 \bar{P}_{xx} + \gamma \bar{P}(1 - \bar{P}) = 0$$

(a) With $\bar{Q} \equiv \bar{P}_x$, we have $\nu^2 \bar{Q}_x + \gamma \bar{P}(1 - \bar{P}) = 0$. Locate all the critical points, that is, equilibrium states of this dynamical system (with x taking on the role of time).

(b) Classify the critical points by a linear stability analysis.

(c) Sketch some typical trajectories in the phase plane.

(d) For the fishery problem, we are only interested in the solution which satisfies the boundary conditions, $\bar{P}_x(0) \equiv \bar{Q}(0) = 0$ and $\bar{P}(l) = 0$. Sketch such a trajectory in the phase diagram.

7. In a circular lake of radius a (and of uniform depth), the polarly symmetric diffusion-growth process is governed by the PDE
$$p_t = \gamma p(1 - p) + \nu^2 \left(\frac{1}{r} p_r + p_{rr} \right) \qquad (0 < r < a, t > 0)$$

where p is the (normalized) fish population density and r is the radial distance from the lake center. Evidently, we have $p_r(a, t) = 0$ since there is no flux across the edge of the lake and $p(r, t)$ must be bounded, in particular at $r = 0$.

(a) Show that both $\bar{p}_0 = 0$ and $\bar{p}_1 = 1$ are equilibrium solutions of the PDE.

(b) Show that $\bar{p}_0$ is unstable according to a linear stability analysis.

(c) Show that $\bar{p}_1$ is asymptotically stable by a linear stability analysis.

8.
$$p_t = \gamma p(1 - p)(2 - p) + \nu^2 p_{xx} \qquad (0 < x < l, t > 0)$$

$$p_x(0, t) = 0 \qquad p(l, t) = 0$$

(a) Find all the spatially uniform equilibrium solutions.

(b) Perform a linear stability analysis of all the equilibrium solutions obtained in part (a).

9. Find a spatially nonuniform equilibrium solution of the reaction-diffusion problem in Exercise 8.

10. Sketch the yield-effort curve (similar to Figure 11.5b) for a typical (noncritical) depensation growth model (see Figure 11.4a). Justify your answer.

11. For the logistic growth model, obtain an upper and lower bound on the eigenvalues $\{\bar{\lambda}_n^2\}$, as determined by (11.38), by noting that $q - 1 = -2U(x)/u_c$ and $0 < U(x) < 1$. [*Hint:* Determine the eigenvalues for the two extreme cases $U(x) \equiv 0$ and $U(x) \equiv 1$ and then apply Sturm's comparison theorem.]

Control and Optimization

It is often said that scientific enquiries are just attempts to understand our own environment. However, with understanding, we can exercise control and make improvements. Soon after the discovery of the semiconducting property of silicon and germanium came the transistor. From the development of the theory of transonic and supersonic fluid flows came more speedy aircraft. A usable superconducting material will undoubtedly be used to improve many aspects of our life, from faster trains to cheaper electricity. It seems to be human nature to never stop short of the optimum. Even as the speed of our serial computer is seen to be limited by the speed of light, our computer scientists are well on their way toward circumventing that speed barrier by parallel processing.

In the past, control and optimization were exercised through qualitative considerations on the part of an experienced manager or designer. Today, these processes have become much more quantitative and deductive. They have become a companion part of our scientific activities. As we cannot possibly test all alternatives, control and optimization are sought through mathematical analyses; they are now two major areas of research effort in applied mathematics. In the next few chapters, these two processes will be discussed through several mathematical models from different areas of economic analysis.

Economics is a study of the use of resources for production and consumption and related issues. Loosely speaking, the general effort may be divided into two broad areas. Microeconomics is concerned with the behavior of individual firms and consumers that determine the (relative) price level of various commodities, the allocation of resources among alternative productions, etc. In contrast, macroeconomics deals with economic aggregates such as the national income (or the gross national product, GNP), the level of employment, etc., and how they are influenced by fiscal and monetary policy changes. The analyses for these two broad classes of economic problems require a variety of mathematical techniques. The study of several mathematical models in microeconomics and macroeconomics in the next few chapters will provide us with a convenient vehicle for an excursion into the realm of mathematics for control and optimization.

In keeping with the earlier part of this book, we will limit ourselves here to a few selected topics in each area which reinforce the development of the previous chapters. In Chapter 12, we use the recent residential land theories to review classical programming techniques. Through Solow's model of congestion cost of transportation in land use, we are reminded once more of the usefulness and broad applicability of perturbation and asymptotic methods. In Chapter 13, we begin our discussion on growth theory with the simplest neoclassical growth model. The related optimal growth model illustrates the important role of optimal control theory, the modern version of the calculus of variations (first introduced in Chapter 5), in economics. Growth models with exhaustible resources in Chapter 14 show that the many useful techniques for differential equations also find applications in economics. Optimal control problems with binding inequality constraints will be illustrated with some recent forest rotation models in Chapter 15.

12

Suburbs Are for the Affluent

The Structure of the Residential District

A good working knowledge of classical optimization with equality constraints is assumed. Perturbation method is applied to a boundary-value problem for nonlinear ODE. The matched asymptotic expansions technique is introduced for the same problem in a different range of parameter values.

12.1 Residential Area in a Monocentric Circular City

Kepler and Newton had their eyes on heaven and order emerged from their scientific scrutiny of the seemingly complex motion of the planets. Most of us mere mortals, however, are earthbound in body and spirit. Much of our time is spent in making a living and in living itself. One essential element in these rather mundane activities is land. We need agricultural and timber land to grow food and lumber, residential land for dwelling and recreation, and a commercial area for transacting business. There is only a finite amount of land area on Earth, considerably less than the total surface area of our planet. Sooner or later, the appropriate use of a particular parcel of land becomes an issue, sooner in places such as Hong Kong and Singapore and perhaps later in Australia and Alaska. How it is to be resolved will depend on the local social order; it may be by individual preference and the forces of economics or by central planning, or perhaps a combination of both. As our society becomes more and more complex, a comparably large body of knowledge on land economics has evolved. It may be loosely grouped into three interrelated parts: (1) the use of urban land for economic activities, (2) the use of residential land for housing and streets, and (3) the use of fertile "agricultural" land for farm goods and timber. [In most land use models, the land used to improve our environment such as wildlife conservation is usually lumped with (2); "housing" is interpreted in a broad sense here to include recreational space such as city, state and national parks.] The theory for any one part would be too rich in content to be thoroughly discussed in a chapter of this book. On the other hand, much of the theory of land use is optimization in action. The fundamental aspects of the theory offer a novel illustration of the optimization process not seen in other modeling texts. In this chapter, we will use some very crude models in residential land theory to show how the forces of economics and household consumption preferences shape the structure of residential areas as in most cities in North America today. About 80 percent of all privately developed land in major cities in the United States is directed toward residential use. Thus, the theory of residential land use is particularly relevant to our daily life. From the perspective of the mathematical model, it is also of interest that many of the mathematical techniques originally developed for mechanics are also useful in land use problems.

The simplest idealized model in conventional residential land theory considers a featureless, monocentric, circular city consisting of a circular *central business district* (CBD) surrounded by an annular region of roads and houses. The land beyond this annular region is for the production of agricultural goods. Because we pay no attention to natural geographical or topographical features, one unit of land differs from another only in its distance from the CBD. We denote the radius of the CBD by R_c (miles) and the radius of the outer boundary of the residential district by R_e. A population of N_0 *identical* families lives in the city; each household uses a certain amount of *housing space s* (square miles) at a distance X from the CBD and consumes a certain amount c (bags) of a simple composite *consumption good* per annum. Each household sends one commuter traveling to work in the CBD to earn an identical *annual wage Y* (dollars). Our standard model is concerned only with traveling to and from the CBD by the workers, which results in a *transportation cost* of t dollars per annum for each household; t depends on the distance traveled and hence the location X. Traveling within the CBD is assumed to be free, so that t $(X \le R_c) = 0$. The residential land values are determined (primarily) by differential accessibility to the CBD in a manner to be described below. The radial road system is densely and evenly distributed in the circumferential direction, so that we effectively have complete polar symmetry. (Alternatively, we may consider only a pie-shaped sector of the city with a small sectoral angle θ.)

Let $r(X)$ (in dollars/square mile) be the land value or *rent* per unit area of housing space and p (in dollars/bag) be the *price* of a bag of consumption goods. Each household uses its annual wage to pay for housing, consumption goods, and travel cost (to and from the CBD), so that

$$pc + rs + t = Y \tag{12.1}$$

Each of the N_0 families chooses a location X, an amount of land s, and an amount of consumption goods c, which give the household the most *satisfaction* subject to the *budget constraint* (12.1). As we shall see, the land rent per unit area, which varies with location, is determined by the household decisions on location, housing space, and the amount of consumption goods.

12.2 Household Optimum

The satisfaction of a household is measured by a utility index U. In the absence of natural geographical features and socially determined neighborhood characteristics, it is difficult to know whether it is better to live near the CBD or far away from it, given that the costs of transportation (including commuting time) have already been included in the budget equation (12.1). For simplicity and a lack of information, we will assume that location has no *direct* effect on the utility of a household [though it definitely has an indirect effect through (12.1)].

The modern theory of utility is an indispensable tool for the social sciences. Though not without serious limitations, the theory developed by von Neumann and Morgenstern and a modified version suggested by Luce (1957) provide a starting point for our analysis of the household optimum. We assume here that the utility index function has already been determined experimentally, and limit ourselves to listing a few of its basic properties relevant to our analyses of the residential land values.

The *utility function* $U(c, s)$ assigns a real number to each point in the c,s plane (the commodity plane) to reflect household preferences. A point (c, s) in the commodity plane corresponds to a specific combination of c amount of consumption goods and s amount of land for housing; such a combination is called a *commodity bundle*. The numerical values of $U(c, s)$ give the ranking by the household preferences, with $U(c, s) \geq U(c', s')$ whenever the bundle (c, s) is at least as satisfactory or desirable as (c', s'). To the extent that the utility function reflects only preference rankings, it is ordinal and is determined only up to a one-one monotone transformation.

A curve in the commodity plane along which U is a constant is called an *indifference curve*. The household is indifferent to all commodity bundles along an indifference curve in the sense that it considers each as desirable as all the others. A household always prefers more of each commodity if the other is not reduced, that is, $\partial U/\partial c > 0$ and $\partial U/\partial s > 0$. We also expect $U_{cc} < 0$ and $U_{ss} < 0$, as the additional satisfaction to be derived from an extra amount of a commodity diminishes with the amount of the commodity already in possession. The assumptions $U_{cc} < 0$, $U_{ss} < 0$, and $U_{cc}U_{ss}$

$- U_{cs}^2 > 0$ are usually made, so that U is concave with a negative definite Hessian matrix in the entire first quadrant of the commodity plane.

Along an indifference curve, we have

$$dU = U_c \, dc + U_s \, ds = 0 \tag{12.2}$$

with $U_z \equiv \partial U / \partial z$, so that

$$\frac{dc}{ds} = - \frac{U_s}{U_c} < 0 \tag{12.3}$$

The ratio U_s / U_c is called the *marginal rate of substitution*. An extra unit of space may be substituted for U_s / U_c amount of consumption goods without changing the utility. We see from (12.3) that indifference curves are convex to the origin of the commodity plane as shown in Figure 12.1. This property is invariant with respect to a one-one monotone transformation. [The applicability of the figure, however, is limited by the fact that each household needs a subsistent level of each commodity.]

Given a household utility index function $U(c, s)$, an annual wage Y, and the total transportation cost per annum $t(X)$, individual households want to choose c and s to maximize $U(c, s)$ subject to the budget constraint (12.1) as well as the nonnegative constraints $c \geq 0$ and $s \geq 0$ (as we cannot have negative consumption or space):

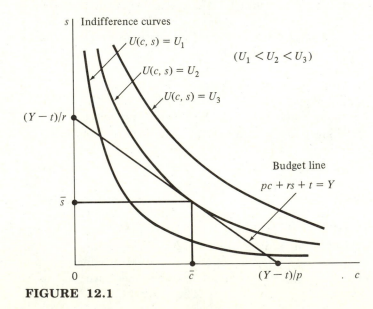

FIGURE 12.1

$$\max_{\{c,s\}} \{ U(c, s) \mid \Gamma(c, s) \equiv Y - t(X) - r(X)s - pc = 0, c \geq 0, s \geq 0 \}$$

$$(12.4)$$

Geometrically, the solution of this problem is straightforward. Given r, t, p, and Y, look for the highest indifference curve for U which is still in contact with the negatively sloped budget line $\Gamma = 0$. The bundle $(\bar{c}, \bar{s})$ which maximizes U, subject to the budget constraint, would be the point $(\bar{c}, \bar{s})$ on a particular $U =$ constant curve where the budget line is tangent to that curve. From Figure 12.1, we see that any indifference curve with a lower utility would intersect the budget line at two places and any one with a higher utility would not intersect it at all.

Considerably more information can be obtained from the mathematical conditions for a stationary point. At such a stationary point $(\bar{c}, \bar{s})$, we have $dU(\bar{c}, \bar{s}) = 0$ with $\Gamma(\bar{c}, \bar{s}) = 0$. We may solve $\Gamma(c, s) = 0$ for c in terms of s and eliminate c from $U(c, s)$, leaving us with a one-variable maximization problem with no constraints. Instead, we observe that we have also $d\Gamma(\bar{c}, \bar{s}) = 0$, which can be combined with $d\bar{U} = 0$ by a Lagrange multiplier to get $d\bar{U} + \lambda \, d\bar{\Gamma} = 0$, or

$$(\bar{U}_c + \lambda \bar{\Gamma}_c)dc + (\bar{U}_s + \lambda \bar{\Gamma}_s)ds = 0 \qquad (12.5)$$

where $\bar{f} \equiv f(\bar{c}, \bar{s})$. The Lagrange multiplier λ, arbitrary up to this point, will now be chosen so that

$$\bar{U}_c + \lambda \bar{\Gamma}_c = 0 \quad \text{or} \quad \bar{U}_c = \lambda p \qquad (12.6)$$

In that case, $d\bar{U} + \lambda d\bar{\Gamma} = 0$ implies

$$\bar{U}_s + \lambda \bar{\Gamma}_s = \bar{U}_s - \lambda r = 0 \quad \text{or} \quad \bar{U}_s = \lambda r(X) \qquad (12.7)$$

The two equations (12.6) and (12.7) are to be solved along with (12.1) for λ, $\bar{c}$, and $\bar{s}$ in terms of r, p, and $I \equiv Y - t$. The results can then be used to express $\bar{U}$ in terms of I, r, and p so that $\bar{U} = U(\bar{c}, \bar{s}) \equiv V(I, r, p)$.

The two equations (12.6) and (12.7) themselves immediately give

$$\frac{\bar{U}_s}{\bar{U}_c} = \frac{r}{p} \qquad (12.8)$$

so that

12A: At household optimum, the marginal rate of substitution [between consumption goods and housing space as defined through (12.3)] equals the price ratio p/r.

Upon differentiating the budget equation (12.1) with respect to X, we get

$$-t'(X) = r'\bar{s} + r\bar{s}' + p\bar{c}' = r'\bar{s} \qquad (12.9)$$

where we have used the *Alonso condition* $r\bar{s}' + p\bar{c}' = 0$ which follows from (12.2) along $d\bar{U} = 0$ and (12.8). Since $s > 0$ and $t'(x) > 0$ (as it costs more to travel a longer distance), we have $r' < 0$ and hence

> **12B:** At household optimum, land rent decreases with increasing distance from the CBD.

With the Alonso condition $rs' + pc' = 0$ along $d\bar{U} = 0$, $\bar{s}'$ and $\bar{c}'$ must be of opposite sign. It will be left as an exercise to show that $\bar{s}' > 0$ and $\bar{c}' < 0$. Observe that with $\bar{s} = \hat{s}(I, r, p)$, we *expect* that $\hat{s}_I > 0$ and $\hat{s}_r < 0$; the housing space of a household normally increases with its net income (after transportation cost) and decreases with increasing rent.

To illustrate, we consider a Cobb-Douglas type utility function often used in urban land economics for households in residential districts:

$$U(c, s) = U(\xi) \qquad \xi = s^\sigma c^{1-\sigma} \qquad (12.10)$$

where $0 < \sigma < 1$ and $U(\cdot)$ is monotone increasing and strictly concave. The first-order necessary conditions for a maximum U yield the stationary point

$$\bar{c} = \frac{1 - \sigma}{p}(Y - t) \qquad \bar{s} = \frac{\sigma}{r}(Y - t) \qquad (12.11)$$

and the corresponding stationary value of U is

$$U(\bar{c}, \bar{s}) = U\left[\frac{(1 - \sigma)^{1-\sigma}\sigma^\sigma}{p^{1-\sigma}r^\sigma}(Y - t)\right] \equiv V(Y - t(X), r(X), p) \tag{12.12}$$

The monotone increasing and strictly concave properties of $U(\cdot)$ ensure that the stationary value is in fact a maximum. From (12.11), we see that $\bar{c}' < 0$ and hence $\bar{s}' > 0$ from $r\bar{s}' + p\bar{c}' = 0$.

12.3 Competitive Equilibrium

Because all N_0 families are identical in our model city, they have the same taste and therefore share the same utility function. For them to settle into an equilibrium dwelling pattern, all households must have attained the same utility (value); otherwise the less satisfied households would move to a more satisfactory location with a higher utility, bidding up the land rent there in the process, etc., to reach equilibrium. When all households are in *competitive equilibrium,* we may write

$$U(\bar{s}, \bar{c}) = V(I(X), r(X), p) = V(I_c, r_c, p) \tag{12.13}$$

where $I_c = I(R_c)$ and $r_c = r(R_c)$ with R_c being the edge of the CBD, that is, the inner boundary of the residential area. The second half of (12.13) may be solved for $r(x)$ to get

$$r(X) = \tilde{r}(I(X); r_c, I_c, p) \qquad I_c = Y - t(R_c) = Y \tag{12.14}$$

[as transportation is free within the CBD so that $t(R_c) = 0$]. All quantities in the model are now determined up to the unknown constant r_c once we know the transportation cost $t(X)$. The quantity $V(I, r, p)$ is sometimes called an *indirect utility index.*

For the Cobb-Douglas type utility function of (12.10), the equi-utility requirement (12.13) implies

$$\frac{(1 - \sigma)^{1-\sigma} \sigma^{\sigma}}{p^{1-\sigma} r^{\sigma}} (Y - t) = c_0$$

where c_0 is an unknown constant, or

$$r^{\sigma} = c_1 [Y - t(X)] = \frac{r_c^{\sigma}}{Y} [Y - t(X)] \tag{12.15}$$

where c_1 is another unknown constant which we can express in terms of the unknown rent at the edge of the CBD, r_c.

To determine the remaining unknown constant r_c, we let $N(X)$ be the number of households living outside the circle of radius X. Evidently, we must have

$$N(R_c) = N_0 \tag{12.16a}$$

$$N(R_e) = 0 \tag{12.16b}$$

because the entire population lives outside the CBD, and no family lives beyond the city limits. In an annular ring in the residential district extending from X to $X + dX$, we have

$$2\pi X \, dX = -\bar{s} \, dN \tag{12.17}$$

since the amount of land area occupied by the households in the ring (of area $2\pi X\, dX$) must equal the total amount of land area in the ring (all available for housing for the time being). From (12.17) and (12.16a), we get

$$N(X) = N_0 - 2\pi \int_{R_c}^{X} \frac{z}{\bar{s}}\, dz \qquad (12.18)$$

Note that $\bar{s}(I(X), r(X), p)$ is a known function of the distance from the city center up to a constant r_c (see Equation 12.14 or 12.15).

Finally, the condition (12.16b) requires

$$N_0 = 2\pi \int_{R_c}^{R_e} \frac{z\, dz}{\bar{s}} \qquad (12.19)$$

which determines r_c once we prescribe $t(X)$.

For the Cobb-Douglas utility, we have $\bar{s} = \sigma(Y - t)/r$ or

$$\bar{s} = \frac{\sigma Y}{r_c} \left(1 - \frac{t}{Y} \right)^{(\sigma-1)/\sigma} \qquad (12.20)$$

so that

$$N_0 = \frac{2\pi r_c}{\sigma Y} \int_{R_c}^{R_e} \left[1 - \frac{t(z)}{Y} \right]^{(1-\sigma)/\sigma} z\, dz$$

or

$$r_c = \frac{\sigma Y N_0}{2\pi} \left\{ \int_{R_c}^{R_e} \left[1 - \frac{t(z)}{Y} \right]^{(1-\sigma)/\sigma} z\, dz \right\}^{-1} \qquad (12.21)$$

12.4 Two Income Classes

The basic model for the households in the residential area is crude and unrealistic, but tractable. Having completed the analysis for that model, we can now make it more realistic in several directions. We begin in this section by allowing for income differentials among households. It suffices to consider the case of two income classes; the method of analysis extends to multiple income classes in a straightforward manner. For the same setting as in the basic model, we now allow two income groups, N_h households with an annual wage Y_h and N_l households with an annual wage $Y_l < Y_h$. It is easy to see immediately:

12C: The two income groups do not mix.

Suppose this is not so and that a rich household and a poor household both reside at location X. In equilibrium, we have then

$$V(I_h(X), r(X)) = V(I_{hc}, r_c) \qquad V(I_l(X), r(X)) = V(I_{lc}, r_c)$$
$$(12.22)$$

omitting the dependence on the constant price p. In general, they give two different values for $r(X)$ which is not acceptable.

Another line of argument for the same result makes use of the fact that land goes to the highest bidder. At a given location X, a high-income household (if it should choose to live there) can always offer a higher bid rent than the low-income households and still buy more land and consumption goods to achieve a higher utility.

Let R_i be the boundary which separates the two income groups. It is not difficult to conclude that

12D: The land rent profile $r(X)$ is continuous; the rent gradient $r'(X)$ is discontinuous at R_i and is steeper on the low-income side of R_i.

If $r(X)$ were not continuous, then a household on the high-rent side could get a finite rent saving by moving an infinitesimal distance to the low-rent side (and thereby achieving a finite increase in its utility). This contradicts the condition of equilibrium.

We take $t'(X)$ to be continuous across R_i (as there is no reason to assume otherwise). Then, in a small neighborhood of R_i, the rich occupy more dwelling space than the poor (as space is a *normal good* so that utility is an increasing function of the good). However close they are to R_i, the difference remains finite because of the differential income. It follows from the "Alonso condition" $r' = -t'/\bar{s}(X)$ that the rent gradient is steeper on the low-income side of R_i. (A similar argument shows that the gradient becomes increasingly steeper as we move toward the CBD.) We now show that

12E: The low-income group lives closer to the CBD; there cannot be an alternating multizone pattern.

If the low-income group were further away from the CBD, then a household on either side of R_i could achieve a higher utility by moving into the other zone, because a continuation of the rent profile (which has a steeper gradient on the low-income side) for that class into the other zone always gives a lower land rent. This induces a relocation and contradicts the condition of equilibrium (see Figure 12.2). The same argument disallows a multizone pattern.

It remains to determine the boundary radius R_i and the land rent at the edge of the CBD r_c to complete the solution. Now, we have for the range $R_c < X < R_i$

$$V(I_l(X), r, p) = V(I_{lc}, r_c, p) \tag{12.23a}$$

and for $R_i < X < R_e$

$$V(I_h(X), r, p) = V(I_{hi}, r_i, p) \tag{12.23b}$$

for some yet unknown rent r_i at R_i. We invert these to get $r(X)$ in terms of the other quantities:

$$r(X) = \begin{cases} \hat{r}(I_l(X), I_{lc}, r_c, p) \equiv r_l(X) & (R_c < X < R_i) \\ \hat{r}(I_h(X), I_{hi}, r_i, p) \equiv r_h(X) & (R_i < X < R_e) \end{cases} \tag{12.24}$$

Continuity requires $r_i = r_l(R_i) = r_h(R_i)$, leaving only two unknown constants r_c and R_i. They are determined by the two conditions on the size of the two income classes:

$$2\pi \int_{R_c}^{R_i} \frac{X}{s_l(X)} \, dX = N_l \qquad 2\pi \int_{R_i}^{R_e} \frac{X}{s_h(X)} \, dX = N_h \tag{12.25}$$

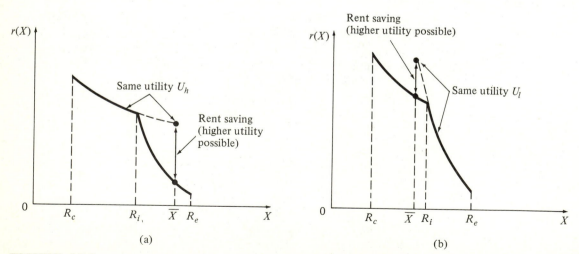

FIGURE 12.2

where

$$s_l(X) = \hat{s}(I_l(x), r_l(X), p) \qquad s_h(X) = \hat{s}(I_h(x), r_h(X), p)$$

$$(12.26)$$

12.5   Other Improvements for the Basic Model

Two Utility Classes

A realistic model must allow for differences in household preference. We consider here the case of two distinct classes of households with a different utility index but otherwise identical. The treatment of multi-utility classes can be handled similarly.

Consider the case of N_m households which require more dwelling space than the remaining N_l households to achieve the same utility. That is, if $U_m(c, s_m) = U_l(c, s_l)$, then $s_m > s_l$. All $N_m + N_l$ households have the same income and are *assumed* to achieve the same utility in equilibrium. An immediate observation on this modified model is that

12F: The two utility groups do not mix.

Otherwise, we have $V_m(I(X), r(X), p) = V_0$ and $V_l(I(X), r(X), p) = V_0$; they give two different rent profiles. Also, the following can be shown to be true:

12G: Rent profile is continuous and rent gradient is continuous except at R_i, the boundary separating the two groups. Those preferring more dwelling space live further away from the CBD where the rent gradient is not as steep as in the other zone.

The arguments are similar to those for the two income classes with $s_m > s_l$ adjacent to R_i being a consequence of the different household preferences.

It remains to determine r_c and R_i to complete the solution. This is done by the same process as in the two-income-class problem.

An Expanding City

Up to now, we have assumed that the city has a fixed boundary R_e for geographical or political reasons. Cities which are surrounded by farm lands have the option to expand outward by converting farm land into residential area (and we see this happening all the time). If the city is *closed* in the sense of no change of population through immigration or emigration, the city reaches its equilibrium size with an outer boundary R_e determined by

$$r(R_e) = r_A \qquad (12.27)$$

where r_A is the *agricultural rent* (or worth) of the farm land. For simplicity, we take r_A to be given (except in an exercise later) as we do not assume any knowledge of the theory of agricultural land economics. Farm land will remain for agricultural production as long as the bid rent for housing does not exceed the worth of the land for producing agricultural goods. The condition (12.27) on the residential land rent at R_e and the condition $N(R_c) = N_0$ (for the case of a single class of identical households) determine r_c and R_e simultaneously.

An Open City

If movement of households in and out of the city is allowed, we have what is called an *open city*. We illustrate the type of analysis required for open cities by working out the case of a "small" city. A city is (open and) *small* if population movement in or out of the city does not affect the economy outside the city; for example, the commodity price p and the indirect utility of the world beyond the city limit V_0 remain unchanged by such movements.

What will be affected by the movement of households is the annual wage rate Y for the households in the city; Y will rise when the number of households decreases through migration and will fall when there is a net gain of population by the city. We denote this wage-population relation by

$$Y = \hat{Y}(N_0) \qquad (12.28)$$

Normally, $\hat{Y}(\cdot)$ is a monotone decreasing function which is convex to the origin. Movement of households in and out of the city will continue until there is no gain by such a move, that is, the utility attained by the households in the city is the same as that outside the city:

$$V(I(X), r(X); p) = V(I_c, r_c; p) = V_0 \qquad (12.29)$$

The wage-population relation (12.28), the equi-utility condition (12.29), and the population size condition $N(R_c) = N_0$ are three simultaneous equations for r_c, Y, and N_0. [If the city is free to expand, then we have also a fourth condition $r(R_e) = r_A$ and a fourth unknown R_e.]

Some remarks on the analysis of cities which are open but not small can be found in (Kanemoto, 1980).

Absentee-Landlord or Public Ownership

Implicit in the previously discussed models on residential land use is an assumption that residential (and the adjacent farm) lands are owned by many landlords who do not live in the city. They collect rent from the households and spend it outside the city. The income of the households in the city in this case is simply its annual wage. For most purposes, the assumption of *absentee landlords* is an adequate and convenient device to simplify the necessary analysis with a minimal loss in generality. For a planned economy, however, the social welfare of all constituents must be taken into consideration, including those of the landlords. The problem becomes much more complicated if we have to compare the utilities of the absentee landlords and those of the tenants. An artificial institutional arrangement can be made to circumvent the need to consider the landlords' utilities; we will now explain this so-called *public-ownership* arrangement.

City residents form a government; it rents land from rural landlords for the price of the agricultural rent r_A. (There are too many small landlords to have much bargaining power with the city government.) The city government in turn rents the land to the households at the competitively determined rent $r(X)$. The net revenue is then divided equally among the households as dividends. The annual income of each household under this arrangement is wage plus dividends, $Y + D$. The net revenue to the city each year is $N_0 D$ given by

$$\int_{R_c}^{R_e} [r(X) - r_A] 2\pi X \, dX = N_0 D \tag{12.30}$$

The income of each household per annum is therefore

$$y = Y + \frac{2\pi}{N_0} \int_{R_c}^{R_e} [r(X) - r_A] X \, dX \tag{12.31}$$

which may be written as

$$N_0 Y = 2\pi \int_{R_c}^{R_e} \left(\frac{pc + t}{\bar{s}} + r_A \right) X \, dX \tag{12.32}$$

with the help of the budget equation $y - t = pc + rs$ and (12.19). Under this setup, the *social dividend D* depends on the land rent profile; consequently, the annual income of a household is not known until we have solved the problem. With $r = r(I(X); r_c, I_c, p)$ where $I(X) = y - t(X) = Y + D - t(X)$, the condition $N(R_c) = N_0$ and any one of the three (equivalent) integrated conditions above determine D and r_c simultaneously.

12.6   Other Criteria for Optimality

Social Optimum

Up to now, the basic model of Section 12.2 and its variations in Sections 12.4 and 12.5 all allow the individual households to freely choose the amount of land and goods to maximize their own utility. Instead, suppose a city government established by the N_0 identical households of the closed city is charged with the responsibility to allocate land and consumption goods to maximize the aggregate welfare of its N_0 constituents. To formulate the corresponding mathematical model for this new setting, we introduce a household density function $n(X)$ for a particular location X with

$$n(X) \equiv \frac{2\pi X}{s(X)} \tag{12.33}$$

so that we do not have to interpret the combination on the right every time it appears. Evidently, we have

$$dN(X) = -n(X)dX = -\frac{2\pi X}{s(X)}\, dX \tag{12.34}$$

as the decrease in population from X to $X + dX$.

With $n(X)$, we can now write down the aggregate utility of all households in the city:

$$W = \int_{R_c}^{R_e} U(c, s)n(X)dX \tag{12.35}$$

which is one measure of the social welfare of the city. The government may wish to choose an appropriate allocation of housing space s and consumption c to maximize W for a *social optimum*. The maximization is subject to three constraints:

$$2\pi X = n(X)s(X) \qquad (R_c \le X \le R_e) \tag{12.33}$$

$$\int_{R_c}^{R_e} n(X)\,dX = N_0 \qquad\qquad (12.36)$$

$$C_0 = \int_{R_c}^{R_e} [n(pc + t) + 2\pi r_A X]\,dX \qquad\qquad (12.37)$$

The last condition gives the total resources available to the households in the city (previously distributed equally to all households as wages and dividends) for consumption, transportation, and the leasing of land from the landlords.

Mathematically, we have now a more complicated problem in the calculus of variations (encountered earlier in Chapter 5) with both equality constraint (12.33) and isoperimetric constraints (12.36) and (12.37). It is possible to obtain necessary conditions for the stationarity of W in the form of Euler differential equations and boundary conditions by the method introduced in Section 5.4 with the help of Lagrange multipliers. We will limit ourselves to the observation that the problem can also be put into the form of an optimal control problem to be analyzed later in Chapters 13–15. This is accomplished by writing (12.33), (12.36), and (12.37) equivalently as

$$\frac{dN}{dX} = -\frac{2\pi X}{s(X)} \qquad\qquad N(R_c) = N_0 \qquad (12.38)$$

$$\frac{dC}{dX} = \frac{2\pi X}{s(X)}(pc + t) + 2\pi r_A X \qquad C(R_c) = 0 \qquad (12.39)$$

and the two boundary conditions

$$N(R_e) = 0 \qquad C(R_e) = C_0 \qquad\qquad (12.40)$$

The optimization problem may now be stated as one of maximizing (12.35) subject to (12.38)–(12.40). The method of solution for such a problem will be discussed later in Chapters 13–15.

Economically, the present optimality criterion for the N_0 households allows for utility differentials among the households if they lead to a larger social welfare index W. In this centrally planned economy, individual households may be sacrificed for a greater social good (or average household utility). If all households attain the same utility, then $W = N_0 U$ and individual household utility is identical to the average utility. One consequence of the new optimality criterion which can be deduced from the necessary conditions for the stationarity of W is that U *varies* with location for social optimum. In fact, it can be shown (Kanemoto, 1980) that for $R_c \le X \le R_e$, we have $dU/dX > 0$ so that

12H: At social optimum, household utility increases with distance from the CBD.

It is rather ironic that the less-well-off constituents in our model always live near the city center whether you have a capitalistic or socialistic government!

Rawlsian Welfare and the Max-Min Criterion

The social welfare index (12.35) is not the only possible measure of city residents' welfare. In fact, it is in some sense an unfair measure because, at optimum, households at different locations attain different levels of utility. A related type of welfare index is

$$W(\nu(X)) = \int_{R_c}^{R_e} \nu(X) U(c, s) n \, dX \qquad (12.41)$$

for some preassigned weight function $\nu(X)$ [with $\nu(X) = 1$ for (12.35)]. For the land use problem, this kind of welfare index is difficult to justify; why should residents in one location be assigned a more favorable weight than another?

A different type of welfare index is

$$W_\alpha = \left\{ \frac{1}{N_0} \int_{R_c}^{R_e} [U(c, s)]^\alpha n \, dX \right\}^{1/\alpha} \qquad (12.42)$$

for some number α, with $W_1 = W/N_0$. In general, the corresponding optimal allocation program still leads to unequal utilities at different locations. The only exception is the limiting case as $\alpha \to -\infty$ favoring the least satisfied household(s). [At the opposite extreme, we have the "royalist welfare index" as $\alpha \to \infty$; it favors the very-well-off household(s)!]

An optimal allocation program which maximizes the "limiting case" $W_{-\infty}$ effectively chooses s and c to maximize the minimum household utility:

$$\max_{(c,s)} \left\{ \min_X U(c(X), s(X)) \right\} \qquad (12.43)$$

subject to the various resource constraints. To see that this optimization problem may be considered as a special case of a generalized social optimum, consider the following discrete analog of (12.42)

$$W_\alpha^{(m)} \equiv \left(\frac{\Delta X}{N_0} \sum_{k=1}^{m} U_k^\alpha \right)^{1/\alpha} \qquad U_k \equiv U(c(X_k), s(X_k)) \qquad (12.44)$$

with $X_k = R_c + k \, \Delta X$, $\Delta X = (R_e - R_c)/m$. It is a little easier to see here that only the smallest component in the sum survives in the limit as $\alpha \to -\infty$. With resource constraints satisfied, a feasible allocation $[c_k(X), s_k(X)]$ generally leads to a minimum utility value at some location X_k. The optimization process seeks that feasible allocation $[\bar{c}(X), \bar{s}(X)]$ which gives the maximum of the "minimum utility values."

The criterion (12.43) for optimal social welfare is effectively a quantitative realization of the concept of social justice advocated by John Rawls (1971). It is egalitarian because

12I: At the max-min optimum, U is uniform for all $R_c \le X \le R_e$.

If $U(c, s)$ were not uniform in location, then there would be another feasible allocation program which would increase the utility of the worst household(s), say by reassigning some consumption goods from high-utility households to the less fortunate.

As a consequence of **12I**, the max-min criterion is equivalent to maximizing the uniform utility U_c subject to the resource constraints. It can be shown that a resource allocation program at the max-min optimum is identical to the corresponding program for household optimum in competitive equilibrium (Kanemoto, 1980).

12.7   Land for Roads and the Distance Cost of Transportation

Up to now, we have assumed that transportation does not need land input. (This is the case for a subway system.) In a modern metropolis where automobiles owned by individual households constitute a significant mode of urban transport, land assigned to roads is often a nontrivial fraction of the total land area available for housing. In that case, to add more roads for improving transportation requires a cost-benefit analysis. At least, the loss of revenue from the land, which would be available for other uses, should be more than offset by the benefits from its use for roads. In the last few sections of this chapter, we analyze the effect of different land allocation policies.

Along the circle at location X, let $b(X)$ be the fraction of land allocated for housing in the thin ring sector $(X, X + dX)$. Then the land constraint $2\pi X = sn$ in (12.34) should be modified to read

$$ns = 2\pi X b(X) \qquad (12.45)$$

and the population constraint becomes

$$N_0 = \int_{R_c}^{R_e} n(X)\,dX = \int_{R_c}^{R_e} \frac{2\pi b(X)X}{s}\,dX \tag{12.46}$$

In competitive equilibrium, the household optimum (for the absentee-landlord case) requires

$$\bar{s} = \tilde{s}(I; r_c, I_c, p) \qquad \bar{c} = \tilde{c}(I; r_c, I_c, p) \qquad I = Y - t(X) \tag{12.47}$$

Then the population constraint (12.46) with $s = \bar{s}$ given by (12.47) determines r_c, the unit land rent at the edge of the CBD.

It should be noted that $I_c \equiv I(R_c) = Y - t(R_c)$ with the travel cost within the CBD being either known or zero. To the extent that we are not concerned with the coupling between the residential area and the CBD, we take $t(R_c) = 0$ in the subsequent development with no loss in generality. If $t(R_c) \neq 0$, we simply take $Y - t_c$ as the annual household income and $I(X) = Y - t = (Y - t_c) - (t - t_c)$ with $t - t_c$ taken to be the transportation cost.

For a Cobb-Douglas household utility function,

$$U = U_0 s^\sigma c^{1-\sigma} \qquad (0 < \sigma < 1)$$

we have from an earlier calculation

$$\bar{s} = \frac{Y}{(\alpha+1)r_c}\left[\frac{Y - t(X)}{Y}\right]^{-\alpha} \qquad \sigma \equiv \frac{1}{\alpha+1} \tag{12.20}$$

but now (12.21) is modified by a factor b in the integrand so that

$$r_c = \frac{\sigma N_0 Y/2\pi}{\int_{R_c}^{R_e}(1 - t/Y)^\alpha b(X)X\,dX} \tag{12.48}$$

Correspondingly, we have

$$r = r_c\left(1 - \frac{t}{Y}\right)^{\alpha+1} \qquad r\bar{s} = \sigma[Y - t(x)]$$

$$p\bar{c} = (1-\sigma)Y\left(1 - \frac{t}{Y}\right) \qquad U(\bar{s}, \bar{c}) = \frac{U_0\sigma^\sigma(1-\sigma)^{1-\sigma}}{Yr_c^\sigma p^{1-\sigma}} \tag{12.49}$$

so that all quantities of interest are determined once we have the annual household travel cost profile $t(X)$ and the land allocation profile (for housing) $b(X)$. From the expressions for $r\bar{s}$ and $p\bar{c}$ in (12.49), we see that σ is the fraction of income net (meaning after taking away) transportation cost spent by a household on housing and $(1 - \sigma)$ is the fraction spent on consumption goods. With this interpretation, the numerical value for σ is usually taken to be around one-fourth in North America.

Until the early seventies, the cost of transportation was taken to depend only on the distance from the CBD. From available data, we can calculate the distance cost per mile per annum τ_d, from which we get the total distance cost of transportation per annum

$$t(X) = t_d(X) \equiv \int_{R_c}^{X} \tau_d \, dX \qquad (12.50)$$

because transportation is free within the CBD. Note that we have $t_d = \tau_d(X - R_c)$ if the travel cost per mile per annum τ_d is a constant.

There is a rather peculiar feature of this basic model for residential land use which can be seen from (12.48). Suppose the fraction of land allocated for housing is the same at every ring so that $b(X) \equiv b_c$. The model gives a well-defined rent profile for $0 < b_c < 1$ and continues to do so as b_c tends to unity. But with $b(X) \equiv 1$, the entire residential district is allocated for housing, and there is no road to get to and from the CBD! For b_c sufficiently close to (but still less than) unity, gigantic traffic congestion should have sent rent for lands near the CBD skyrocketing. Evidently, the assumption that only distance (and fixed) cost of transportation are important in calculating the travel cost is inappropriate for heavy-traffic situations. The effect of traffic congestion must be included in the model if it is to give useful insight to an admittedly complicated economic problem [see Solow (1972)].

12.8 Congestion Cost of Transportation

When traffic is extremely heavy, the commuting cost to and from the CBD is no longer a function of the distance from the CBD alone. Traffic jams cause automobiles to burn more gas and to have more wear and tear. It takes more time to drive the same distance, and time is money. There are also the physical and emotional costs due to the stress and aggravation experienced by the commuters driving in traffic congestion. All these suggest that the transportation cost of a household should have (at least) two components: a distance cost $t_d(X)$ which is taken to be known, and a congestion cost $t_c(X)$ which varies with traffic density. In this chapter, we follow the conventional treatment by taking them to be additive so that

$$t(X) = t_d(X) + t_c(X) \qquad (12.51)$$

To relate t_c to the traffic density, we recall that $(1-b)2\pi X$ is the total road width at ring X and $N(X)$ is the number of households living outside the ring at X (and therefore also the number of commuters crossing the ring each day to go into the CBD). Hence, the quantity $N(X)/2\pi X(1-b)$ is a measure of the traffic density at X and the congestion cost $t_c(X)$ should depend on this quantity. More specifically, we take the transportation cost per unit distance (per annum) of a household located at X to be a function of this quantity, that is,

$$\frac{dt_c}{dX} \equiv \tau_c(X) = f\left(\frac{N(X)}{2\pi X[1-b(X)]}\right) \tag{12.52}$$

where $f(\cdot)$ is to be specified by suitable field studies. With $\tau_d \equiv dt_d/dX$, we have from (12.51) and (12.52)

$$\frac{dt}{dX} = \tau_d(X) + \tau_c(X) = \tau_d(X) + f\left(\frac{N}{2\pi X(1-b)}\right) \tag{12.53}$$

so that the annual household transportation cost is no longer a known function of X as it depends on the unknown household distribution $N(X)$.

A second equation for N and t comes from the land constraint $ns = 2\pi Xb(X)$ which may be written as

$$-n \equiv \frac{dN}{dX} = -\frac{2\pi Xb(X)}{s} \tag{12.54}$$

In competitive equilibrium, the household optimum gives $\tilde{s} = \tilde{s}(I; r_c, p, I_c)$ with $I = Y - t$ and $I_c = Y - t_c = Y$. In that case, (12.54) becomes

$$\frac{dN}{dX} = -\frac{2\pi Xb}{\tilde{s}(Y - t; r_c, p, Y)} \tag{12.55}$$

The two first-order ODE for t and N (12.53) and (12.55) are supplemented by two initial conditions

$$N(R_c) = N_0 \tag{12.56a}$$

$$t(R_c) = 0 \tag{12.56b}$$

This initial-value problem (IVP) determines a one-parameter family of solutions, with the unknown parameter r_c determined by the condition

$$N(R_e) = 0 \tag{12.57}$$

The solution for N, t, and r_c may be obtained by numerical integration of the IVP coupled with Newton's method for root finding. Alternatively, we may introduce a new ODE $dr_c/dX = 0$ and solve the problem as a boundary-value problem for the third-order system (of three first-order ODEs) by a number of existing methods.

For explicit solutions of our model problem for residential land use with congestion cost of transportation, we will work with a Cobb-Douglas utility function and the particular congestion cost function

$$f\left(\frac{N}{2\pi X(1-b)}\right) = \frac{aN}{2\pi X(1-b)} \tag{12.58}$$

where $a > 0$ is a known constant. The ODE for N and t may be written as

$$\frac{dw}{dX} = -\frac{\tau_d}{Y} - \frac{aN}{Y\,2\pi X(1-b)} \qquad \frac{dN}{dX} = -\frac{r_c 2\pi X b(X)}{\sigma Y}\,w^\alpha \tag{12.59}$$

where $w \equiv 1 - t/Y$ is the fraction of household income after transportation cost. With $\sigma = 1/(\alpha + 1)$, the above system is linear ($\alpha = 1$) only if housing expenditure is half of the household income net transportation cost. Few cities in North America are in that category. For the (unlikely) case of $\alpha = 1$, the solution of the problem when $b(X)$ is a constant may be expressed in terms of modified Bessel functions.

For $\alpha \neq 1$, the ODE is nonlinear and an exact solution does not seem possible. In the next two sections, we show how asymptotic methods (developed originally for problems in mechanics) may be useful for an accurate approximate solution which clearly delineates the effect of various parameters in the problem.

12.9   Perturbation Solutions

For the purpose of applying asymptotic methods, we put the ODE (12.59) and the auxiliary conditions (12.56a, b) and (12.57) in dimensionless form. To do so, we introduce the dimensionless quantities

$$x = \frac{X}{R_c} \qquad u = \frac{N}{N_0} \qquad \nu_c = \frac{2\pi r_c R_c^2}{\sigma Y N_0}$$

$$\tau_d(X) = \tau_{do}\,g(x) \qquad \epsilon_d^2 = \frac{\tau_{do}R_c}{Y} \qquad \epsilon_c^2 = \frac{aN_0}{2\pi Y} \tag{12.60}$$

$$\epsilon^2 = \epsilon_d^2 + \epsilon_c^2 \qquad \eta = \frac{\epsilon_d^2}{\epsilon^2} \qquad R = \frac{R_e}{R_c} \qquad (>1)$$

where $g(x)$ is a dimensionless unit distance cost function not greater than unity in magnitude. Note that u is the fraction of the households living outside location X and η is a measure of the distance cost as a fraction of total transportation cost. The two governing ODEs may then be written as

$$\frac{dw}{dx} = -\epsilon^2\left[\eta g(x) + (1-\eta)\frac{u}{x(1-b)}\right] \qquad (12.61)$$

$$\frac{du}{dx} = -\nu_c x b w^\alpha \qquad (12.62)$$

and the auxiliary conditions become

$$u(1) = 1 \qquad (12.63a)$$

$$w(1) = 1 \qquad (12.63b)$$

$$u(R) = 0 \qquad (12.63c)$$

In (12.62), the parameter ν_c is an unknown constant to be determined in the solution process. The dimensionless combination

$$\epsilon^2 \frac{R_e - R_c}{R_c} = \frac{\tau_{do}(R_e - R_c)}{Y} + \frac{aN_0(R_e - R_c)}{2\pi R_c Y}$$

is a measure (or of the order of magnitude) of the travel cost-to-income ratio which is certainly much less than unity. Therefore, we have $\epsilon^2 \ll 1$ as long as $(R_e - R_c)/R_c$ is not large compared to unity. In that case, the structure of the differential equations for u and w suggest a perturbation series solution in powers of ϵ^2. (For integer values of α, the solution is known to be analytic in the parameter ϵ^2.)

For a perturbation solution, we let

$$\{u(x;\epsilon), w(x;\epsilon), \nu_c(\epsilon)\} = \sum_{n=0}^{\infty} \{u_n(x), w_n(x), \nu_n\}\epsilon^{2n} \qquad (12.64)$$

similar to what we did in Chapters 3 and 11. The two equations for u and w and the auxiliary conditions may be written as

$$w_0' + \epsilon^2\left[w_1' + \frac{1-\eta}{x(1-b)}u_0 + \eta g\right] + \epsilon^4\left[w_2' + \frac{1-\eta}{x(1-b)}u_1\right]$$

$$+ \cdots + \epsilon^{2n}\left[w_n' + \frac{1-\eta}{x(1-b)}u_{n-1}\right] + \cdots = 0$$

$$(u_0' + v_0 xbw_0^\alpha) + \epsilon^2 (u_1' + v_0 xba w_1 w_0^{\alpha-1} + v_1 xbw_0^\alpha) + \epsilon^4 (\cdots) + \cdots = 0$$

$$[u_0(1) - 1] + \epsilon^2 [u_1(1)] + \cdots + \epsilon^{2n}[u_n(1)] + \cdots = 0$$

$$[w_0(1) - 1] + \epsilon^2 [w_1(1)] + \cdots + \epsilon^{2n}[w_n(1)] + \cdots = 0$$

$$u_0(R) + \epsilon^2 u_1(R) + \epsilon^4 u_2(R) + \cdots + \epsilon^{2n} u_n(R) + \cdots = 0$$

$$(12.65)$$

For these equations to be satisfied identically in ϵ, the coefficient of each power of ϵ^2 must vanish identically. This requires the leading term coefficients w_0, u_0, and v_0 to satisfy

$$w_0' = 0 \qquad w_0(1) = 1 \qquad u_0' = -v_0 xbw_0^\alpha \qquad u_0(1) = 1 \qquad u_0(R) = 0$$

$$(12.66)$$

The solutions of these two linear problems are

$$w_0(x) = 1 \qquad u_0(x) = 1 - v_0 \int_1^x xb(x) w_0^\alpha \, dx \qquad v_0 = \frac{1}{\int_1^R xb(x)\,dx}$$

$$(12.67)$$

The quantities $w_0(x)$, $u_0(x)$, and v_0 are completely determined once we specify $b(x)$. This leading term solution corresponds to the $t(X) \equiv 0$ case. To obtain the effect of nontrivial transportation cost, especially the effect of congestion cost, we must continue the solution process to get $\{w_1, u_1, v_1\}$, $\{w_2, u_2, v_2\}$, etc. This process involves only the solution of a sequence of *linear* boundary-value problems.

To illustrate, we consider the case of a uniform fraction of land allocation for housing with $b(x) \equiv b_1$ and a constant unit distance cost per annum so that $g(x) \equiv 1$. The expressions for v_0 and $u_0(x)$ in (12.67) for this case become

$$v_0 = \frac{2/b_1}{R^2 - 1}$$

$$(12.68)$$

$$u_0(x) = \frac{R^2 - x^2}{R^2 - 1}$$

$$(12.69)$$

Correspondingly, the linear BVP governing $w_1(x)$, $u_1(x)$, and v_1 takes the form

$$w_1' = -\eta - \frac{1-\eta}{1-b_1} \frac{u_0}{x} \qquad w_1(1) = 0$$

$$(12.70)$$

$$u_1' = -v_0 \alpha b_1 xw_1 - v_1 b_1 x \qquad u_1(1) = u_1(R) = 0$$

$$(12.71)$$

The solution for (12.70) is

$$w_1(x) = -\eta(x - 1) - \frac{1 - \eta}{2(1 - b_1)(R^2 - 1)} [R^2 \ln x^2 - (x^2 - 1)] \qquad (12.72)$$

The ODE and first auxiliary condition of (12.71) give

$$\frac{u_1(x)}{v_0 b_1 \alpha} = -\frac{v_1}{2\alpha v_0}(x^2 - 1) + \eta[\tfrac{1}{3}(x^3 - 1) - \tfrac{1}{2}(x^2 - 1)]$$

$$+ \frac{1 - \eta}{8(1 - b_1)(R^2 - 1)} \{2R^2[x^2 \ln x^2 - (x^2 - 1)] - (x^4 - 1) + 2(x^2 - 1)\}$$

$$(12.73)$$

The condition $u_1(R) = 0$ then gives

$$\frac{R^2 - 1}{2\alpha} \frac{v_1}{v_0} = \eta[\tfrac{1}{3}(R^3 - 1) - \tfrac{1}{2}(R^2 - 1)]$$

$$+ \frac{1 - \eta}{8(1 - b_1)(R^2 - 1)} \{2R^2[R^2 \ln R^2 - (R^2 - 1)] - (R^4 - 1) + 2(R^2 - 1)\}$$

$$(12.74)$$

It is not difficult to show that $w_1(x) < 0$ for $1 < x \leq R$ and $w_1(x)$ is monotone decreasing as x increases. Hence, we have the not particularly surprising result that there is less income available for consumption and housing if the effect of congestion is also included in the total cost of transportation. It can also be shown that $v_1 > 0$ and $u_1(x) < 0$ for $1 < x \leq R$ so that for a sufficiently small ϵ^2, we have the following conclusions:

12J: Unit land rent at the edge of the CBD is higher and the rent profile is steeper if congestion cost is included in the cost of transportation.

and

12K: One effect of congestion cost is to keep more households closer to the edge of the CBD.

Higher-order terms in the perturbation series for w, u, and v_c may be obtained similarly. However, a constant fraction land allocation function,

$b(x) \equiv b_1$, is not a particularly realistic policy. With more traffic near the CBD, a larger fraction of available land should be allocated to roads there. A linear allocation function $b(x) = b_1 + (b_R - b_1)(x + 1)/(R - 1)$ with $b_R < b_1$ or a hyperbolic profile $b(x) = 1 - (1 - b_1)/x$ would be more appropriate. The solutions for these land allocation functions will be left as exercises [see also Wan (1977)].

If the boundary of the outer city limit R_e is not prescribed but determined by the economic condition of $r(R_e) = r_A$, we have a *free-boundary problem* with R as an additional unknown. The solution for R_e will also depend on ϵ and, for a regular perturbation series solution, we should also take

$$R(\epsilon) = \sum_{n=0}^{\infty} R_n \epsilon^{2n} \tag{12.75}$$

with R_n determined successively just as the coefficients of the other series (Wan, 1977).

We deduced earlier the expression $r\bar{s} = \sigma(Y - t)$ and concluded from this that σ is the fraction of net income (after travel cost) spent on housing. In that case, we may take it to be about one-fourth. To the extent that rs is actually the expenditure on land alone, we have effectively assumed a certain complementary relationship between the size of dwelling space and the kind of house built on that lot. While the correct analysis requires a separate accounting of land and capital expenditures on housing, we may consider for our simple model values of σ which truly reflect the expenditure on land alone. In a typical North American city, the fraction of net income spent on land for housing is about one-tenth or less. For $\sigma \ll 1$, the regular perturbation method of solution is not practical unless we have also $\epsilon^2/\sigma \ll 1$; otherwise too many terms would be required for an accurate approximate solution. A more useful approximate solution has been obtained in Wan (1977 and 1983) for $\sigma \ll 1$ by the method of *matched asymptotic expansions* [see Kevorkian and Cole (1981)].

12.10   Allocation of Land for Roads

In all the preceding sections, we have assumed that there is a known land allocation policy $b(x)$ in the residential area for streets and housing. But how does one arrive at such a policy? When traffic is congested, transportation cost could be reduced by building more roads. But there would also be an attendant loss in rental revenue and/or household utility from giving up housing space for these roads. One policy for land allocation would be for the city to continue to build more roads as long as the benefit from doing so outweighs the loss in rental revenue (Kanemoto, 1977; Arnott and MacKinnon, 1978). Suppose an additional fraction Δb of land at location X is to be used for new roads. To a first approximation, the cost in lost rent would be $r(X) 2\pi X \Delta b \, dX$ while the benefit would be N times the differential congestion cost for the distance dX at that location. The condition of cost equaling benefit for a congestion cost function of the type given in (12.58) may be taken in the form:

$$r(X) 2\pi X \Delta b \, dX = N \, dX \left[\frac{aN}{2\pi X (1 - b)} - \frac{aN}{2\pi X (1 - b + \Delta b)} \right]$$

$$= \frac{aN^2 \Delta b \, dX}{2\pi X (1 - b)(1 - b + \Delta b)} \simeq \frac{\alpha N^2 \Delta b \, dX}{2\pi X (1 - b)^2}$$

or

$$r(X) = \frac{1}{a} \left[\frac{aN}{2\pi X (1 - b)} \right]^2 \tag{12.76}$$

(A similar relation can be derived for a general unit congestion cost τ_c.) This condition determines $b(X)$.

For a Cobb-Douglas utility function, $r(X) = r_c w^{\alpha+1}$ and the condition (12.76) for $b(X)$ may be written as

$$\frac{u}{x(1 - b)} = \frac{1}{\epsilon} \left(\frac{\sigma \nu_c}{1 - \eta} w^{\alpha+1} \right)^{1/2} \tag{12.77}$$

Fortuitously, this single relation enables us to eliminate both u and b from the first ODE leaving us with a single equation for w

$$\frac{dw}{dx} = -\epsilon^2 \eta g(x) - \epsilon [(1 - \eta) \sigma \nu_c w^{\alpha+1}]^{1/2} \tag{12.78}$$

Together with auxiliary condition $w(1) = 1$, it determines $w(x)$ up to the unknown parameter ν_c. When τ_d is a constant, we have $g(x) \equiv 1$ and the first-order ODE for w is separable and a solution in the form of a quadrature is immediate. Once we have $w(x; \nu_c)$, the algebraic condition (12.77) relating w, b, and u and the ODE (12.62) determines u and b with the only constant of integration specified by $u(1) = 1$. Finally, ν_c is determined by the condition $u(R) = 0$. The ODE (12.78) is separable for $g(x) \equiv 1$ and an exact solution is always possible (Wan, 1979).

Unfortunately, the quadratures involved in the solution process cannot be expressed in terms of elementary or special functions in general. For $\epsilon \ll$

1, a perturbation solution is more useful. For such a solution, we see from the relation (12.77) among u, w, and b that $b = 1 + O(\epsilon)$ since u, w, and ν_c are all of order unity. With

$$b(x) = 1 + \epsilon b_1(x) + \epsilon^2 b_2(x) + \cdots \tag{12.79}$$

[i.e., $b_0(x) \equiv 1$], we have immediately from (12.77)

$$b_1(x) = -\frac{u_0}{x}\left(\frac{1-\eta}{\sigma\nu_0 w_0^{\alpha+1}}\right)^{1/2} \tag{12.80}$$

where

$$w_0' = 0 \qquad w_0(1) = 1 \tag{12.81}$$

and

$$u_0' = -\nu_0 x b_0(x) w_0^\alpha \qquad u_0(1) = 1 \qquad u_0(R) = 0 \tag{12.82}$$

After some calculations, we get

$$w_0(x) \equiv 1 \qquad \nu_0 \equiv \frac{2}{R^2 - 1} \qquad u_0(x) = \frac{R^2 - x^2}{R^2 - 1} \tag{12.83}$$

so that

$$b(x) = 1 - \epsilon\left[\frac{R^2 - x^2}{x}\sqrt{\frac{(\alpha+1)(1-\eta)}{2(R^2-1)}}\right] + O(\epsilon^2) \equiv b_{CB}(x) \tag{12.84}$$

While higher-order terms may be calculated for the various unknowns (Wan, 1979 and 1983), we merely note that the perturbation series is now in powers of ϵ (instead of ϵ^2 as in the case when b is prescribed). We note also that the costs and benefits are calculated from the market rent, available housing space, and household distributions before the new road construction. In actual fact, the households would redistribute themselves to reach a new equilibrium configuration (and hence a new rent profile) as soon as additional land is allocated for roads. Thus, the road width (or the allocation of land for roads) determined by the cost-equals-benefit criterion may not be optimal. Either one of the two criteria of optimality discussed in Section 12.6 would be more appropriate for determining an optimal allocation of s, c, and b.

Still another optimal land allocation criterion is to choose b to maximize $V(I_c, r_c, p)$ [see (12.13)] which depends on b through r_c and I_c. This so-called *second-best allocation* has been analyzed in Kanemoto (1977) and Wan

(1979 and 1983). A perturbation solution for the case of a Cobb-Douglas utility function gives (Wan, 1979 and 1983):

$$b_{SB}(x; \epsilon) = 1 - \epsilon \left[\frac{R^2 - x^2}{x} \sqrt{\frac{\alpha(1 - \eta)}{2(R^2 - 1)}} \right] + O(\epsilon^2) \qquad (12.85)$$

so that

$$\frac{1 - b_{SB}(x)}{1 - b_{CB}(x)} = \sqrt{\frac{\alpha}{\alpha + 1}} \, [1 + O(\epsilon)] = \sqrt{1 - \sigma} \, [1 + O(\epsilon)] \qquad (12.86)$$

The ratio (12.86) indicates that except for higher-order terms in ϵ,

> **12L:**   The cost-benefit criterion using market land price allocates more land for roads than the second-best allocation resulting in a lower common utility for the households.

EXERCISES

1. $$U(s, c) = U_0 s^\sigma c^{1-\sigma} \qquad (0 < \sigma < 1)$$

(a) Show that at the household maximum,

$$\bar{s} = \frac{\sigma}{r}(Y - t) \qquad \bar{c} = \frac{1 - \sigma}{p}(Y - t)$$

(Thus, the parameter σ is the fraction of net income after-transportation-cost that the household spends on housing.)

(b) In equilibrium, show that

$$\left(\frac{r}{r_c} \right)^\sigma = \frac{Y - t(x)}{Y - t_c} \qquad t_c \equiv t(X_c)$$

and therewith

$$\bar{s} = \frac{Y - t_c}{(\alpha + 1)r_c} \left[\frac{Y - t(X)}{Y - t_c} \right]^{-\alpha} \qquad \sigma \equiv \frac{1}{\alpha + 1}$$

(c) Show that the unknown unit land rent at the edge of the CBD is determined by

$$r_c = \frac{\sigma N_0 (Y - t_c)^{\alpha+1}}{\int_{R_c}^{R_e} [Y - t(X)]^\alpha 2\pi X \, dX}$$

(d) Suppose

$$t(X) = \frac{\tau}{2} X^2$$

For the case of a fixed boundary R_e (with $\tau R_e^2 / 2 < Y$ as transportation cost it is only a small fraction of the household income), obtain an explicit solution for r_c in terms of the parameters N_0, τ, R_e, and σ (or α).

(e) For a city free to expand outward, determine its equilibrium outer radius R_e from the condition $r(R_e) = r_A$ for the same quadratic transportation cost function. Here, r_A is taken to be a known constant.

2. At the household optimum, c and s are determined by $I(X) = Y - t(X)$ and $r(X)$ (and, of course, also by p which will not be explicitly stated in this problem as it plays no role in the results to be derived): $\bar{c} = \hat{c}(I, r)$, $\bar{s} = \hat{s}(I, r)$. Correspondingly, we have $U(\bar{s}, \bar{c}) \equiv V(I, r)$.
  (a) Show that $\hat{s} + r\hat{s}_r + p\hat{c}_r = 0$ and $r\hat{s}_I + p\hat{c}_I = 1$.
  (b) Derive Roy's identity $V_r = -V_I \bar{s}$.
  (c) In equilibrium, we have $V(I, r) = V_c \equiv V(I_c, r_c)$ which can be solved to get $r = \tilde{r}(I(X), V_c)$. Show that

$$\tilde{r}_I = \frac{I}{\hat{s}(I, r)} \equiv \frac{I}{\tilde{s}(I, V_c)} \qquad \tilde{r}_{V_c} = -\frac{I}{\tilde{s} V_I(I, r)}$$

3. Let the utility index $U(x, y)$ depend on the amounts x and y of the two commodities X and Y consumed. Both goods are "desirable" so that $U_x > 0$ and $U_y > 0$. For such a utility function, we see that $m \equiv dy/dx < 0$ along an indifference curve.
  (a) For the indifference curves to be convex to the origin ($dm/dx > 0$), show that the "bordered Hessian" of U must be positive, that is,

$$\begin{vmatrix} U_{xx} & U_{xy} & U_x \\ U_{xy} & U_{yy} & U_y \\ U_x & U_y & 0 \end{vmatrix} > 0 \qquad (|B| \equiv \text{determinant of } B)$$

and U is said to be *strict quasi-concave*.
  (b) If $\hat{U}(x, y) \equiv f(U(x, y))$ where f is a one-one, monotone increasing and twice differentiable function and U is strict quasi-concave, show that we have $\hat{U}_x > 0$, $\hat{U}_y > 0$, and $\hat{U}$ is also strict quasi-concave.
  (c) Let p and q be the constant unit price of X and Y, respectively. The first-order necessary condition for a maximum, constrained by the budget equation $px + qy = I$, determine $x = f_1(I)$,

$y = f_2(I)$, and $\lambda = f_3(I)$ where we have suppressed the appearance of p and q in f_k, $k = 1, 2, 3$ (and where λ is the Lagrange multiplier). Show that at the household maximum, we have $\lambda = dU/dI$ and λ is therefore the *marginal utility of income*.

4. Suppose, in the residential land problem, the composite consumption goods are in fact produced by a single (profit-maximizing) farmer who owns all the land in the same region beyond the city limit R_e. Given the knowledge of the unit price function $p = f(Q_0)$, determine the size of the residential zone in competitive equilibrium and the area of cultivated farm land (or alternatively the amount of goods produced) in equilibrium. Assume that goods must be shipped to the city center $X = 0$ and sold there, and that the yield per unit land area is a constant q_l with no labor substitution.

   (a) For a general utility function $U(c, s)$, transportation cost function $t(X)$ for the commuter, unit transportation cost $\tau_f(X)$ for farm products, and price function $f(Q_0)$, obtain two appropriate conditions for the determination of the city outer limit R_e and the outer limit of the (cultivated) farm land R_0.

   (b) Specialize these conditions for $U = U_0 s^\sigma c^{1-\sigma}$, $t(x) = \tau_d X^2/2$, $f(Q_0) = p_0\sqrt{Q^*/Q_0}$, and $\tau_f = $ constant in the transportation cost term for the farmer. Do as much as you can toward an explicit solution for R_e and R_0.

5. (a) With the help of the relation $U(s, c) \equiv V(I, r) = V(I_c, r_c) \equiv V_c$ so that $r = r(I, V_c)$ and Roy's identity, show that $\partial r/\partial I = 1/s > 0$, so that rent increases with income level when utility is kept constant.

   (b) Alternatively, we may solve for $\bar{s}$, $\bar{c}$, and I in terms of r and V_c. Show that $\bar{c}_r \bar{s}_r \le 0$ where $(\ )_r \equiv [\partial(\ )/\partial r]$ with V_c fixed. (In fact, with a little more work, we can show that $\bar{s}_r \le 0$, which is not surprising, and therefore $\bar{c}_r \ge 0$ which is also not surprising.)

   (c) Treating $\bar{s}$ and $\bar{c}$ as functions of X, use $\bar{c}_r \ge 0$ and $\bar{s}_r \le 0$ to show (again with V_c fixed)

$$\frac{d\bar{s}}{dX} \ge 0 \qquad \frac{d\bar{c}}{dX} \le 0$$

6. For a closed city with $\tau = \tau_d = $ constant, a fixed outer boundary so that R_e is prescribed, and public ownership, obtain the social optimum allocation for $U = \sigma \ln(s/s_0) + (1 - \sigma) \ln(c/c_0)$.

7. Suppose one absentee landlord owns all land surrounding the CBD. How large an annular region ($R_c \le X \le R_e$) would the landlord develop

to house the N_0 households if the aggregate rent revenue net the aggregate agricultural rent payment is maximized?

$$\max_{\{R_e \geq R_c\}} \left\{ \int_{R_c}^{R_e} [r(X) - r_A] 2\pi X \, dX \,\Big|\, \int_{R_c}^{R_e} \frac{2\pi X}{\bar{s}} \, dX = N_0 \right\}$$

(The transportation cost is taken to be the distance cost alone.) To be specific, work with a Cobb-Douglas utility function $U = s^\sigma c^{1-\sigma}$, $0 < \sigma < 1$, and interpret your results.

8. When migration is possible, the (single) absentee landlord may increase revenue by attracting more households to the land he/she owns. Suppose that the uniform household utility, annual household income from wages, and the commodity price (for the whole economy in and outside our city) are unaffected by migration. Solve the monopolist's optimum problem for a Cobb-Douglas utility function and (if you wish) a constant transportation cost per unit distance $\tau = \tau_d = $ constant.

9. Suppose $U(s, c)$ is concave (so that $U_{cc}U_{ss} - U_{sc}^2 > 0$) as well as strictly quasi-concave.
   (a) With $\bar{s} = \hat{s}(I, r)$ and $\bar{c} = \hat{c}(I, r)$, show that

   $$\frac{\partial \hat{s}}{\partial I} = \frac{U_c}{pD} (U_{sc}U_c - U_s U_{cc})$$

   where $D \equiv 2U_{sc}U_c U_s - U_s^2 U_{cc} - U_{ss}U_c^2$.
   (b) If U is strictly quasi-concave, show that $U_{sc}U_c - U_s U_{cc} > 0$ when housing is a normal good, that is, $U_s > 0$.

10. For a competitive equilibrium household optimum with $U = s^\sigma c^{1-\sigma}$ and

    $$\tau = \tau_d + \frac{aN}{2\pi X(1-b)} \qquad X = R_c x$$

    where $\tau_d = $ constant and $b = 1 - (1 - b_1)/x$ with $b_1 \equiv b(x = 1)$, obtain the first two terms of the perturbation series solution for w, u, and v_c in the dimensionless BVP for the case of a fixed outer boundary and an absentee landlord formulated in Section 12.9.

11. Show that the incremental change of the aggregate net revenue ΔR^* at the monopolistic maximum of agricultural land use, associated with an incremental change of $\Delta \tau$ of the constant unit transportation cost τ, is identical to that at the (perfectly) competitive maximum.

12. Consider the problem of minimum expenditure to achieve a given utility $\bar{U}$:

$$\min_{\{x=(x_1,\ldots,x_n)\}} \{p \cdot x \,|\, U(x) = \bar{U}\} \qquad p = (p_1, \ldots, p_n)$$

Let $x^*(p, \bar{U})$ be the minimum point and $E(p, \bar{U}) \equiv p \cdot x^*$ be the minimum expenditure. Show that

$$\frac{\partial E}{\partial \bar{U}} = \lambda^* \qquad \frac{\partial E}{\partial p_i} = x_i^*$$

13. For a competitive equilibrium household optimum with $U = s^\sigma c^{1-\sigma}$ and a unit transportation cost function

$$\tau = \tau_d + \frac{aN}{2\pi X(1-b)}$$

where $\tau_d = $ constant and $b = 1 - (1 - b_1)/x$, obtain the first two terms of the perturbation series solution for w, u, v, and $R \equiv R_e/R_c$ if the city expands its outer boundary until $r(R_e) = r_A$ [or $v_c w^{\alpha+1}(R) = v_A$ in dimensionless form].

14.
$$U(s, c) = \sigma \ln \frac{s}{s_0} + (1 - \sigma) \ln \frac{c}{c_0}$$

Write down the dimensionless BVP for the max-min optimum land allocation for a city with a fixed outer boundary and with the unit transportation cost function of exercise 13. Obtain the first two terms of the perturbation series for $b(x)$, that is, $b(x;\epsilon) = b_0(x) + b_1(x)\epsilon + O(\epsilon^2)$.

15. With $U = s^\sigma c^{1-\sigma}$, work out enough terms in the perturbation solution of the second-best allocation of residential land for a city with absentee landlords, the unit transportation cost function of exercise 13, and a fixed outer boundary to get the first two terms of $b(x; \epsilon) = b_0(x) + b_1(x)\epsilon + O(\epsilon^2)$.

13

Pay or Save

Neoclassical Economic Growth Theory

A variational formulation of optimal economic growth gives rise to a nonlinear boundary-value problem in ODE; its solution is deduced by a stability analysis of the critical points in the phase plane. The method of Lagrange multipliers and the Hamiltonian are introduced to cope with equality constraints in calculus of variations problems.

13.1   The One-Sector Neoclassical Growth Model

Macroeconomics is mainly concerned with certain gross features of an economy. In the extreme case, all commodities produced in the economy are aggregated as a single homogeneous good. A certain amount C of this good produced in a given period of time is consumed and the rest invested to accumulate capital assets K (buildings, machines, etc.) for increased production in the next period. One area of macroeconomics, called the theory of economic growth (or simply *growth theory*), is concerned with how consumption and accumulated capital stock in this single-sector economic model as well as in other more sophisticated models change with time. A better understanding of these models is expected to suggest possible regulation of the evolution of these time-dependent quantities to attain certain objectives such as *economic efficiency*.

The mechanism for regulating the trend of the above one-sector model of the economy is to choose between consumption (provision for the present) and investment for capital accumulation (provision for the future). While more consumption is preferable to less at any moment of time, more consumption means less capital accumulation. A smaller capital stock means a smaller future output, and hence a smaller future consumption. Therefore, a choice has to be made at each instant of time among feasible consumption policies. At one extreme is the policy of consuming as much as possible now even though the potential for future consumption is jeopardized: "Let us eat, drink and be merry for tomorrow we shall die." At the other extreme is the Stalinist policy of consuming as little as possible today so as to increase the capital stock and thereby the future consumption (though you may not live to enjoy it).

With a choice of consumption-to-investment ratio at each instant of time, we have then a particular time path $C(t)$ for consumption corresponding to these choices over a time period. Evidently, there are many such consumption time paths. In this chapter, we will study the effect of different consumption paths on the evolution (or growth) of the economy. We will then discuss several different criteria for optimal growth and how each such criterion determines the consumption path.

In the simplest neoclassical growth model, the economy uses two homogeneous inputs, *labor* $L(t)$ and *capital* $K(t)$, to produce a single *homogeneous good*. The output of each time period (or each instant of time) $Q(t)$, depends on the amount of labor and capital available and is given in terms of L and K by a *production function* $F(K, L)$. This output is partly consumed and partly invested:

$$Q(t) = F(K, L) = C(t) + I(t) \qquad (13.1)$$

Note that capital stock K, labor pool L, and consumption rate C are all nonnegative quantities; in fact consumption rate should be no less than a subsistence level. As stipulated at the beginning of this chapter, *investment* is used both to augment the stock of capital currently at the level K_0 and to replace depreciated capital. We assume henceforth that the existing capital stock depreciates at a rate proportional to the current capital stock so that

$$I = K^{\cdot} + \gamma K \qquad K(0) = K_0 \qquad (\ )^{\cdot} \equiv \frac{d(\ )}{dt} \qquad (13.2)$$

While there is no need to do so, it is customary to take the labor force currently at the level L_0 to grow exponentially with a (constant) rate n:

$$L^{\cdot} = nL \qquad L(0) = L_0 \qquad (13.3)$$

Given a particular consumption rate time path $C(t)$, Equations 13.1–13.2 determine $K(t)$, $L(t)$, $Q(t)$, and $I(t)$.

In economic analysis, the production function $F(K, L)$ is conventionally assumed to be smooth and with piecewise continuous second derivatives in both arguments. In addition, F has the following properties:

$$F_K > 0 \qquad F_{KK} < 0 \tag{13.4}$$

$$F_L > 0 \qquad F_{LL} < 0 \tag{13.5}$$

$$\lim_{K \to \infty} F_K = 0 \qquad \lim_{L \to \infty} F_L = 0 \tag{13.6}$$

wherever the derivatives are defined and where $(\ )_x \equiv \partial(\ )/\partial x$ as before. In other words, $F(K, L)$ is monotone increasing, concave to the origin, and asymptotic to a maximum in each factor. These qualitative features reflect in a general way the observed features in actual production (Burmeister and Dobell, 1970). Furthermore, we will assume for convenience that F is *homogeneous of degree one* so that

$$F(\alpha K, \alpha L) = \alpha F(K, L) \tag{13.7}$$

This assumption allows us to eliminate the explicit appearance of L in the problem. For $\alpha = 1/L$, we have

$$\frac{1}{L} F(K, L) = F\left(\frac{K}{L}, \frac{L}{L}\right) = F(k, 1) \equiv f(k) \tag{13.8}$$

where $k \equiv K/L$ is the capital stock per head of labor force. One example of such $F(K, L)$ is the Cobb-Douglas production function $F(K, L) = F_0 K^a L^{1-a}$ for some positive constant F_0 and $0 < a < 1$. With

$$F(0, L) = F(K, 0) = 0 \tag{13.9}$$

this production function has the additional property of "no free lunch".

Along with the per head capital accumulation $k(t)$, we introduce also the per head consumption rate $c(t) = C(t)/L(t)$. With

$$k^\bullet = \frac{K^\bullet}{L} - \frac{K}{L}\frac{L^\bullet}{L} = \frac{K^\bullet}{L} - nk \tag{13.10}$$

Equations 13.1 and 13.2 may be combined to give

$$\frac{1}{L} F(K, L) = c(t) + \frac{K^\bullet}{L} + k = c(t) + k^\bullet + \mu k \qquad k(0) = \frac{K_0}{L_0} \tag{13.11}$$

where $\mu \equiv n + \gamma$. We now use (13.8) to write (13.11) as

$$k^{\cdot} = f(k) - \mu k - c \qquad k(0) = k_0 \tag{13.12}$$

where $k_0 \equiv K_0/L_0$. The IVP (13.12) summarizes the essential features of the neoclassical growth theory. Given $c(t)$, the IVP (13.12) determines $k(t)$ and therefore $K(t)$, $Q(t)$, and $I(t)$.

It is customary in economics to use $c(t)$ in (13.12) as a *control* to steer the economy in a particular direction. For example, we may wish to steer the economy toward a particular capital accumulation k_T at time $t = T$. This may or may not be possible depending on the form of the actual production function and the value of μ and k_0. Consider, for example, the case $f(k) = \beta k$ and $L(t) \equiv L_0$ for which the IVP (13.12) has the exact solution

$$k = \left[k_0 - \int_0^t c(\tau) e^{-(\beta-\mu)\tau} \, d\tau \right] e^{(\beta-\mu)t} \tag{13.13}$$

If $\beta < \mu$, then $k(t)$ is a monotone decreasing function of t as $k^{\cdot} = (\beta - \mu)k - c$ is negative. If the initial stock of capital is less than k_T, it will not be possible for $k(T) = k_T$ no matter what consumption rate we choose (which must be ≥ 0). In general, whether there exists at least one control $\bar{c}(t)$ which steers the economy to k_T at $t = T$ is a question in *control theory* (Burghes and Graham, 1980; Tou, 1964).

13.2   Fixed Fraction Saving and the Golden Rule

To see how a given per head consumption pattern determines the time path of the economy, consider a society which saves a constant fraction s of the production at each instant of time (for investment) and consumes the balance so that

$$c = (1 - s) \frac{Q}{L} = (1 - s)f(k) \tag{13.14}$$

The IVP (13.12) becomes

$$k^{\cdot} = sf(k) - \mu k \qquad k(0) = k_0 \tag{13.15}$$

This determines $k(t)$ completely, and therefore also $c(t)$, $Q(t)$, etc., once the production function $f(k)$ is specified.

For a Cobb-Douglas production function, $f(k) = f_0 k^a$ $(0 < a < 1)$, the solution of (13.15) is (see exercises)

$$k(t) = k_0 \left\{ e^{-(1-a)\mu t} + \frac{sf_0}{\mu k_0^{1-a}} \left[1 - e^{-(1-a)\mu t} \right] \right\}^{1/(1-a)} \tag{13.16}$$

It follows that the economy evolves toward an equilibrium (or steady) state as

$$\lim_{t \to \infty} k(t) = k_\infty \equiv \left(\frac{sf_0}{\mu} \right)^{1/(1-a)}$$

$$\lim_{t \to \infty} c(t) = c_\infty \equiv f_0(1 - s) \left(\frac{sf_0}{\mu} \right)^{a/(1-a)} \tag{13.17}$$

With the parameter s still unspecified, we may choose it to maximize the steady state per head consumption rate. It is not difficult to show that the maximum value of the per head equilibrium consumption rate, denoted by $\bar{c}_\infty$, is attained at $s = a \equiv \bar{s}$ with

$$\bar{c}_\infty = f_0(1 - a) \left(\frac{af_0}{\mu} \right)^{1/(1-a)} \tag{13.18}$$

and the corresponding equilibrium capital stock is

$$\bar{k}_\infty = \left(\frac{af_0}{\mu} \right)^{1/(1-a)} \tag{13.19}$$

For other production functions, it is not always possible to obtain an exact solution of the IVP (13.15) in terms of elementary or special functions. On the other hand, we can still obtain the equilibrium states of the economy by the graphical method. At steady state, we have $k^\cdot = 0$ so that the ODE in (13.15) becomes

$$sf(k) - \mu k = 0 \tag{13.20}$$

For a Cobb-Douglas-like production function with $f(0) = 0$, $f'(k) > 0$, $f''(k) < 0$ and $f'(\infty) = 0$, Equation 13.20 has two real solutions, $k = 0$ and $k = k_\infty > 0$. It is evident from a plot of $k^\cdot$ versus k that $k = 0$ is unstable while $k = k_\infty$ is asymptotically stable (Figure 13.1).

Now, Equation 13.20 determines k_∞, and therefore $c_\infty = (1 - s)f(k_\infty)$, in terms of s (as well as μ and other parameters in production functions).

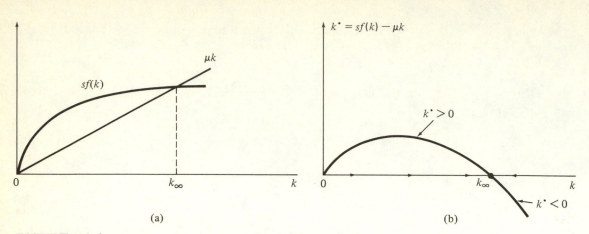

(a)   (b)

FIGURE 13.1

We can choose s to maximize the steady state per head consumption $c_\infty(s)$. Suppose c_∞ attains its maximum at $\bar{s}$, then $dc_\infty/ds = 0$ at $s = \bar{s}$ where

$$\frac{dc_\infty}{ds} = \frac{d}{ds}[(1-s)f(k_\infty)] = (1-s)f'(k_\infty)\frac{dk_\infty}{ds} - f(k_\infty) \qquad (13.21)$$

For an expression for dk_∞/ds, we differentiate (13.20) with respect to s to get

$$f(k_\infty) + sf'(k_\infty)\frac{dk_\infty}{ds} = \mu \frac{dk_\infty}{ds}$$

or

$$\frac{dk_\infty}{ds} = \frac{f(k_\infty)}{\mu - sf'(k_\infty)} \qquad (13.22)$$

Upon using (13.22) to eliminate dk_∞/ds from (13.21), we get

$$\frac{dc_\infty}{ds} = \frac{f(k_\infty)}{\mu - sf'(k_\infty)}[f'(k_\infty) - \mu] \qquad (13.23)$$

Since $f(k_\infty) \neq 0$, we must have

$$f'(k_\infty(\bar{s})) = \mu \qquad (13.24)$$

for dc_∞/ds to vanish at $s = \bar{s}$. Equation 13.24 determines $\bar{s}$ and is called the *Golden Rule of (Capital) Accumulation*. For a Cobb-Douglas production function $f(k) = f_0 k^a$, $0 < a < 1$, Equation 13.20 gives $k_\infty(s) = (sf_0/\mu)^{1/(1-a)}$ which together with Equation 13.24 requires $\bar{s} = a$. These results are in agreement with what we found earlier from the explicit solution of the IVP (13.15).

Graphically, we have the picture in Figure 13.2. For simplicity, we take $f(0) = 0$ so that the curve $f(k_\infty)$ starts at the origin. For $s = 1$, the straight line μk_∞ intersects the monotone increasing concave curve $sf(k_\infty)$ only at the origin and another point $k_\infty^{(0)}$. There is no consumption for this case since we save all output of each production period. For $s = s_1 < 1$, the second intersection is at $k_\infty^{(1)}$ and the consumption rate is $c_1 = f(k_\infty^{(1)}) - s_1 f(k_\infty^{(1)}) = (1 - s)f(k_\infty^{(1)})$ as indicated in Figure 13.2. For s_1 close to 1, we have $f'(k_\infty^{(1)}) < \mu$; but for a much smaller value of s, say $s_2 \ll s_1 < 1$, we have $k_\infty(s_2) \equiv k_\infty^{(2)}$ and $f'(k_\infty^{(2)}) > \mu$. As s decreases to zero, $c_\infty(s)$ also decreases to zero. Given the monotone increasing and concave property of $f(k)$, there is a unique $\bar{s}$ for which c_∞ attains a maximum, denoted by $\bar{c}_\infty \equiv c_\infty(\bar{s})$. The Golden Rule (13.24) tells us that this unique optimal saving rate $\bar{s}$ locates $\bar{k}_\infty = k_\infty(\bar{s})$ at a position where the slope of the tangent of f at $\bar{k}_\infty$ is equal to μ.

If $n = 0$, then we have $\mu = \gamma$; the Golden Rule merely tells us that we should save at a rate to accumulate a capital stock k_∞ so that the additional

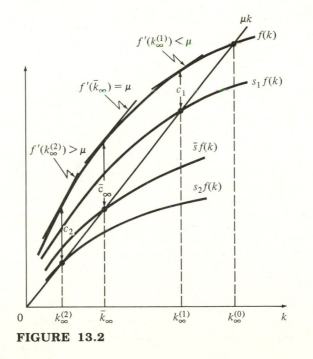

FIGURE 13.2

accumulation from another unit of capital would just balance out the depreciation rate γ per unit of capital. If there is no depreciation but $n > 0$, then the additional accumulation from an extra unit of capital would balance the per head growth rate of the labor population. Hence,

13A: In stable equilibrium, the Golden Rule of Capital Accumulation gives the maximum per head consumption rate, attained at a per head savings rate for which the additional gain from an extra unit of available per head capital just balances the loss through capital depreciation and the (per head) growth rate of the labor population.

The optimal savings rate can also be taken from the steady-state relation (13.20) with μ given by (13.24):

$$\bar{s} = \frac{\mu k_\infty(\bar{s})}{f(k_\infty(\bar{s}))} = \frac{\bar{k}_\infty f'(k_\infty(\bar{s}))}{f(k_\infty(\bar{s}))} \qquad (13.25)$$

which may be interpreted as

13B: The optimal savings rate $\bar{s}$ equals the profit share of the national income.

Upon differentiating (13.23) once more with respect to s, we get

$$\left.\frac{d^2 c_\infty}{ds^2}\right|_{s=\bar{s}} = \left[f(k_\infty)\, \frac{dk_\infty}{ds}\, f''(k_\infty) \right]_{s=\bar{s}} \qquad (13.26)$$

We know $f(\bar{k}_\infty) > 0$ and $f''(\bar{k}_\infty) < 0$ from the stipulated properties of the production function. By (13.22) and (13.24), we have also $dk_\infty/ds > 0$ at $s = \bar{s}$. It follows that $(d^2 c_\infty/ds^2) < 0$ at $s = \bar{s}$. Therefore, the extremum of c_∞ attained at $s = \bar{s}$ is in fact a maximum.

13C: For a fixed fraction saving policy, the Golden Rule of Capital Accumulation is in fact optimal.

More elaborate savings functions and the corresponding golden rules are discussed in Burmeister and Dobell (1970).

13.3 Social Optimum

In the last section, we prescribed a particular savings rule and saw how the economy evolves under this rule. In this section, we will prescribe instead a certain social goal and see what savings (or consumption) rule we should adopt to achieve this goal. One possible social objective is the social optimum introduced in Chapter 12. We consider again a *utility* function $U(c)$ as a measure of consumer satisfaction. Evidently, satisfaction increases with consumption so that $U'(c) > 0$ but with diminishing return so that $U''(c) < 0$. We also define a *social welfare* index W which is now the sum of utility over time weighted in favor of the present,

$$W = \int_0^T e^{-\delta t} U(c(t)) \, dt \qquad (13.27)$$

where δ is the constant *discount rate* and T is the final time of the planning period. We want to choose a consumption rate $c(t)$ to maximize this social welfare index subject to the given initial capital stock and the rule for production and capital accumulation characterized by the IVP (13.12).

As we indicated earlier, a capital stock cannot be negative and the consumption rate must be no less than some subsistence level (taken to be zero for simplicity) so that

$$c(t) \geq 0 \quad \text{and} \quad k(t) \geq 0 \qquad (13.28)$$

Without the nonnegativity constraints (13.28), optimization problems may be formulated in the framework of the classical calculus of variations of Chapter 5. With the inequality constraints, the problem would be more conveniently treated by the modern theory of optimal control. We will not introduce the elaborate mathematical machinery of optimal control theory to analyze our problem but will work informally with the variational methods already encountered in this book. More specifically, we will ignore the constraints (13.28) and find the solution of our problem by way of the Euler differential equation of (13.27). At the same time, we will keep an eye on whether the inequality constraints are satisfied. If they are not, we will stop and try to patch things up.

Suppose $\bar{c}(t)$ maximizes W and $\bar{k}(t)$ is the corresponding $k(t)$ from (13.12). Consider another $k(t) = \bar{k}(t) + \epsilon \eta(t)$, with $\eta(0) = 0$ since $k(0) = \bar{k}(0) = k_0$, and the corresponding $c(t) = \bar{c}(t) + \epsilon \xi(t)$. For this class of $k(t)$, the welfare index becomes

$$W = \int_0^T e^{-\delta t} U(f(k) - \mu k - k^\cdot) \, dt$$

$$= \int_0^T e^{-\delta t} U(f(\bar{k} + \epsilon \eta) - \mu \bar{k} - \mu \epsilon \eta - \bar{k}^\cdot - \epsilon \eta^\cdot) \, dt \equiv W(\epsilon) \qquad (13.29)$$

where we have used the ODE in (13.12) to eliminate c from $U(c)$. Since $\bar{c}$ (and hence $\bar{k}$) maximizes W, we must have $dW/d\epsilon = 0$ at $\epsilon = 0$ for an interior maximum. But

$$\left.\frac{dW}{d\epsilon}\right|_{\epsilon=0} = \int_0^T e^{-\delta t} U'(\bar{c})\left\{[f'(\bar{k})-\mu]\eta - \eta^{\cdot}\right\}dt$$

$$= -\bar{v}(T)\eta(T)e^{-\delta T}$$

$$+ \int_0^T \left\{\bar{v}^{\cdot} + [f'(\bar{k})-\mu-\delta]\bar{v}\right\}e^{-\delta t}\eta(t)\,dt \qquad (13.30)$$

where $\bar{v}(t) \equiv U'(\bar{c}(t)) = (dU/dc)_{c=\bar{c}(t)}$, and use has been made of the condition $\eta(0) = 0$. A prime in this chapter indicates differentiation with respect to the argument of the function involved. For $W(\epsilon)$ to attain an interior maximum at $\epsilon = 0$, we must have (as in Chapter 5)

$$\bar{v}^{\cdot} + [f'(\bar{k})-(\mu+\delta)]\bar{v} = 0 \qquad (13.31)$$

and

$$e^{-\delta T}\bar{v}(T)\eta(T) = 0 \qquad (13.32)$$

We know from Chapter 5 that the ODE (13.31) is the *Euler differential equation* for our variational problem and the system (13.31), (13.32), and (13.12) determines $\bar{k}(t)$ and $\bar{c}(t)$ simultaneously. (For the optimal control problem, the inequality constraints on k and c must also be satisfied by the solution.) To see that this system is effectively a two-point boundary-value problem in ODE, we now look at the implications of the terminal conditions (13.32). There are altogether three different situations:

1. For the case of a finite T and no constraint on the economy at $t = T$, $\eta(T)$ is arbitrary and $e^{-\delta T} > 0$. Therefore, we must have

$$\bar{v}(T) \equiv U'(\bar{c}(T)) = 0 \qquad (13.33)$$

   In calculus of variations, this condition on $\bar{c}$ is called the *Euler (or natural) boundary condition* for the problem. In optimal control, it is sometime called the *transversality condition*. For this case, (13.12), (13.31), and (13.33) constitute a two-point boundary value problem for the determination of $\bar{c}$ and $\bar{k}$.

2. A second case involves a specific constraint imposed on the economy at a finite terminal time T, usually in the form

$$\bar{k}(T) = k_T \qquad (13.34)$$

   as we may want to leave a specified amount of capital stock for the next planning period. In this case, we must have $\eta(T) = 0$ [since $\bar{k}(T) = k(T) = k_T$] and (13.32) is automatically satisfied. But now, the system (13.31), (13.34), and (13.12) forms a two-point boundary value problem for $\bar{c}(t)$ and $\bar{k}(t)$.

3. Finally, if the planning is done for the entire future (usually called *infinite horizon* in growth economics) so that $T = \infty$, the condition (13.32) is satisfied if we have $\eta(t)$ bounded and

$$\lim_{t \to \infty} |\bar{v}(t)| = \lim_{t \to \infty} |U'(\bar{c}(t))| < \infty \qquad (13.35)$$

It is for this case that we need $\delta > 0$ to ensure the convergence of the improper integral (13.29). [Note that (13.32) is also satisfied when $\eta \bar{v}$ grows exponentially with a rate $\nu < \delta$. But the improper integral (13.29) may not converge in that case. In any event, it is still an open question whether (13.35) is necessary for optimality.] Now, (13.31), (13.35), and (13.12) form a two-point boundary value problem for $\bar{k}$ and $\bar{c}$.

In all cases where the terminal condition is prescribed in terms of $\bar{v}$, it may be more convenient to think of the BVP as one for $\bar{k}$ and $\bar{v}$ (instead of $\bar{k}$ and $\bar{c}$) since $\bar{v}$ appears in a rather neat form in (13.31). With $v = U'(c) > 0$ and $U''(c) < 0$, we can always invert the relation to get c as a function of v and write the ODE in (13.12) as an equation for $\bar{k}$ and $\bar{v}$.

In a later section, we will examine the solution of the optimal growth problem for a specific $U(c)$ and a specific $f(k)$. But before we leave the general problem, it should be noted that the Euler differential equation and the Euler boundary condition are only necessary conditions for an (interior) extremum. The extremum is in fact a (local) maximum in our case because

$$\frac{d^2 W}{d\epsilon^2}\bigg|_{\epsilon=0} = \int_0^T e^{-\delta t} \{ U'(\bar{c}) f''(\bar{k}) \eta^2 + U''(\bar{c}) [f'(\bar{k})\eta - \mu\eta - \eta^{\bullet}]^2 \} dt \quad (13.36)$$

is negative since f'' and U'' are always negative and U' is positive. Thus,

13D: For cases where both $U(c)$ and $f(k)$ are increasing concave functions, any (interior) extremum of the social welfare index is also (at least) a local maximum.

13.4 Lagrange Multipliers and the Hamiltonian

For more complicated problems, it is too cumbersome to use the constraints to eliminate some of the auxiliary quantities, as we did by using (13.12) to eliminate c from the expression (13.27) for W. Instead, it is customary to incorporate the constraints into the optimization process by way of *Lagrange multipliers*. We illustrate this procedure with the same example discussed in the last section.

Write (13.12) as $f(k) - \mu k - c - k^{\cdot} = 0$ and observe that (13.29) can be written as

$$W = \int_0^T \{e^{-\delta t} U(c) + \lambda(t) [f(k) - \mu k - c - k^{\cdot}]\} dt$$

$$= \left[-\lambda k\right]_0^T + \int_0^T \{e^{-\delta t} U(c) + \lambda^{\cdot} k + \lambda[f(k) - \mu k - c]\} dt \qquad (13.37)$$

for any differentiable function $\lambda(t)$, the *Lagrange multiplier*. Again, suppose $\bar{c}(t)$ maximizes W and $\bar{k}$ the corresponding capital stock time path. The variational procedure of setting $c = \bar{c}(t) + \epsilon \eta(t)$ and $k = \bar{k}(t) + \epsilon \xi(t)$, etc., gives

$$\frac{dW}{d\epsilon}\bigg|_{\epsilon=0} = \left[-\lambda \xi(t)\right]_0^T + \int_0^T \{[e^{-\delta t} U'(\bar{c}) - \lambda(t)] \eta(t)$$

$$+ [\lambda f'(\bar{k}) - \mu \lambda + \lambda^{\cdot}] \xi(t)\} dt \qquad (13.38)$$

As $\lambda(t)$ is still not specified, we choose it to make the second bracketed quantity in the integrand of (13.38) equal to zero:

$$\lambda^{\cdot} = -\lambda[f'(\bar{k}) - \mu] = -\lambda \frac{\partial}{\partial \bar{k}} [f(\bar{k}) - \mu \bar{k} - \bar{c}] \qquad (13.39)$$

For $\eta(t)$ and $\xi(t)$ to be arbitrary except for $\xi(0) = 0$ [and possibly $\xi(T) = 0$ if $k(T) = k_T$], we have $dW/d\epsilon = 0$ at $\epsilon = 0$ only if

$$\lambda = e^{-\delta t} U'(\bar{c}) \qquad (13.40)$$

and, when there is no terminal condition specified at $t = T$,

$$\lambda(T) = 0 \qquad (13.41)$$

The Euler B.C. (13.41) would be replaced by a terminal condition on $k(T)$ [such as $k(T) = \bar{k}_T$] if such a condition should be prescribed at $t = T$.

The ODE (13.12), (13.39), and the algebraic equation (13.40) form a second-order system for $\bar{k}$, $\bar{c}$, and λ. This system is supplemented by the initial condition $\bar{k}(0) = k_0$ and either the transversality condition (13.41) or a terminal condition at $t = T$. Together, they determine $\bar{k}(t)$, $\bar{c}(t)$, and $\lambda(t)$.

More generally, suppose we want to choose $c((t)$ and $r(t)$ to maximize

$$J \equiv \phi(k(T), w(T), r(T), c(T)) + \int_0^T F(c, r, k, w)\,dt \qquad (13.42)$$

subject to constraining *equations of state*

$$k^{\cdot} = f(k, w, c, r) \qquad w^{\cdot} = g(k, w, c, r) \qquad (13.43)$$

and the initial conditions

$$k(0) = k_0 \qquad w(0) = w_0 \qquad (13.44)$$

The two functions $c(t)$ and $r(t)$ are the controls for the problem.

Let $\lambda(t)$ and $\Lambda(t)$ be the Lagrange multipliers associated with first and second ODE in (13.43), respectively. The necessary conditions for an interior maximum W deduced by variational methods may be summarized in terms of a Hamiltonian H defined by

$$H \equiv F(c, r, k, w) + \lambda(t)f(k, w, c, r) + \Lambda(t)g(k, w, c, r) \qquad (13.45)$$

The necessary conditions themselves are the Euler equations

$$\lambda^{\cdot} = -\frac{\partial H}{\partial k} = -\left(\frac{\partial F}{\partial k} + \lambda\frac{\partial f}{\partial k} + \Lambda\frac{\partial g}{\partial k}\right) \qquad (13.46)$$

$$\Lambda^{\cdot} = -\frac{\partial H}{\partial w} = -\left(\frac{\partial F}{\partial w} + \lambda\frac{\partial f}{\partial w} + \Lambda\frac{\partial g}{\partial w}\right) \qquad (13.47)$$

$$\frac{\partial H}{\partial c} = \frac{\partial F}{\partial c} + \lambda\frac{\partial f}{\partial c} + \Lambda\frac{\partial g}{\partial c} = 0 \qquad (13.48)$$

$$\frac{\partial H}{\partial r} = \frac{\partial F}{\partial r} + \lambda\frac{\partial f}{\partial r} + \Lambda\frac{\partial g}{\partial r} = 0 \qquad (13.49)$$

and the Euler boundary conditions

$$\lambda(T) = \frac{\partial \phi}{\partial k(T)} \qquad \Lambda(T) = \frac{\partial \phi}{\partial w(T)} \qquad (13.50)$$

where we have omitted the overhead bar for the optimal solution. The solution of the boundary value problem for k, w, c, r, λ, and Λ, defined by (13.43),

(13.44), (13.46)–(13.50) is called the *interior solution* of the optimal control problem if all the inequality constraints (if any) are satisfied. Otherwise, one or more constraints will be binding and replace (13.48) or (13.49) or both.

Further generalization of the problem (13.45) to more states and more controls is straightforward. Applications of this more general class of problems will be described in Chapter 14.

13.5 The Turnpike Solution

For a utility function of the form

$$U(c) = \frac{c^{1-\sigma}}{1-\sigma} \qquad (\sigma > 0) \tag{13.51}$$

we have $U' = c^{-\sigma} \equiv v$ or $c = v^{-1/\sigma}$. The two-point BVP for optimal growth of Section 13.3, with an infinite horizon, becomes (after omitting the overhead bars)

$$\begin{aligned} k^{\cdot} &= f(k) - \mu k - v^{-1/\sigma} & k(0) &= k_0 \\ v^{\cdot} &= -v[f'(k) - (\mu + \delta)] & \lim_{t \to \infty} v e^{-\delta t} &= 0 \end{aligned} \tag{13.52}$$

Alternatively, we may express the same problem in terms of k and c. In that case, we have

$$\begin{aligned} k^{\cdot} &= f(k) - \mu k - c & k(0) &= k_0 \\ c^{\cdot} &= \frac{c}{\sigma}[f'(k) - (\mu + \delta)] & \lim_{t \to \infty}(c^{-\sigma} e^{-\delta t}) &= 0 \end{aligned} \tag{13.53}$$

The solution of either BVP can be described qualitatively by the phase-plane trajectories. For illustrative purposes, we further specialize $f(k)$ to

$$f(k) = k^a \qquad (0 < a < 1) \tag{13.54}$$

In that case, the three critical points of the ODE (13.53) in the k,c plane are $(0, 0)$, $(k_\infty, 0)$, and $(\bar{k}, \bar{c})$ where

$$k_\infty = \left(\frac{1}{\mu}\right)^{1/(1-a)} \qquad \bar{k} = \left(\frac{a}{\mu+\delta}\right)^{1/(1-a)} \qquad \bar{c} = \frac{\bar{k}}{a}[\mu(1-a)+\delta]$$

$$\tag{13.55}$$

(The use of overhead bars here is not to be confused with those in previous sections.) It is not difficult to verify that, as critical points (see Chapter 4), the two equilibrium solutions $(0, 0)$ and $(k_\infty, 0)$ both behave as a node, the former is a source and the latter a sink. In contrast, the critical point $(\bar{k}, \bar{c})$ is a saddle point with separatrices near the critical point characterized by the slopes:

$$\frac{dc}{dk}\bigg|_{(\bar{k},\bar{c})} = \frac{1}{2}\left[\delta \pm \sqrt{\delta^2 + \frac{4(1-a)(\mu+\delta)(\mu+\delta-a\mu)}{\sigma a}}\right] \qquad (13.56)$$

one positive and one negative. The phase-plane portrait is shown qualitatively in Figure 13.3.

An important observation is that trajectories heading toward the c axis get there in finite time for we have $c \gg 0$ as $k \to 0$ so that $k^{\cdot}$ remains finite as $k \to 0$. With $k = 0$, no further production (and consumption) is possible. Such a time path (with $c \equiv 0$ for $t > T$ for some T) may not be desirable as it calls for *extinction* at some finite T. If extinction is allowed in optimal growth, then we should reformulate the problem as a free end-point problem.

Next, trajectories which do not head toward the c axis will eventually end up at $(k_\infty, 0)$. The only exceptions are the two (asymptotically stable) separatrices $(k_{\text{opt}}(t), c_{\text{opt}}(t))$ which head toward the saddle point $(\bar{k}, \bar{c})$; this critical point offers the highest steady-state (equilibrium) consumption rate possible. For optimal growth with any given k_0, there is a unique c_0 which lands the point (k_0, c_0) on a stable separatrix; the dynamics then drives the economy from (k_0, c_0) toward $(\bar{k}, \bar{c})$ along the separatrix. For the trajectories

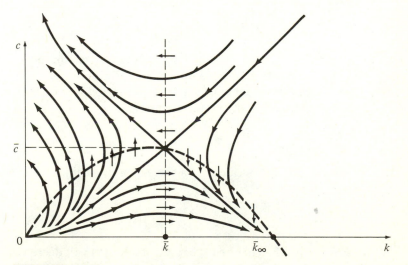

FIGURE 13.3

which do not head toward the c axis, it is not difficult to see that $c_{\mathrm{opt}}(t) \geq c(t)$ for $t \geq T_0$, for some $T_0 \geq 0$.

For the case of a finite planning horizon T and a prescribed capital stock at both $t = 0$ and $t = T$, it is not possible to choose c_0 to get on the stable separatrix. Instead, we have to pick c_0 to be on a trajectory which gets to the prescribed $k(T)$ at $t = T$. As T becomes longer and longer, this trajectory is known to arch more and more toward the separatrix so that the system generally spends a substantial fraction of the total period in the neighborhood of the saddle point $(\bar{k}, \bar{c})$ (Burmeister and Dobell, 1970). This tendency to be near the stable separatrix, especially the saddle point, is called a *consumption turnpike* phenomenon in economic growth theory. The term turnpike refers to the economic efficiency associated with the planned economic growth along the stable separatrix leading to the highest (steady-state) consumption possible. The optimal growth path near or along the separatrix is therefore called the turnpike solution.

13.6   The Max-Min Optimum Growth

Still another criterion for choosing the time path of consumption rate (and of capital accumulation) is the more egalitarian max-min policy of Section 12.6. We found there that, at the max-min optimum, the utility index $U(c(t))$ should be identical for the whole planning interval $0 \leq t \leq T(\leq \infty)$. The optimization problem is to choose $c(t)$ to maximize the temporally uniform utility $U(c(t)) \equiv \hat{U}$. With U being a monotone increasing function of c (with $U' > 0$ and $U'' < 0$), it follows from $U(c) \equiv \hat{U}$ that

$$c(t) \equiv \hat{c} \tag{13.57}$$

and the problem is reduced to finding the maximum admissible constant value for $\hat{c}$. We do this below for the infinite horizon planning problem.

For a constant (per head) consumption rate $\hat{c}$, the dynamics governing the accumulation of capital becomes

$$k^{\boldsymbol{\cdot}} = f(k) - \mu k - \hat{c} \qquad k(0) = k_0 \tag{13.58}$$

For a given k_0, we calculate a corresponding *equilibrium value* for $\hat{c}$, denoted by $\hat{c}_0$ with

$$\hat{c}_0 = f(k_0) - \mu k_0 \tag{13.59}$$

For $\hat{c} > \hat{c}_0$, the (per head) capital stock $k(t)$ will decrease with time; for $\hat{c} < \hat{c}_0$, $k(t)$ increases with time, at least for awhile. Let $\bar{k}$ be the equilibrium capital stock per head at the maximum constant per head consumption rate determined by $f'(\bar{k}) = \mu$. Two situations should be analyzed separately:

1. $k_0 < \bar{k}$: For this case, $\hat{c}_0$ is the maximum constant per head consumption rate possible. If $\hat{c} > \hat{c}_0$, the per head capital stock would dwindle to zero in finite time and there would be no further production (or consumption) possible beyond that point. On the other hand, any value $\hat{c} < \hat{c}_0$ cannot be optimal since we can always sustain a larger per head consumption rate. Hence,

> **13E:**   The insistence on intertemporal equity in our single sector model would keep an economy with a small initial capital endowment at a low level of per head consumption rate forever.

A lack of flexibility for improving future welfare is a major weakness of the insistence on (intertemporal) fairness and equity.

2. $k_0 > \bar{k}$: In this case take $\hat{c} = \bar{c} = \hat{c}(\bar{k})$. With this choice of $\hat{c}$, $k(t)$ decreases initially, reducing k_0 to $\bar{k}$. At this point $k(t) = \bar{k}$, the economy is in equilibrium since $f(\bar{k}) - \mu\bar{k} - \bar{c} = 0$, and the per head consumption rate $\bar{c}$ will be sustained indefinitely. It is clear from the definition of $\bar{c}$ that anything less than $\bar{c}$ would not be

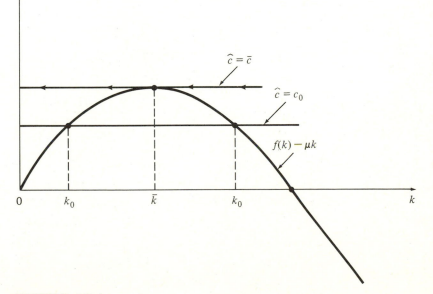

FIGURE 13.4

optimal. On the other hand, the equilibrium position is rather precarious; a slight miscalculation may be catastrophic.

The above two situations can be seen from Figure 13.4.

EXERCISES

1. (a) For economic growth with a fixed fraction savings s so that $c = (1 - s)f(k)$ and with a Cobb-Douglas production function, $f(k) = f_0 k^a$, $0 < a < 1$, solve the initial value problem governing the (per head) capital accumulation:

    $$k^{\cdot} \equiv \frac{dk}{dt} = sf(k) - \mu k \qquad k(0) = k_0 \qquad (0 < s < 1)$$

    (b) Deduce from the above solution

    $$k_\infty \equiv \lim_{t \to \infty} k(t) = \left(\frac{sf_0}{\mu}\right)^{1/(1-a)} \qquad c_\infty \equiv \lim_{t \to \infty} c(t) = \frac{\mu(1 - s)k_\infty}{s}$$

    (c) Find the value of s which maximizes c_∞.
    (d) Find the value of s which maximizes the welfare index W

    $$W = \int_0^\infty e^{-\delta t} U(c(t))\,dt \qquad U(c) = U_0 c^\sigma$$

    with $0 < \sigma < 1$ and $a\sigma/(1 - a) = 1$.

2. (a) Show that the production function

    $$F(K, L) = \alpha L(1 - e^{-\alpha K/L}) + \beta K \qquad (\alpha > 0, \beta > 0)$$

    may be written as $F(K, L) = Lf(k)$, where $k = K/L$.
    (b) Graph $f(k)$ versus k.
    (c) Show on a suitable graph all equilibrium solutions of $k^{\cdot} = sf(k) - \mu k$ with $0 \le \mu - \beta < \alpha^2$ and discuss their stability. [*Hint:* Do not try to obtain an exact solution for $k(t)$.]
    (d) Show that the Golden Rule savings rate $\hat{s}$ and the corresponding capital/labor ratio $\hat{k}$ are given by

    $$\hat{k} = \frac{1}{\alpha} \ln \gamma \qquad \hat{s} = \frac{\mu \ln \gamma}{\alpha^2 - (\mu - \beta) + \beta \ln \gamma}$$

    where $\gamma \equiv \alpha^2/(\mu - \beta)$. What is maximized by the Golden Rule?
    (e) Discuss the consequences of $\alpha^2 < \mu - \beta$.

3. Until recently, the theory of growth economics has been dominated by *fixed coefficients* production functions. For our simple one-sector growth model, such a production function takes the form

$$F(K, L) = \min[\alpha K, \beta L] = \begin{cases} \alpha K & (\alpha K \le \beta L) \\ \beta L & (\beta L \le \alpha K) \end{cases}$$

(a) Show that F is homogeneous of degree one, that is, $F(\lambda K, \lambda L) = \lambda F(K, L)$, and, therefore,

$$\frac{Q}{L} = \min[\alpha k, \beta] \equiv f(k)$$

(b) Graph $f(k)$ versus k.

(c) Find all the equilibrium solutions of the fundamental equation of neoclassical growth theory

$$k^{\cdot} = -sf(k) - nk$$

for the fixed coefficients production function of part (a) with $\alpha > n$.

(d) Discuss the stability of all the equilibrium solutions of part (c).

(e) What is the largest equilibrium consumption (per worker) possible $(\alpha > n)$?

(f) What is the economic implication of the case $n > \alpha$?

14

Justice for All

Exhaustible Resources and Intergenerational Equity

*Familiarity with the solution technique for a Bernoulli equation is
assumed. Treatment of infinity in a critical point analysis is
discussed. Nonisolated critical points appear in a specific example.*

14.1 Exhaustible Resource Essential to Production

When the production of the single composite good in a one-sector growth
model requires a certain natural resource as a third input, we have $Q = F(K, L, R)$ where $R(t)$ is the flow of resource input (quantity per unit time).
We are interested here only in the case of a resource such as fossil fuel with
a *finite* deposit $\bar{D}_0$. Any amount of the resource used in the production of
the composite good will be lost forever. The replenishment of the resource,
if any, is such a small fraction of consumption that it can be taken to be
zero. For such a resource, we have the following *finite deposit constraint* on
its extraction rate:

$$\int_0^\infty R(t)dt \le \bar{D}_0 \tag{14.1}$$

For the problem to be nontrivial, we stipulate $F(K, L, 0) = 0$ so that the
resource input is *essential* to the production. Otherwise, we may exhaust the
finite resource deposit in a finite time period and shift the economy into a
new phase of production and growth without the natural resource (as in
Chapter 13). We also take F to be an increasing function of R with diminishing returns:

$$F_R > 0 \qquad F_{RR} < 0 \qquad \lim_{R \to \infty} F_R = 0 \qquad (14.2)$$

As before, the output Q is used for (1) current consumption, (2) replacement for depreciated capital stock, and (3) investment to increase the capital stock for future production. In addition, a part of the output may now have to be used to pay for the cost of extracting the resource. The disposition of output for economic growth with an essential resource input for production is summarized by

$$F(K, L, R) = Q + K^{\cdot} + \gamma K + \theta R + C \qquad (14.3)$$

where θ is the cost of extracting a unit of resource. In general, θ is a function of t, $K(t)$, and $R(t)$. However, we consider in this chapter (mainly) the case of a constant unit extraction cost so that θ is just a number.

We again limit our discussion to either F being homogeneous of degree one or a stable labor force so that $L(t) \equiv L_0$. For either class of problems, we may simplify the above mathematical model by working with $k \equiv K/L$, $r \equiv R/L$, and $c \equiv C/L$. In terms of these quantities, the dynamics of economic growth is governed by

$$k^{\cdot} = f(k, r) - \mu k - \theta r - c \qquad k(0) = k_0 \qquad (14.4)$$

where $k(t)$, $c(t)$, and $r(t)$ are all nonnegative quantities, $f(k, r) \equiv F(K, L, R)/L$, and $\mu = \gamma + n$ as in Chapter 13.

For a given initial capital endowment and labor population and for any chosen combination of per head consumption rate $c(t)$ and resource flow $r(t)$, the above initial value problem determines the per head capital accumulation $k(t)$. The planning problem is to choose $r(t)$ and $c(t)$ to steer the economy toward some desirable direction. Similar to growth without essential resource input, we may analyze the consequences of the following planning policies:

1. A given savings rule, such as the fixed fraction saving $c = (1 - s)f(k, r)$, with s and $r(t)$ chosen to maximize (per head) consumption rate or some other measure of social welfare.

2. Optimal growth over an infinite horizon which maximizes the conventional social welfare measure

$$W = \int_0^{\infty} e^{-\delta t} U(c(t)) dt \qquad (14.5)$$

3. The max-min program which maximizes a uniform utility index $U(c)$. [For a positive, increasing, and concave utility function $U(c)$,

this is equivalent to finding the largest constant consumption rate $c(t) \equiv \hat{c}$ which can be sustained by the initial stock of capital.]

For these problems, the inequality constraint (14.1) is replaced by the corresponding equality constraint. If the resource were not depleted, we can always extract more to produce a little more to increase consumption as long as θ is assumed to be constant (or to remain the same order of magnitude). In that case, we can take the equality constraint on the resource deposit in the form

$$D^{\cdot}(t) = -r(t) \qquad D(0) = \bar{D}_0 \qquad D(\infty) = 0 \tag{14.6}$$

14.2 Social Optimum

For the conventional optimal growth problem, we maximize

$$W = \int_0^\infty e^{-\delta t} U(c)\, dt \tag{14.7}$$

subject to

$$k^{\cdot} = f(k, r) - \mu k - \theta r - c \tag{14.8}$$

$$D^{\cdot} = -r \tag{14.9}$$

$$k(0) = k_0 \tag{14.10}$$

$$D(0) = \bar{D}_0 \tag{14.11}$$

$$D(\infty) = 0 \tag{14.12}$$

$$r(t) \geq 0 \qquad k(t) \geq 0 \qquad c(t) \geq 0 \tag{14.13}$$

As we mentioned at the end of Section 14.1, the properties specified for $U(c)$ and $f(k, r)$ imply that a program of growth which does not exhaust the resource deposit cannot be optimal. With the Lagrange multipliers (also called the *adjoint or costate variables* in optimal control theory) λ and Λ, we have the following expression for the Hamiltonian of the problem introduced in Chapter 13:

$$H \equiv e^{-\delta t} U(c(t)) + \Lambda[-r] + \lambda[f(k, r) - \mu k - \theta r - c] \tag{14.14}$$

The first-order necessary conditions for optimality require (see Section 13.4)

$$\Lambda^{\cdot} = -\frac{\partial H}{\partial D} = 0 \qquad \lambda^{\cdot} = -\frac{\partial H}{\partial k} = -(f_k - \mu)\lambda \tag{14.15}$$

$$\lambda = e^{-\delta t}U'(c) \equiv e^{-\delta t}\frac{dU}{dc} \tag{14.16}$$

$$\Lambda = \lambda(f_r - \theta) \tag{14.17}$$

and

$$\lambda(\infty) = 0 \tag{14.18}$$

The four differential equations involving $k^{\cdot}$, $D^{\cdot}$, $\Lambda^{\cdot}$, and $\lambda^{\cdot}$ along with the two algebraic equations and four boundary conditions on $k(0)$, $D(0)$, $D(\infty)$, and $\lambda_k(\infty)$ define a two-point boundary value problem for the six unknowns k, D, R, c, λ, and Λ. For the types of $U(c)$ and $f(k, r)$ considered, it can be verified that the inequality constraints are satisfied by the solution of this BVP so that the *interior solution* prevails. The solution of the BVP for specific choices of f, U, and parameter values will be discussed later. For the rest of this section, we deduce two general consequences of the first-order conditions.

We begin by noting that Λ is a constant since $\Lambda^{\cdot} \equiv 0$. It follows upon differentiating (14.17)

$$\lambda^{\cdot} = \left[\frac{\Lambda}{f_r - \theta}\right]^{\cdot} = -\frac{\Lambda[f_r]^{\cdot}}{(f_r - \theta)^2} \qquad \frac{\lambda^{\cdot}}{\lambda} = -\frac{[f_r]^{\cdot}}{f_r - \theta} \tag{14.19}$$

so that we get from the second equation of (14.15)

$$\frac{[f_r]^{\cdot}}{f_r - \theta} = f_k - \mu \tag{14.20}$$

The relation (14.20) is known as the *Hotelling rule* for an efficient program of economic growth which gives the following characterization of the optimal growth.

14A: For optimal growth, the economy must be operated in such a way that the planner is indifferent to choosing between the potential gain from holding a dollar's worth of resource deposit and the gain from a dollar's worth of capital stock for future production.

To see this, we note that the output flow from an increment of resource flow Δr is $q = (f_{,r} - \theta)\Delta r$. Investing q gives an incremental capital accumulation flow $(f_{,k} - \mu)q\,\Delta t = (f_{,k} - \mu)(f_{,r} - \theta)\Delta r\,\Delta t$. A unit of resource is worth holding in the ground if it increases in value with time. In terms of the output flow, this means $q^{\cdot}\,\Delta t = f_{,r}^{\cdot}\,\Delta r\,\Delta t$ is positive. Hotelling's rule tells us that for the optimal policy, this increase in value is equal to the incremental capital accumulation. If it is higher, we should hold the resource. If it is lower, we should have been extracting at a higher rate.

If we start with $\lambda = e^{-\delta t}U'(c)$ instead and go through the same calculation leading to the Hotelling rule, we get

$$\frac{\dot\lambda}{\lambda} + \delta = -\eta(c)\frac{\dot c}{c} = -(f_k - \mu - \delta) \tag{14.21}$$

$$\eta(c) \equiv -c\frac{U''(c)}{U'(c)} \tag{14.22}$$

or

$$-\frac{U''(c)}{U'(c)} = f_k - \mu - \delta \tag{14.23}$$

The relation (14.23) is known as *Ramsey's rule*. Its interpretation is discussed in some detail in Burmeister and Dobell (1970). In terms of the preference for an extra unit of output for investment or consumption, the rule says roughly that,

> **14B:**   At optimal growth, the planner should be indifferent to a choice between using the extra unit of output flow for producing more capital and for a higher consumption rate.

Suppose the output flow q is available over a time interval Δt and set $\delta = 0$ for simplicity. The loss of utility from investing $q\,\Delta t$ for capital accumulation is $\Delta U = U'(c)\Delta c = q\,\Delta t U'(c)$. The capital increment from investing $q\,\Delta t$ is $\Delta k = q\,\Delta t[f_{,k} - \mu]$. It gives rise to additional future utility

$$\Delta U = \int_t^\infty q\,\Delta t[f_{,k} - \mu]U'(c)\,dt$$

The conclusion **14B** follows from (14.23) upon differentiating the two different expressions for ΔU with respect to t.

14.3   Constant Return to Scale

For a production function $f(k, r)$ which is homogeneous of degree one [and $L(t) \equiv L_0$], we have

$$\frac{1}{r}f(k, r) = f\left(\frac{k}{r}, 1\right) \equiv g(x) \qquad x \equiv \frac{k}{r} \tag{14.24}$$

$$f_k = [rg(x)]_k = g'(x) \qquad (\ )' \equiv \frac{d(\ )}{dx} \tag{14.25}$$

$$f_r = [rg(x)]_r = g(x) - xg'(x) \tag{14.26}$$

Hotelling's rule may be written in this case as a first-order ODE for $x \equiv k/r$,

$$x^{\cdot} = \frac{(g' - \mu)(xg' + \theta - g)}{xg''} \tag{14.27}$$

Since $g(x) \equiv f(k/r, 1) = f(k, r)/r$ is a known function, Hotelling's rule is now a first-order *separable* ODE for $x(t)$ and an exact solution of such an equation is immediate. Note that this result is independent of the form of $U(c)$.

With $x \equiv k/r$ determined as a function of t up to a constant $x_0 \equiv k_0/r_0$ where $r_0 \equiv r(0)$, we obtain $\lambda(t)$ from the second ODE in (14.15) and the end condition (14.18), now written as

$$\frac{\lambda^{\cdot}}{\lambda} = g'(x) - \mu \qquad \lambda(\infty) = 0 \tag{14.28}$$

Next, $c(t)$ is determined from (14.16). Also, upon writing $k(t) = x(t)r(t)$, we transform the equation for capital accumulation $k^{\cdot} = f(k, r) - \mu k - \theta r - c$ into

$$xr^{\cdot} + [x^{\cdot} - g(x) + \theta + \mu x]r = -c(t) \tag{14.29}$$

An exact solution of this *linear* first-order ODE is immediate as $c(t)$ and $x(t)$ are both known. Finally, we have $k(t) = x(t)r(t)$ and the constants of integration r_0 and Λ are determined by $k(0) = k_0$ and

$$\int_0^\infty r(t)dt = \bar{D}_0 \tag{14.30}$$

Two specific examples have been worked out for

$$U(c) = \frac{c^{1-\sigma}}{1 - \sigma} \tag{14.31}$$

$$f(k, r) = f_0 k^a r^{1-a} \tag{14.32}$$

1. *$\theta = \mu = 0$:* For this case, we have $g(x) \equiv f_0 x^a$ and $x^{\cdot} = f_0 x^a$, etc. It is not difficult to obtain (see exercises) for this case

$$c(t) = (c_0^{b\sigma/a} + F_0 t)^{a/b\sigma} e^{-\delta t/\sigma} \tag{14.33}$$

where $b = 1 - a$, $F_0 = (bf_0/\Lambda^b)^{1/a}$, and $c_0^{b\sigma/a} = (bf_0/\Lambda)^{b/a}(k_0/r_0)^b$. The two unknown constants r_0 and Λ are determined by (14.30) and $r(\infty) = 0$ [which is necessary for the improper integral in (14.30) to exist as $r(t) \geq 0$].

Even without an explicit solution for r_0 and Λ, we see that $c(t)$ tends to zero as $t \to \infty$ for any positive discount rate $\delta > 0$. For $\delta = 0$ (with $\sigma > 1$ and $a < \sigma$), we have $c(t) \to \infty$ as $t \to \infty$ instead. This rather undesirable knife-edge situation with regard to the value of δ is quite typical for maximum welfare growth.

2. *$\theta = 0$, $\mu \neq 0$:* For this case, we have again $g(x) \equiv f_0 x^a$. But now $x^{\cdot} = (f_0 a x^a - \mu x)/a$ is a Bernoulli equation and can be solved exactly. Straightforward calculations (see exercises) give

$$c(t) = e^{-\delta t/\sigma} \left\{ \left[c_0^{\sigma b/a} - \frac{a}{\mu b} \left(\frac{bf_0}{\Lambda b} \right)^{1/a} \right] e^{-b\mu t/a} + \frac{a}{\mu b} \left(\frac{bf_0}{\Lambda b} \right)^{1/a} \right\}^{a/\sigma b} \tag{14.34}$$

etc. with $b = 1 - a$. Again, we have the knife-edge situation with $c(t) \to 0$ as $t \to \infty$ if $\delta > 0$ while $c(t) \to (a/\mu b)^{a/\sigma b} (bf_0/\Lambda b)^{1/\sigma b}$ as $t \to \infty$ if $\delta = 0$.

From these two examples, we see that

14C: With the usual social rate of time preference in favor of the present ($\delta > 0$), the growth path which maximizes the conventional social welfare index (W) dictates eventual extinction.

Eventual extinction for $\delta > 0$ is certain for the case $\theta = \mu = 0$. It is also programmed for the $\mu \neq 0$ case even with $\delta = 0$ if $\bar{D}_0$ and k_0 are not large enough so that $c(\infty)$ is well below the subsistence level. As long as the destined-to-become-defunct last generation cannot be present to defend its own interest, it behooves those of us who control its destiny to reexamine and perhaps to reformulate our concept of (and approach to) optimal economic growth. Hopefully, the Earth's exhaustible resources may be exploited in a manner that is fair to future generations, at least in the context of economic growth theory.

14.4 The Max-Min Optimum

The results in the last section for the conventional optimal economic growth under the maximum social welfare criterion are disconcerting for at least two reasons. First, there is the knife-edge situation with respect to the discount rate. Second, the optimal program for $\delta > 0$ in fact calls for a per head consumption rate which eventually falls below the subsistence level (and can therefore be taken to mean the eventual extinction of the population). Extinction is also possible for the $\mu \neq 0$ case even for $\delta = 0$.

The cause of possible extinction is clear. The conventional utilitarian approach defines social welfare as the sum of the utilities of individual members of the society (which are the different generations in the present context). It allows for the possibility that a loss of utility to one or more generations can be more than offset by a positive increment to others. This offsetting process is further weighted by the preference for present and near future utilities. It has been argued (Rawls, 1971) that unequal distribution of utilities (or wealth) is justified only if it is necessary for the improvement of the poorest member of the society. This concept of social welfare, originally proposed for contemporary (intragenerational) members of a society, may be formulated as a quantitative statement in the form of the max-min principle introduced earlier in Chapter 12 and applied to an intertemporal society (Solow, 1974).

In the intergenerational setting, the constituent members of the society are the entire populations living at different instances in time. The utilities of all these constituents at the max-min optimum have been found to be necessarily identical (see Section 12.6). It follows from the concavity of $\bar{U}(c)$ that the per head consumption rate must remain constant for all time $c(t) \equiv c_0$. The optimization problem for economic growth with exhaustible resources is now to choose a (per head) extraction rate $r(t)$ to maximize the constant (per head) consumption rate c_0 subject to the dynamics of capital accumulation

$$k^{\cdot} = f(k, r) - \theta r - c_0 \qquad (14.35a)$$

$$k(0) = k_0 \qquad (14.35b)$$

and the finite resource deposit constraint formulated as

$$D^{\cdot} = -r \qquad D(0) = \bar{D}_0 \qquad D(\infty) = 0 \qquad (14.6)$$

In the ODE for capital accumulation, the $-\mu k$ term for capital depreciation has been omitted to simplify our discussion; the quantity $f(k, r)$ is now the actual output net capital depreciation (if any).

The above optimization problem for optimal economic growth under the max-min criterion is not a conventional optimal control problem. In his landmark paper, the 1987 Nobel Laureate of Economics, Robert Solow (1974) reformulated the problem as one of choosing $r(t)$ to minimize its integral over the infinite horizon in (14.30) and then choosing c so that the value of the integral equals the prescribed value $\bar{D}_0$. By now it is known that the original problem may be cast in the form of a Mayer problem in the conventional optimal control theory (Wan, 1980). This is done by regarding the constant c_0 as the initial value of a time function $c(t)$, with $c(t)$ defined by the first-order ODE

$$c^{\cdot} = 0 \qquad (14.36)$$

and the initial condition $c(0) = c_0$. In that case, the optimization problem for the max-min program becomes

$$\max_{r(t) \geq 0} \{ c(0) \} \qquad (14.37)$$

subject to (14.35a, b), (14.36), and (14.6), as well as the inequality constraints

$$k \geq 0 \qquad r \geq 0 \qquad D \geq 0 \qquad (14.38)$$

The Hamiltonian for this problem is

$$H \equiv \lambda_k [f(k, r) - \theta r - c] + \lambda_c [0] + \lambda_D [-r] \qquad (14.39)$$

and the first-order necessary conditions for a maximum are the three Euler ODEs for the Lagrange multipliers $\lambda_k(t)$, $\lambda_c(t)$, and $\lambda_D(t)$,

$$\lambda_k^{\cdot} = -\frac{\partial H}{\partial k} = -\lambda_k f_k \qquad (14.40a)$$

$$\lambda_D^{\cdot} = -\frac{\partial H}{\partial D} = 0 \qquad (14.40b)$$

$$\lambda_c^{\cdot} = -\frac{\partial H}{\partial c} = \lambda_k \qquad (14.40c)$$

the condition for an interior maximum

$$\frac{\partial H}{\partial r} = \lambda_k (f_r - \theta) - \lambda_D = 0 \qquad (14.41)$$

and the Euler boundary conditions

$$\lambda_k(\infty) = 0 \tag{14.42a}$$

$$\lambda_c(0) = -1 \tag{14.42b}$$

$$\lambda_c(\infty) = 0 \tag{14.42c}$$

These first-order conditions in (14.40) and (14.41) and the equations of state in (14.35), (14.6), and (14.36) form a two-point BVP for a sixth-order system of ODE for the seven unknown functions k, r, D, c, λ_k, λ_D, and λ_c.

  An immediate consequence of the above ODEs is the Hotelling efficiency condition

$$\frac{(f_r)^{\cdot}}{f_r - \theta} = f_k \tag{14.43}$$

As before, this is obtained by differentiating the expression for $\partial H/\partial r$ in (14.41) with respect to time and using the ODE (14.40a, b) to eliminate λ_k and λ_D from the resulting expression. It is important to note that the Hotelling condition (14.43) is effectively a first-order ODE of the form

$$r^{\cdot} = g(k, r; \theta, c) \tag{14.44}$$

with

$$g(k, r; \theta, c) = \frac{f_k(f_r - \theta) - f_{rk}(f - \theta r - c)}{f_{rr}} \tag{14.45}$$

This ODE and the three ODEs in (14.35a), (14.36), and (14.6) form a fourth-order system for k, D, r, and c. Supplemented by $k(0) = k_0$, $D(0) = \bar{D}_0$, $D(\infty) = 0$, and $r(\infty) = 0$ (which is needed for the convergence of the improper integral of r over the infinite horizon[1]), this system determines the four functions without any reference to the adjoint variables λ_k, λ_c, and λ_D. These adjoint variables may be determined subsequently by a single quadrature, namely,

$$\left(\frac{\lambda_c}{\lambda_D}\right)^{\cdot} = \frac{\lambda_k}{\lambda_D} \tag{14.46a}$$

$$\frac{\lambda_c(\infty)}{\lambda_D} = 0 \tag{14.46b}$$

---

[1] We are interested here only in piecewise differentiable solutions of the BVP and ignore pathological situations such as $r(t)$ nonvanishing on a set of measure zero beyond some finite time (which incidentally may, in fact, be a feasible solution).

with λ_k / λ_D known from

$$\frac{\lambda_k}{\lambda_D} = \frac{1}{f_r - \theta} \qquad (14.47)$$

and with the unknown constant λ_D determined by the only remaining boundary condition $\lambda_c(0) = -1$ so that

$$\lambda_D = \left. \frac{-1}{\lambda_c / \lambda_D} \right|_{t=0} \qquad (14.48)$$

14.5   A Cobb-Douglas Production Function

For the total resource extracted over an infinitely long planning period to remain finite, we must have $r(t) \to 0$ as $t \to \infty$. Given that the exhaustible resource is an essential input to production so that $f(k, 0) = 0$, the capital stock $k(t)$ would be drawn down by a negative $k^{\cdot}$ unless $f(k, r)$ remains finite and no smaller than c_0 as $r \to 0$. It is therefore not obvious that a constant per head consumption rate can be sustained for all $t > 0$ in the face of a dwindling essential resource reserve. To investigate the feasibility of a maxmin program, we consider the special case of a *Cobb-Douglas production function* $f(k, r) = f_0 k^a r^b$ $(0 < b < a, a + b < 1)$. Appropriate values for a and b for different industries in different countries have been compiled in Walters (1963).

We assume for the moment that there is no extraction cost so that $\theta = 0$. For this case, the Hotelling condition may be written as

$$r^{\cdot} = \frac{ac_0 r}{k(1 - b)} \qquad (14.49)$$

with $r(\infty) = 0$, while the ODE for capital accumulation becomes

$$k^{\cdot} = f_0 k^a r^b - c_0 \qquad (14.50)$$

with $k(0) = k_0$. In both equations, we have made use of $c^{\cdot} = 0$ and set $c(t) \equiv c_0$, an unknown constant to be determined. It follows immediately from the form of f that $k(t)$ must become unbounded as $t \to \infty$; otherwise we would have $k^{\cdot} \to -c_0$ as $t \to \infty$ so that k also dwindles to zero in finite time which is not acceptable. On the other hand, we do not need k to be too large at any given instant; in particular, we do not need $k^{\cdot}$ itself to become unbounded in the limit. At any t, what we need is just enough output to cover the current consumption and to increase the capital stock to make up for the resource reduction. A limiting behavior such as $f_0 k^a r^b$ tending to a constant $\phi_0 > c_0$ would do. For such a limiting behavior we have $k \to (\phi_0 - c_0)t$ for large t. This suggests that we try the solution

$$k(t) = (k_0 + \psi_0 t) \qquad r = R_0 k^{-a/b} \tag{14.51}$$

for some $\psi_0 > 0$ as a feasible max-min program. It is not difficult to verify that the total resource consumed by such a program is finite as long as $a > b > 0$, and that for a suitable choice of ψ_0, R_0, and c_0, this feasible program is in fact the actual max-min optimal program.

Instead of making the educated guess (14.51), we can actually deduce the same optimal program directly from the first-order necessary conditions for the max-min optimum. By combining (14.49) and (14.50), we get

$$\frac{dk}{dr} = \frac{k^\cdot}{r^\cdot} = \frac{k(f_0 k^a r^b - c_0)}{rac_0/(1-b)} \tag{14.52}$$

Equation (4.52) is a Bernoulli equation and can be solved exactly. The general solution may be taken in the form

$$k^a r^b = r \left[A_0 + \frac{f_0}{c_0}(1-b)r \right]^{-1}$$

In order for $k^a r^b$ not to tend to zero as $t \to \infty$ (and $r \to 0$), the constant of integration A_0 must be zero so that

$$(1-b)f_0 k^a r^b = c_0 \tag{14.53}$$

which we will call the *optimal trajectory* in the r,k plane. We use this relation in the ODE for capital accumulation to simplify it to

$$k^\cdot = f_0 k^a r^b - c_0 = \frac{c_0}{1-b} - c_0 = \frac{bc_0}{1-b} \tag{14.54}$$

giving

$$k = k_0 + \frac{bc_0 t}{1-b} \quad \text{and} \quad r = \frac{[c_0/f_0(1-b)]^{1/b}}{k^{a/b}} \tag{14.55}$$

Finally, the only unknown constant c_0 is determined by

$$\bar{D}_0 = \int_0^\infty r\, dt = \left[\frac{c_0}{f_0(1-b)}\right]^{1/b} \int_0^\infty \left(k_0 + \frac{bc_0 t}{1-b}\right)^{-a/b} dt$$

$$= \frac{1-b}{c_0(a-b)}\left[\frac{c_0}{f_0(1-b)}\right]^{1/b} k_0^{(b-a)/b} \qquad (14.56)$$

where we have made use of the fact that $0 < b < a < 1$. From (14.56), we get

$$c_0 = f_0^{1/b}(a-b)\bar{D}_0^{b/(1-b)} k_0^{(a-b)/(1-b)}(1-b) \qquad (14.57)$$

Note that the improper integral in (14.56) diverges if $b \ge a$.

Given that $\Delta Q/Q \simeq (f_k\,\Delta k + f_r\,\Delta r)/f = a(\Delta k/k) + b(\Delta r/r)$, we have the following conclusion for a Cobb-Douglas (production) technology and a finite deposit of essential resources:

14D: To sustain a constant per head consumption rate, it is necessary that a fractional decrease in resource input can be made up by a smaller fractional increase in capital input with the output fraction unaffected by the input changes.[2]

A direct proof of this observation is given in an appendix at the end of this chapter.

It is of some interest to see the qualitative features of the optimal trajectory in the r,k plane directly from the ODE for k and r. Equations (14.49) and (14.50) have no finite critical point. The direction field in the r,k plane is horizontal along the hyperbola $f_0 k^a r^b = c_0$, slanting upward (in the northwest direction) to the right and slanting downward (in the southwest direction) to the left. Along the (vertical) k axis, the direction field is vertical downward. Any trajectory close to the (horizontal) r axis rapidly heads toward the origin without crossing the r axis (Figure 14.1a). Now, not all trajectories in the r,k plane cross the $f_0 k^a r^b = c_0$ curve and turn downward. To see this, we note that the two ODEs may be written in terms of r and $u \equiv 1/k$:

$$u^\cdot = -u(f_0 u^{1-a} r^b - c_0 u) \qquad r^\cdot = \frac{urac_0}{1-b} \qquad (14.58)$$

---

[2] In the language of the economists, the economic content of $a > b$ is that the *elasticity* of output with respect to reproducible capital exceeds that with respect to the exhaustible resource.

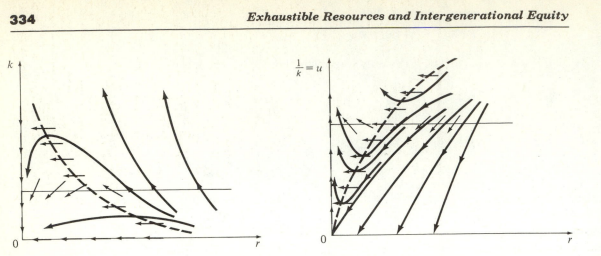

(a)   (b)

FIGURE 14.1

It follows that every point along the r axis in the $(r, 1/k \equiv u)$ plane is a *nonisolated* critical point. Many trajectories end up heading toward one of these critical points. The only possible solution for the max-min optimum is the trajectory which heads toward the origin in the r,u plane since r must tend to zero as $t \to \infty$ (Figure 14.1b). That there is only one such trajectory is confirmed by the exact solution obtained earlier.

14.6 The Effect of Extraction Cost

Consider now the case $\theta = \theta_0 > 0$ (a constant) in the ODE for capital accumulation (14.35a). The first-order necessary conditions for a max-min optimum are similar to those for the $\theta = 0$ case, with the Lagrange multipliers (adjoint variables) λ_k, λ_c, and λ_D satisfying the three ODEs in (14.40), the boundary conditions (14.41), and, for an interior solution, the algebraic conditions (14.42).

It is again an immediate consequence of these first-order conditions and the ODE for capital accumulation that k, c, and r satisfy the Hotelling efficiency condition (14.43). This condition and the ODE for capital accumulation form a pair of first-order autonomous ODE for k and r with $c \equiv c_0$ as a parameter. Once we specify $f(k, r)$, methods for qualitative and quantitative solutions may be applied. We again illustrate this with a Cobb-Douglas production function $f(k, r) = f_0 k^a r^b (a > b > 0, a + b < 1)$.

For the Cobb-Douglas case, the Hotelling condition may be written as

$$r^\bullet = -\frac{ac_0 r}{(1-b)k}\left[1 - \frac{\theta_0(1-b)}{bc}r\right] \tag{14.59}$$

and the equation for capital accumulation takes the form

$$k^{\cdot} = f_0 k^a r^b - \theta_0 r - c_0 \tag{14.60}$$

In the r,k plane, these two autonomous ODE for k and r have a finite critical point at the intersection of $r = bc_0/\theta_0(1-b)$ (vertical) line and the curve $f_0 k^a r^b - \theta_0 r = c_0$. The critical point is an unstable improper node (a source point). The system also has nonisolated critical points at $k = \infty$ for all (nonnegative) values of r. A phase portrait can be constructed as in the $\theta = 0$ case to find that there is only one possible trajectory which does not exhaust either the capital endowment or the resource deposit (or both) in finite time (see Figure 14.2).

In the remainder of this section, we discuss in some detail a method for a quantitative solution which was found important for the analysis of the multigrade resource deposit cases (Solow and Wan, 1976; Ascher and Wan, 1980). The first step in the solution process is to form

$$\frac{dk}{dr} = \frac{dk/dt}{dr/dt} = \zeta(r) k^{a+1} - \eta(r) k \tag{14.61}$$

where

$$\zeta(r) = \frac{1-b}{ac_0} \frac{f_0 r^{b-1}}{\bar{\theta}_0 r - 1} \tag{14.62a}$$

$$\eta(r) = \frac{1-b}{ac_0 r} \frac{\theta_0 r + c_0}{\bar{\theta}_0 r - 1} \tag{14.62b}$$

$$\bar{\theta}_0 = \frac{\theta_0(1-b)}{bc_0} \tag{14.62c}$$

This is again a Bernoulli equation; its general solution is

$$k^{-a} r^{1-b} = \frac{f_0 b}{\theta_0} + A_0 \left[r - \frac{bc_0}{(1-b)\theta_0} \right] \tag{14.63}$$

where A_0 is a constant of integration. We rewrite the solution in the form

$$f_0 b k^a r^{b-1} = \frac{\theta_0}{\{1 - [A_0 c_0/f_0(1-b)]\} + (A_0 \theta_0/f_0 b) r} \tag{14.64}$$

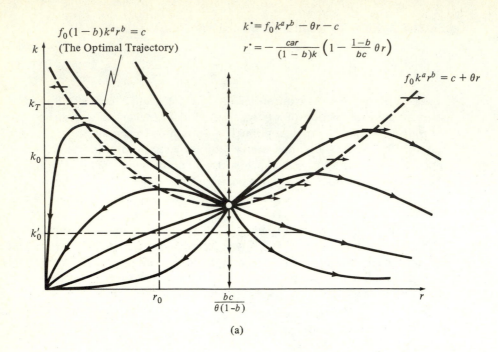

$f_0(1-b)k^a r^b = c$
(The Optimal Trajectory)

$k^{\bullet} = f_0 k^a r^b - \theta r - c$

$r^{\bullet} = -\dfrac{car}{(1-b)k}\left(1 - \dfrac{1-b}{bc}\,\theta r\right)$

$f_0 k^a r^b = c + \theta r$

k

k_T

k_0

k_0'

r_0 $\dfrac{bc}{\theta(1-b)}$ r

(a)

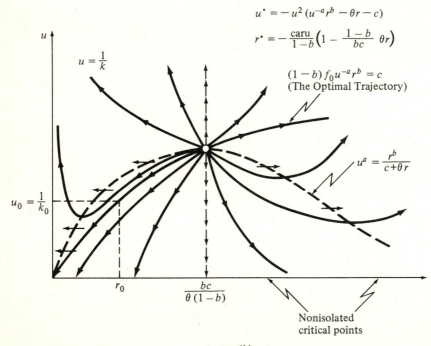

$u^{\bullet} = -u^2(u^{-a} r^b - \theta r - c)$

$r^{\bullet} = -\dfrac{caru}{1-b}\left(1 - \dfrac{1-b}{bc}\,\theta r\right)$

$u = \dfrac{1}{k}$

$(1-b)f_0 u^{-a} r^b = c$
(The Optimal Trajectory)

$u^a = \dfrac{r^b}{c+\theta r}$

$u_0 = \dfrac{1}{k_0}$

r_0 $\dfrac{bc}{\theta(1-b)}$

Nonisolated
critical points

(b)

FIGURE 14.2

and use it in the expression for λ_k given by (14.47) to get

$$\lambda_k = \frac{\dfrac{\lambda_D}{\theta_0}\left\{\dfrac{A_0\theta_0}{f_0 b}\,r + \left[1 - \dfrac{A_0 c_0}{f_0(1-b)}\right]\right\}}{1 - \left\{\left[1 - \dfrac{A_0 c_0}{f_0(1-b)}\right] + \dfrac{A_0\theta_0}{f_0 b}\,r\right\}} \tag{14.65}$$

where λ_D is a yet undetermined constant (from $\lambda_D^{\boldsymbol{\cdot}} = 0$). The transversality condition $\lambda_k(\infty) = 0$ and the condition $r(\infty) = 0$, needed for the convergence of the improper integral (14.30), may be combined into a single condition $\lambda_k(r = 0) = 0$. Since $\lambda_D \neq 0$, this condition can only be satisfied by setting

$$A_0 = \frac{f_0}{c_0}\,(1 - b) \tag{14.66}$$

so that the unique solution for the Bernoulli equation which satisfies the condition $\lambda_k(\infty) = 0$ is

$$f_0(1 - b)\,k^a r^b = c_0 \tag{14.67}$$

It is not difficult to verify that the corresponding trajectory in the phase plane passes through the critical point of the autonomous system of ODE for k and r. It is the only trajectory in the phase plane for which $r \to 0$ as $k \to \infty$, and is therefore the optimal trajectory.

To get k, r, and D as explicit functions of t, we use the relation (14.67) in the ODE for capital accumulation to get

$$k^{\boldsymbol{\cdot}} = \frac{b c_0}{1 - b} - \theta_0 r \tag{14.68}$$

Upon integrating both sides of (14.68) from 0 to t, we get

$$k(t) = k_0 + \frac{b c_0 t}{1 - b} - \theta_0 \int_0^t r(t)\,dt$$

$$= k_0 + \frac{b c_0 t}{1 - b} - \theta_0 [D(t) - \bar{D}_0]$$

$$= \bar{k}_0 + \frac{b c_0 t}{1 - b} + \theta_0 D(t) \tag{14.69}$$

with

$$\bar{k}_0 = k_0 - \theta_0 \bar{D}_0 \tag{14.70}$$

Correspondingly, we have along the optimal trajectory

$$r = \left[\frac{c_0}{f_0(1-b)}\right]^{1/b} k^{-a/b} = \left[\frac{c_0}{f_0(1-b)}\right]^{1/b}\left[\bar{k}_0 + \frac{bc_0 t}{1-b} + \theta_0 D\right]^{-a/b} \qquad (14.71)$$

It follows that the ODE $D^\cdot = -r$ may be written as a single ODE for D alone:

$$D^\cdot = -r = -\left[\frac{c_0}{f_0(1-b)}\right]^{1/b}\left[\bar{k}_0 + \frac{bc_0 t}{1-b} + \theta_0 D\right]^{-a/b} \qquad (14.72)$$

Together with $D(0) = \bar{D}_0$ and $D(\infty) = 0$, it determines $D(t)$ and c_0, and therewith the solution of the problem.

For completeness, we note that k, r, and λ_k/λ_D can be obtained from $D(t)$ by way of the algebraic relations derived earlier. If we wish, we may solve

$$\left(\frac{\lambda_c}{\lambda_D}\right)^\cdot = \frac{\lambda_k}{\lambda_D} \qquad \lambda_c(0) = -\lambda_D \qquad \lambda_c(\infty) = 0 \qquad (14.73)$$

to determine $\lambda_c(t)$ and λ_D.

14.7   Investment Rule for Intergenerational Equity

The max-min program with a Cobb-Douglas production function follows the optimal trajectory

$$f_0(1-b)k^a r^b = c \qquad (14.67)$$

For the case $\theta = 0$, we may write the above relation as

$$f(k, r) - c = bf_0 k^a r^b = rf_r \qquad (14.74)$$

But, from the accounting equation (production = consumption + investment), the left-hand side is just $k^\cdot$. Thus, along the optimal trajectory, capital is accumulated according to the following investment rule:

14E: Without cost for resource extraction the max-min optimum reinvests the share of output attributable to the resource input, that is, invest the rents from the resource input to production for capital accumulation.

When θ is a nonzero constant, the accounting equation takes the form

$$rf_r = f(k, r) - c = k^{\cdot} + \theta r \quad \text{or} \quad k^{\cdot} = r(f_r - \theta) \tag{14.75}$$

so that the investment rule is now modified to read:

14F: When there is a constant cost per unit extraction rate, the maxmin optimum invests the resource rents net (i.e., after deducting) extraction cost to increase the capital stock.

The above observation for a Cobb-Douglas production function suggests that the same investment rule may hold for a more general class of production functions. The following theorem (Hartwick, 1977) confirms our expectation:

The Hotelling efficiency condition $(f_r - \theta)^{\cdot} = (f_r - \theta)(f_k - \mu)$ and the Hartwick investment rule $k^{\cdot} = rf_r - \theta r$ implies $c^{\cdot} = 0$ for a constant unit extraction cost θ.

The theorem follows from differentiating the accounting equation where $k^{\cdot}$ is replaced by $r(f_r - \theta)$ repeatedly and the Hartwick rule is observed whenever appropriate:

$$c^{\cdot} = [f(k, r) - \mu k - \theta r - k^{\cdot}]^{\cdot} = [f(k, r) - \mu k - rf_r]^{\cdot}$$
$$= f_k k^{\cdot} + f_r r^{\cdot} - \mu k^{\cdot} - r^{\cdot} f_r - r(f_r)^{\cdot} = (f_k - \mu)(f_r - \theta)r - r(f_r)^{\cdot} = r[(f_k - \mu)(f_r - \theta) - (f_r)^{\cdot}]$$

Of course, investing according to the Hartwick rule does not necessarily guarantee that the constant per head consumption rate can be sustained forever for production functions other than the Cobb-Douglas type.

EXERCISES

1. Let $U(c) = c^{1-\sigma}/(1 - \sigma)$, $f(k, r) = f_0 k^a r^{1-a}$ and $\theta = \mu = 0$. Use the approach of Section 14.3 to obtain the following management program for social optimum:

$$c(t) = (c_0^{b\sigma/a} + F_0 t)^{a/b\sigma} e^{-\delta t/\sigma} \qquad F_0 = \left(\frac{bf_0}{\Lambda^b}\right)^{1/a} \quad (b = 1 - a)$$

$$k(t) = \left(\frac{\Lambda}{f_0 b}\right)^{1/a} (F_0 t + c_0^{b\sigma/a})^{1/b} r(t) \qquad c_0^{b\sigma/a} = \left(\frac{bf_0}{\Lambda}\right)^{b/a}\left(\frac{k_0}{r_0}\right)^b$$

$$r(t) = r_0 - \left(\frac{bf_0}{\Lambda}\right)^{1/a} \int_0^t (F_0 \tau + c_0^{b\sigma/a})^{(a-\sigma)/b\sigma} e^{-\delta\tau/\sigma} d\tau$$

Give conditions for the determination of any unknown constants in this solution.

2. *Diminishing return to scale, no discounting and social optimum:* If f is not homogeneous of degree one, we can no longer reduce the BVP to one for a single first-order separable ODE which determines $k/r \equiv x$ as a function of t. Nevertheless, we can still reduce the problem to solving another kind of first-order ODE, provided $\delta = 0$.

   (a) Use (14.17) in (14.15) to get after some rearrangement $f_{rr}r^{\cdot} = (f_k - \mu)(f_r - \theta) - f_{rk}(f - \mu k - \theta r - c)$.

   (b) Obtain a first-order ODE for k as a function r with Λ as a parameter

   $$\frac{dk}{dr} = \frac{f_{rr}[f(k, r) - \mu k - \theta r - c]}{(f_k - \mu)(f_r - \theta) - f_{rk}(f - \mu k - \theta r - c)}$$

   Explain how you would express c in terms of r, k, and Λ.

   (c) With $k(r_0) = k_0$ where $r_0 = r(0)$, the above ODE determines k (and hence c) in terms of r with Λ and r_0 as parameters. How do you determine Λ and r_0?

3. Apply the procedure outlined in Exercise 2 to the case

   $$f(k, r) = f_0 k^a r^b \qquad U(c) = \frac{c^{1-\sigma}}{1 - \sigma} \qquad (\sigma > 1)$$

   to get

   (a) $$c = \left(\frac{f_0 b}{\Lambda}\right)^{1/\sigma} k^{a/\sigma} r^{(b-1)/\sigma}$$

   (b) $$r^{\cdot} = \frac{car}{k(b - 1)}$$

   (c) $$\frac{dk}{dr} = -\left(\frac{b - 1}{a}\right)\frac{k}{r} \frac{(b - 1)f_0}{a}\left(\frac{\Lambda}{f_0 b} k^{a(\sigma-1)+\sigma} r^{-(1-b)(\sigma-1)}\right)^{1/\sigma}$$

   (d) $$k^a = \left(\frac{(1 - b)(\sigma - 1)f_0}{\sigma}\left(\frac{\Lambda}{bf_0}\right)^{1/\sigma} r^{1-b}\right)^{\sigma/(1-\sigma)}$$

   [*Hint:* The ODE in part (c) is a Bernoulli equation.]

   (e) $$c = \left[\frac{b\sigma}{\Lambda(1 - b)(\sigma - 1)} r^{-1}\right]^{1/(\sigma-1)}$$

   and hence $c \to \infty$ as $t \to \infty$ if $a > 0$, $b > 0$, and $a + b < 1$.

   (f) Indicate how you would determine r as a function of t.

Appendix:

Nonexistence of a Max-Min Optimum for a Cobb-Douglas Technology with $b > a$

Suppose $c_0 > 0$, we have from the equation for capital accumulation $k^{\cdot} = f_0 k^a r^b - c_0$

$$\left(\frac{k^{1-a}}{1-a}\right)^{\cdot} = f_0 r^b - c_0 k^{-a} < f_0 r^b$$

so that

$$\frac{k^{1-a} - k_0^{1-a}}{1-a} < f_0 \int_0^t r^b(\tau) d\tau = f_0 \int_0^t r^b(\tau) \cdot [1]^{1-b} d\tau$$

By Hölder's inequality, we have (for $0 < b < 1$)

$$\int_0^t r^b(\tau) [1]^{1-b} d\tau \le \left[\int_0^t r(\tau) d\tau\right]^b \left(\int_0^t d\tau\right)^{1-b} < t^{1-b} \bar{D}_0^b$$

so that

$$k^{1-a} < K_0 t^{1-b} + k_0^{1-a}$$

It follows that there is a positive constant K_1 such that

$$k^{1-a}(t) < K_1 t^{1-b}$$

for all $t \ge 1$.

But for $t > 1$, the equation for capital accumulation gives

$$k^{\cdot} = f_0 k^a r^b - c_0 < f_0 (K_1 t^{1-b})^{a/(1-a)} r^b - c_0$$

After integration from 1 to t, we get by Hölder's inequality again

$$k(t) - k(1) < K_2 \int_1^t (\tau^{a/(1-a)})^{1-b} r^b(\tau) d\tau - c_0(t-1)$$

$$\le K_2 \left(\int_1^t t^{a/(1-a)} d\tau\right)^{1-b} \left[\int_1^t r(\tau) d\tau\right]^b - c_0(t-1)$$

It follows that

$$k(t) < K_2 \bar{D}_0^b \left(\int_1^t \tau^{a/(1-a)} d\tau \right)^{1-b} - c_0(t-1) + k(1)$$

$$= K_3(t^{1/(1-a)} - 1)^{1-b} - c_0(t-1) + k(1)$$

$$< K_3 t^{(1-b)/(1-a)} - c_0(t-1) + k(1)$$

Now, $b > a$ implies $1 - b < 1 - a$ so that $k(t)$ is dominated by the $-c_0 t$ term. Therefore, $k(t)$ becomes negative for sufficiently large (but finite) t.

15

Economically Optimal Forest Harvesting Schedule

A solution process is described for an optimal control problem with binding inequality constraints. Some knowledge of monotone sequences is required near the end of the chapter.

15.1 The Fisher Age

In Chapters 12 to 14, we have seen a number of dynamic optimization problems for which the optimal program is determined for a period of time. Some of these problems involve inequality constraints and are problems in optimal control theory. From a pedagogical viewpoint, they do not fully illustrate the solution process for optimal control problems. The inequality constraints play only a passive role in these problems; the constraints are automatically satisfied by the solution of the Euler differential equations and boundary conditions. We have then what is known as the *interior solution* for these optimal control problems. Not all solutions to optimal control problems are interior solutions; the inequality constraints are not always satisfied by the solution of the BVP. For some problems, these constraints must be enforced. The enforcement of inequality constraints changes the solution process and gives what is called a *corner solution* of the optimization problem. It is important to see the process for constructing a corner solution. In this chapter, we illustrate this process by way of a renewable resource problem: What is the (economically) optimal schedule for logging trees?

Forestry is a major industry in many parts of the world and proper forest management is of major concern to governments as well as to the lumber companies, not necessarily for the same reasons. Private companies are interested in maximizing the financial return for their capital investment. Governments are interested in maximizing tax revenues from lumber industry and employment for the labor force. An economic analysis is useful for both.

To an economist, a stand of trees is a stock of capital which increases in value with tree age. For example, a typical stand of 110-year-old British Columbia douglas fir was worth $1000 (as usable timber) after harvesting and shipping costs in 1967, while a 30-year-old stand had no net commercial value (Pearse, 1967). It is a biological fact that the growth of a tree increases rapidly during its early years and declines steadily with age. At some advanced age, biological decay sets in: Not only does the tree stop growing, its (usable) timber content eventually decreases. The sketch in Figure 15.1 captures the qualitative features of the variation of usable (commercial) timber content $w(A)$, with w measured in units of usable tree biomass or volume and A being the age of the tree. The age of maximum usable timber content is typically more than 100 years old for most commercial timber species in the northern hemisphere. Given this typical profile of $w(A)$ and a positive real discount rate (net inflation) for future income, reflecting the time preference of our society (or the forest owners), the "when to cut a tree" question has long been a fundamental issue in forestry economics and management. In the first three sections of this chapter, we will summarize the important classical results for this problem. In the rest of the chapter, the more realistic models developed in the last 10 years (especially those pertaining to limited harvesting capacity) will be described.

To the extent that they constitute the main thrust of theoretical research in forest economics, the results presented in this chapter are conspicuous in their preoccupation with profit. This private sector orientation might well have been tied to an effort on the part of the economists to gain the acceptance of forest managers [see Samuelson (1976)]. Nevertheless, it has the unfortunate consequence that researchers in forest economics have devoted very little attention to the environmental issues related to forest logging. An attempt (however inadequate) will be made in this chapter

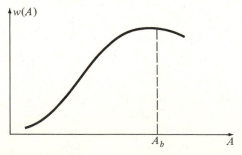

FIGURE 15.1

to incorporate some environmental factors in the conventional economic analysis (see Section 15.2 and the appendix).

Let p and c be the unit price and harvesting cost of trees, respectively. The net commercial timber value (also called the *stumpage value*) V of a tree is given by $V = pw - c$. As a function of the tree age A, it is monotone increasing (at least before decay sets in) and concave (after the age of maximum growth rate). Suppose δ is the constant real discount rate for discounting future revenue. Then, the optimal cutting time of the tree is determined by

$$\max_{t>0} \{ e^{-\delta t} V(A) \,|\, A = t - T_0 \} \tag{15.1}$$

where $T_0 \leq 0$ is the *planting* (or *germination*) *time* of the tree. The solution of this simple optimization problem is the *Fisher* (also Jevons, Wicksell and Thuneu) *age* A_F defined as the root of the equation

$$\frac{\dot{V}(A_F)}{V(A_F)} = \delta \tag{15.2}$$

obtained from $d[e^{-\delta t} V(A)]/dt = 0$ at $A = A_F$. In this chapter, a dot on top indicates differentiation with respect to the argument. From (15.2) we see that

15A: The present value of (discounted) future revenue of a tree is maximized at its Fisher age when the percentage increase of the stumpage value equals the discount rate.

If $A_F + T_0 < 0$ so the tree is already older than its Fisher age at $t = 0$, the best we can do is to log it immediately.

It is clear that logging should take place before the onset of biological decay, A_b. Logging should also be after the age of maximum growth rate. Hence, we need only to consider a range of tree age A over which $V(A)$ is a monotone increasing concave function. In that case, A_F is geometrically the intersection of the graph of the monotone increasing concave function $\delta V(A)$ and the graph of the monotone decreasing function $\dot{V}(A)$. Hence A_F is unique. Furthermore, $V(A_F) e^{-\delta(T_0 + A_F)}$ is a maximum because we have

$$\frac{d^2}{dt^2} [V(t - T_0) e^{-\delta t}] \Big|_{t = T_0 + A_F} = e^{-\delta(T_0 + A_F)} [\ddot{V} - \delta^2 V]_{t = T_0 + A_F} < 0$$

$$\tag{15.3}$$

given $\ddot{V}(T_0 + A_F) \leq 0$. (Note that $\dot{V}$ is positive before the age of maximum growth rate.)

For a homogeneous forest of N_0 trees, we only have to replace V by $N_0 V$ in the optimization problem and get the same Fisher age as the optimal cutting time.

We can also extend the analysis of this section to forests which are inhomogeneous in tree age, as trees may be planted at different times and would therefore be of different vintages. Suppose there are n_i trees of age A_i in the forest at time t. Then $T_i = t - A_i$ is the planting time of the A_i-year-old trees. The total timber content of the forest at time t is

$$w(t) = \sum_{i=0}^{m} n_i w(t - T_i) \tag{15.4}$$

if there are $m + 1$ different tree vintages. The net revenue in (15.1) is now $V(t) = pw(t) - c$. The stumpage value $V(t)$ of a forest with a continuous age distribution can be similarly characterized [with the sum in the expression for $w(t)$ above replaced by an integral]. Optimization problems for non-uniform age distributions will be investigated more thoroughly later.

15.2 Opportunity Cost

In the preceding calculation of the present value of net future revenue, it was tacitly assumed that the land on which trees are planted has no other economic value. Otherwise, the *opportunity cost* of not having the land available for other income production should be taken into account. Suppose the land has an alternative use which produces a revenue of $I(t)$ per unit time. The expression for the present value of the future net revenue $P(t)$ (for profit) takes the form

$$P(t) \equiv e^{-\delta t} V(t - T_0) - \psi(t) \qquad \psi(t) = \int_0^t I(\tau) e^{-\delta \tau} \, d\tau$$

$$\tag{15.5a}$$

The optimization problem is now

$$\max_{t > 0} \left\{ P(t) \equiv e^{-\delta t} V(t - T_0) - \int_0^t I(\tau) e^{-\delta \tau} \, d\tau \right\} \tag{15.5b}$$

Assuming for simplicity that $I(t)$ is nondecreasing (except possibly for a total cutoff after $t = T_c$), it is straightforward to show that the maximum (if it exists) is attained at a root of the equation

$$\frac{\dot{V}(t - T_0) - I(t)}{V(t - T_0)} = \delta \qquad (15.5c)$$

which has the same economic content as the condition for the determination of the Fisher age. The intersection of $\delta V(t - T_0)$ and $\dot{V}(t - T_0) - I(t)$ takes place at a smaller value of A than the case $I(t) \equiv 0$. As expected

15B: The optimal logging time is earlier than A_F when there is an alternative (revenue producing) use of the land.

If logging the forest degrades the environment, it is possible to incorporate the interest for a better environment into this classical economic model of forest harvesting in a number of ways. We do it by way of the opportunity cost $\psi(t)$ now taken in the form

$$\psi(t) = -\int_0^t e^{-\delta\tau} I(\tau) d\tau + \int_t^\infty e^{-\delta\tau} I(\tau) d\tau \qquad (15.6a)$$

The first term may be interpreted as the income from user's fees or government subsidy before logging to reflect the environmental value of the forest. The second term corresponds to the penalty associated with the loss of the use of the forest after logging, assuming again no replanting after harvest. The condition (15.5c) for the optimal harvest time is now replaced by

$$\frac{\dot{V}(t - T_0) + 2I(t)}{V(t - T_0)} = \delta \qquad (15.6b)$$

For sufficiently large $I(t)$, it may be necessary now to remove the simplifying assumption of $V(A)$ being monotone increasing and retain the more representative bell-shape feature of $V(A)$ to include biological decay for large A. In any event

15C: The income stream from the unharvested forest pushes back the optimal harvest time.

15.3   The Faustmann Rotation

If the land is immediately replanted after cutting, and the cut-and-replant process repeats indefinitely, then the income stream $I(t)$ can be calculated more specifically. The present value of discounted future net revenue is evidently

$$P(t_1, t_2, \ldots) = e^{-\delta t_1} V(t_1 - t_0) + e^{-\delta t_2} V(t_2 - t_1) + e^{-\delta t_3} V(t_3 - t_2)$$

$$+ \cdots + e^{-\delta t_k} V(t_k - t_{k-1}) + \cdots$$

$$= \sum_{k=1}^{\infty} e^{-\delta t_k} V(t_k - t_{k-1}) \tag{15.7}$$

assuming for simplicity that the discount rate, price, and cost (in constant dollars) do not change over time. The optimization problem is to choose a sequence of harvest times $t_1 < t_2 \cdots < t_k < \cdots$ to maximize P. Note that t_0 is the planting time of the initial homogeneous forest, and is not a control parameter.

Formally, the first-order necessary conditions require that a maximum P be attained at a stationary point of P which is a solution of the infinite system of equations:

$$\frac{\partial P}{\partial t_k} = [\dot{V}(t_k - t_{k-1}) - \delta V(t_k - t_{k-1})] e^{-\delta t_k} - \dot{V}(t_{k+1} - t_k) e^{-\delta t_{k+1}}$$

$$= 0 \qquad (k = 1, 2, \ldots)$$

Even if we leave alone theoretical questions such as the convergence of the series in (15.7), the determination of a formal solution $(t_1^*, t_2^*, \ldots)$ of the infinite system, exact or approximate, is in itself a formidable task. Fortunately, a little reflection allows us to reduce the original problem to a manageable form.

Suppose we have found the optimal cutting time for the initial harvest, say $t_1 = T + t_0$. To decide on the optimal harvest schedule for the remaining harvests is exactly the same problem as the one before the initial harvest. For an infinite harvest sequence, the situation after any particular harvest is identical to the initial situation. It follows that the tree age must be the same for all harvests of the optimal harvest schedule, that is,

$$t_k - t_{k-1} = T \qquad (k = 1, 2, \ldots) \tag{15.8}$$

In that case, we have

$$P = e^{-\delta t_0} V(T) \sum_{k=1}^{\infty} e^{-\delta kT} = \frac{e^{-\delta(t_0 + T)} V(T)}{1 - e^{-\delta T}} = \frac{e^{-\delta t_0} V(T)}{e^{\delta T} - 1} \tag{15.9}$$

It is not difficult to show now that for a maximum P, trees should be harvested by the so-called *Faustmann rotation* schedule R_F (M. Faustmann, 1849, 1968). This rotation is determined by the condition

$$\frac{\dot{V}(R_F)}{V(R_F)} = \frac{\delta}{1 - e^{-\delta R_F}} \tag{15.10}$$

If we write this condition as

$$\dot{V}(R_F) = \delta V(R_F) + \delta \frac{V(R_F)}{e^{\delta R_F} - 1} \tag{15.11}$$

we see that the first term on the right is just the "interest" earned by the stumpage value for a unit time at the rate δ. The second term on the right side is the interest per unit time of the present value of the stream of future net revenue to be earned from all future harvests, known as the *site value*. Thus, we have from (15.10):

> **15D:** The optimal harvest schedule for an ongoing forest is to cut trees according to the Faustmann rotation when the increase in stumpage value by waiting a unit time longer equals the sum of all opportunity costs of investment for that unit time tied up in the standing trees and in the land of the future tree site.

From the diagram in Figure 15.2, we see that $R_F < A_F$ so that an ongoing forest should be harvested sooner than the once-and-for-all forest. Our anal-

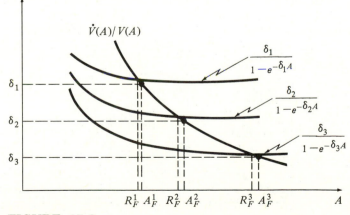

FIGURE 15.2

ysis merely confirms and quantifies the rather obvious fact that the earlier realization of future incomes can more than make up for the revenue lost in the earlier harvest.

15.4 Maximum Sustained Yield

For a variety of reasons (including a strong tradition), the notion of Faustmann's rotation has not gained wide acceptance in forestry management (Samuelson, 1976). Instead, most foresters hold to the view that the goal of a good harvesting policy is to have a sustained forest yield, preferably *maximum sustained yield*. They fully realize that to wait until a tree achieves its top lumber content at the age A_b cannot be optimal. Beyond its best growth rate, a grown tree should be cut to make land available for a faster-growing young tree and thereby earn a higher revenue.

As a measure of the growth rate, we may use $w(A)/A$. Maximizing this average gain per year gives the tree age A_g for the maximum sustained (gross) yield with A_g determined by

$$\dot{w}(A_g) = \frac{w(A_g)}{A_g} \tag{15.12}$$

A geometric interpretation of the above condition gives A_g as the point where the slope of $w(A)$ equals the slope of the straight line, $w = \dot{w}(A_g)A$, through the origin (see Figure 15.3).

For company profits, a more appropriate measure of the growth rate would be the (average) net growth rate $[w(A) - c/p]/A$ with the term c/p being the harvesting cost paid in lumber. Maximizing this average net gain per year gives the tree age A_n for the maximum sustained (net) yield with A_n determined by

$$\dot{w}(A_n) = \frac{1}{A_n}\left[w(A_n) - \frac{c}{p} \right] \tag{15.13}$$

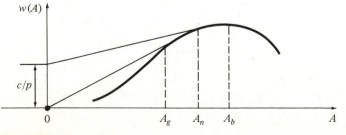

FIGURE 15.3

A geometric interpretation for this condition gives A_n as the point where the slope of $w(A)$ equals the slope of the straight line $w = \dot{w}(A_n)A + c/p$.

Upon writing (15.13) as

$$\frac{\dot{w}(A_n)}{w(A_n) - c/p} = \frac{1}{A_n} \qquad (15.13a)$$

the condition for A_n is of the same form as the condition determining the Fisher age which may be written as

$$\frac{\dot{w}(A_F)}{w(A_F) - c/p} = \delta \qquad (15.2a)$$

Depending on the size of δ, A_F may be greater or smaller than A_n (see Figure 15.4).

On the other hand, the Faustmann rotation R_F is determined by (15.10) which may be written as

$$\frac{\dot{w}(R_F)}{w(R_F) - c/p} = \frac{\delta}{1 - e^{-\delta R_F}} \qquad (15.10a)$$

We know already $R_F < A_F$. With a little calculation, we can show that $R_F < A_n$ for all δ. First, for $\delta \ll 1$, we have

$$1 - e^{-\delta R_F} \simeq \delta R_F(1 - \tfrac{1}{2}\delta R_F) \qquad (\delta \ll 1) \qquad (15.14)$$

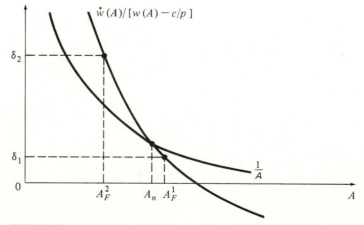

FIGURE 15.4

and therewith

$$\frac{\dot{w}(R_F)}{w(R_F) - c/p} \simeq \frac{1}{R_F(1 - \frac{1}{2}\delta R_F)} \qquad (\delta \ll 1) \qquad (15.15)$$

It follows that R_F tends to A_n from below as $\delta \to 0$.

More generally, the minimum of $\delta/(1 - e^{-\delta R_F})$ as a function of δ with R_F fixed is $(1 + \delta R_F)/R_F$ attained at the single real root of $e^{\delta R_F} = 1 + \delta R_F$. Therefore, the graph of $\delta/(1 - e^{-\delta A})$, now as a function of A for a fixed δ, is always above $1/A$. In that case, the graph of $\delta/(1 - e^{-\delta A})$ intersects the monotone decreasing graph of $\dot{w}(A)/[w(A) - c/p]$ before $1/A$ does.

The results of this section are summarized in the following theorem:

> **15E:** At the sustained net (gross) yield maximum attained at A_n (A_g), the gain in net (gross) increase in lumber content by waiting a unit time longer equals the average net (gross) increase per unit time. Furthermore, we have $R_F < A_n$ with $R_F \to A_n$ from below as $\delta \to 0$. On the other hand, A_F may be greater or less than A_n.

15.5 Finite Harvesting Rate and Ordered Site Access

In the derivation of the Fisher age, Faustmann rotation, maximum sustained yields, etc., there was an implicit assumption on the harvesting capacity of the logging company: A forest (or any part of it) can be clear-cut by the company in a time span much shorter than the time scale for tree growth, timber price fluctuation, etc., so that all harvests may be considered instantaneous. For a very large forest or a firm with limited capacity, we can no longer allow instantaneous harvesting of a finite forest. An appropriate formulation of the optimal harvest schedule problem for firms operating on a bounded harvesting rate would be of interest. Similar to fishery problems, a rather natural formulation of this so-called *blocked harvest* problem would be in terms of a harvesting rate function $h(t)$ at a time t, say the acreage of forest area harvested in unit time. For the more realistic model, we have the inequality constraint $0 \le h(t) \le h_{\max}$ for a company with finite logging capacity working on a very large forest. By choosing a starting time t_s and a particular time path of harvesting rate $h(t)$, a forest of size F_0 is completely logged during the time interval (t_s, t_e) with t_e determined by the "stock constraint"

$$F_0 = \int_{t_s}^{t_e} h(t)\,dt \qquad (15.16)$$

Let $v(t)$ be the net stumpage value per unit area of the forest harvested at time t; it may depend on $h(t)$ to reflect the set-up costs and *overload* costs (such as overtime pay, excessive wear on machines, etc.). We consider in this section the case of a single harvest with no replanting (called a *once-and-for-all forest*) and (for simplicity) with the after-harvest forest site having no commercial value. In that case, the present value of the discounted future net revenue is given by

$$P = \int_{t_s}^{t_e} e^{-\delta t} v(t) h(t)\, dt \qquad\qquad (15.17)$$

where δ is the constant discount rate. The forest manager's optimization problem is to choose $h(t)$ and t_s to maximize P subject to the "stock constraint" (15.16) and harvesting rate constraint $0 \le h(t) \le h_{\max}$ (Heaps and Neher, 1979). This is a difficult problem in optimal control theory as it involves a free end point and a singular control. Even more serious is the corresponding formulation for the *ongoing forest* case (with replanting immediately after each harvest) where the equations of state are differential-difference equations and the conventional theory of optimal control does not apply. For these reasons, only some partial results have been obtained for this problem, (Heaps and Neher, 1979).

The above formulation of the *blocked harvest* problem has another severe limitation. To be able to execute the optimal harvesting schedule, we need all the trees to be of the same age or to be able to log them in any order which may be prescribed by the optimal schedule. In practice, there may be requirements or regulations dictating the order in which trees in the forest are to be cut. It may be physically or legally necessary or economically prudent to cut trees in a certain order from a logging camp. In this chapter, we will take a different approach to the blocked harvest problems which removes this limitation as well as the mathematical difficulties mentioned earlier.

For simplicity, we consider only the situation where a single logging crew is to cut a (uniformly) narrow row of trees along a prescribed path from the logging camp winding through the entire forest. In that case, the position of any tree site can be described by the *arc length s* along the path to the site. For a continuous model of the forest, the discrete tree stands of the forest are smeared out over their respective assigned areas. Except for cases of sharp discontinuities in the initial age distribution, the actual distribution of tree age is replaced by a continuous approximation. The commercial value of the stumpage at different sites may be different because of a nonuniform age distribution, different growth conditions, market price differences at different cutting times, etc. (Wan and Anderson, 1983).

Let $T(s) \ge 0$ be the time at which the tree site at location s along the path is harvested in the future, $T = 0$ being now. The initial age distribution of the trees in the forest is denoted by $-T_0(s)$ with $T_0(s) \le 0$ being the germination time of current trees on site s. At cutting time, the tree stand at s will be $A(s) \equiv T(s) - T_0(s)$ years old. By construction, $T' \equiv dT/ds$ is nonnegative along the path with $T' = 0$ only if instantaneous harvesting is possible (with unlimited harvesting capacity), as T' is a measure of the time consumed in logging a particular tree site and $1/T'$ is therefore a measure of the harvesting rate h at location s. Let $p(s, T, A, T')ds$ and $c(s, T, A, T')ds$ be the commercial price and the harvesting (cutting, shipping, etc.) cost of the timber from the incremental strip $(s, s + ds)$ of the logging path. Evidently, the price per unit site harvested at location s and the harvesting cost per unit site at that location generally depend on tree age A and absolute time. The dependence on absolute time reflects in part the fluctuation of the lumber and labor markets. The possible dependence of p and c on location is not unexpected; trees may grow faster at one site than another, and they may be more difficult to log at some locations because of the geography and topography. Due to fixed setup costs and variable overload costs, harvesting cost per unit site may change with harvesting rate. If the logging company has any degree of monopolistic power, lumber price may also be affected by the rate of harvest.

15.6   The Once-and-for-All Forest with Ordered Site Access

The present value of the discounted future net revenue for the tree stumpage along an incremental path $(s, s + ds)$ is $e^{-\delta T(s)}(p - c)ds$. For a path so normalized that it is of unit length, the present value of the discounted future net revenue for the entire forest is given by

$$P \equiv \int_0^1 (p - c)e^{-\delta T}\, ds \qquad (15.18)$$

The management problem for the logging company is to choose a harvest schedule $T(s)$ for the forest so that this present value is a maximum. The maximization is subject to the constraint on the harvesting rate, which takes the form of $T' \geq \tau \; (\geq 0)$ in our model. At this point, we have effectively completed the formulation of our basic (purely) economic model. Many features not previously analyzed by the classical model can be incorporated by specifying p and c appropriately. Other more sophisticated features such as the environmental factor may also be added by suitable modifications of this basic model (see appendix). In our formulation, the harvesting rate is $F_0\, ds/dT$, where F_0 is the forest area to be cut. A change of the independent variable from s to T transforms (15.18) back to the expression for P used in the conventional model. (We have taken $F_0 = 1$ for convenience and will continue to do so.)

Since p and c may depend on T', we introduce a new control variable $u(s)$ by the defining equation (of state)

$$\frac{dT}{ds} = u \tag{15.19}$$

and write the present value of future net revenue P as

$$P = \int_0^1 e^{-\delta T} V(s, T, A, u)\, ds \tag{15.20}$$

where $V \equiv p - c$ and $A = T - T_0$. The Hamiltonian for this problem is

$$H \equiv V(s, T, A, u)e^{-\delta T} + \lambda u \tag{15.21}$$

In terms of H, the Euler differential equations for our problem are

$$\frac{d\lambda}{ds} = -\frac{\partial H}{\partial T} = -(V_T + V_A - \delta V)e^{-\delta T} \tag{15.22}$$

$$\frac{\partial H}{\partial u} = e^{-\delta T}\frac{\partial V}{\partial u} + \lambda = 0 \tag{15.23}$$

with $(\ )_y \equiv \partial(\ )/\partial y$. The Euler boundary conditions for the problem are

$$\lambda(0) = \lambda(1) = 0 \tag{15.24}$$

The optimal solution must also satisfy the inequality constraints

$$u \geq \tau (\geq 0) \qquad T(0) \geq 0 \tag{15.25}$$

By allowing τ to be positive, we include the possibility of an imposed maximum feasible harvesting rate. If there is no such imposed upper limit, then $\tau = 0$ and the first inequality in (15.25) simply reflects the fact that, in our model, the tree sites are ordered for the purpose of harvesting, as is usually the case in reality.

When the solution of the BVP defined by (15.19), (15.22), (15.23), and (15.24) also satisfies (15.25), we have an *interior solution* for our optimal control problem. In that case, (15.23) and (15.19) may be used to eliminate

λ and u from (15.22) and (15.24) to get a second-order ODE for T and one boundary condition for T at each end of the interval. This two-point boundary value problem may then be solved to get the optimal harvest time for different tree sites denoted by $T^*(s)$. Strictly speaking, $T^*(s)$ satisfies only the necessary condition for a maximum P. For brevity, we call it the *optimal policy*, with the understanding that optimality is still to be demonstrated. In some cases (e.g., when V is strictly concave in u and T), this is straightforward.

When the "interior solution" does not satisfy the bounded harvest rate condition $u \geq \tau$, it evidently requires that $u < \tau$ for the stationary point T^*. Given the properties of V (see Figure 15.5), our experience with classical optimization problems in calculus suggests that we get as close to the T^* as possible by taking

$$u = \tau \tag{15.26}$$

The Pontryagan Maximum principle confirms this rigorously (Bryson and Ho, 1969). In that case, the inequality constraint $u \geq \tau$ is said to be *binding* and the condition (15.26) replaces (15.23) in the boundary value problem for u, T, and λ. The resulting optimal program is called the *corner solution*. With $u = \tau$, Equation 15.19 can be integrated immediately to give

$$T(s) = \tau s + t_0 \tag{15.27}$$

where the constant of integration t_0 will be determined presently. Upon substituting (15.26) and (15.27) into (15.22), we get

$$\lambda(s) = -\int_0^s [(V_T + V_A - \delta V)e^{-\delta T}]\,ds \tag{15.28}$$

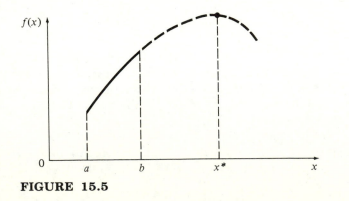

FIGURE 15.5

where the condition $\lambda(0) = 0$ has been used to eliminate a new constant of integration and the relations (15.19), (15.26), and (15.27) are to be used in the integrand. The remaining end condition $\lambda(1) = 0$ becomes

$$\int_0^1 [(V_T + V_A - \delta V)e^{-\delta T}]\,ds = 0 \tag{15.29}$$

and serves as a condition for the determination of the constant t_0. Again, (15.19), (15.26), and (15.27) are to be used in the integrand of (15.29).

It is also possible that the "interior solution" does not satisfy the constraint $T(0) \geq 0$, say $T(s) < 0$ for $s \leq s_0$. As in Section 15.1, we harvest as soon as we can by taking optimal policy to be the corner solution with $T(0) = 0$ and $T'(s) = \tau$ for $0 < s < s_0 \ (\leq 1)$. For $s_0 < s \leq 1$, the optimal policy is determined by the conditions (15.19), (15.22), (15.23), and $\lambda(1) = 0$ with a second auxiliary condition coming from the continuity of T at $s = s_0$. Optimality must still be verified by checking the appropriate concavity conditions or other sufficiency conditions (Bryson & Ho, 1969).

Once we have found the optimal harvest schedule $T^*(s)$, we may integrate (15.22) and use (15.24) to get

$$\int_0^1 (V_T^* + V_A^* - \delta V^*)e^{-\delta T^*}\,ds = 0 \tag{15.30}$$

where $(\quad)^*$ is $(\quad)$ evaluated at $T = T^*(s)$. We rewrite (15.30) as

$$\int_0^1 (V_T^* + V_A^*)e^{-\delta T^*}\,ds = \delta \int_0^1 V^* e^{-\delta T^*}\,ds \tag{15.30a}$$

For a small change in T^*, say ΔT, we have $\Delta V \simeq (dV^*/dT^*)\Delta T = (V_T^* + V_A^*)\Delta T$. We may therefore interpret $dV/dT = V_T + V_A$ as the *incremental* (or *marginal*) revenue resulting from an additional unit of time delay in harvesting the tree stumpage at location s. Even without an explicit solution of the optimal harvest schedule, we see from (15.30a) that

> **15F:** At maximum profit, the incremental gain in discounted net revenue from all sites in the forest by delaying harvesting one more unit of time equals the interest earned on the present value of net income from the total harvest.

15.7   Fisher's Rule and Nonuniform Initial Tree Age Distribution

In this section, we consider the classical situation of an age-dependent revenue function $V = V(A)$ and an unlimited harvesting capacity so that $\tau =$

0. This would allow a comparison with known results (Section 15.1) and the generation of some new results for forests with a nonuniform initial age distribution. For this case, we have $V_u = 0$ and from (15.23) $\lambda \equiv 0$ (which trivially satisfies the two end conditions) for an interior solution. The Euler differential equation (15.22) reduces to (15.2) and we recover the classical Fisher's rule for either a forest of uniform initial tree age distribution or a forest with no ordered site access requirements.

When the initial age distribution $T_0(s)$ is nonuniform and trees are to be cut in a certain order, the situation is more complicated. We distinguish and treat separately the three cases: (1) $T_0'(s) \geq 0$; (2) $T_0'(s) \leq 0$; and (3) T_0' changes sign over the interval $0 \leq s \leq 1$. We will assume $T_0(s)$ to be continuous. The case of $T_0(s)$ having simple jump discontinuities can also be treated with no difficulties (Wan and Anderson, 1983).

1. *A Nondecreasing $T_0(s)$:* The interior (Fisher) solution for this case is $T(s) = A_F + T_0(s)$ so that $T'(s) = T_0'(s) \geq 0$. The inequality constraint $u \geq 0$ is satisfied and $T^*(s) = A_F + T_0(s)$ holds for the whole forest unless $A_F + T_0(s) < 0$ for $0 \leq s \leq s_0 \ (\leq 1)$. When some trees are already too old, we do the best we can and take

$$T^*(s) = \begin{cases} 0 & (0 \leq s \leq s_0) \\ A_F + T_0(s) & (s_0 \leq s \leq 1) \end{cases} \qquad (15.31)$$

The two scenarios are shown in Figure 15.6.

2. *A Nonincreasing $T_0(s)$:* If $A_F + T_0(0) \leq 0$, then we must do the best we can by taking $T^*(s) \equiv 0$ for $0 \leq s \leq 1$. For $A_F + T_0(0) >$

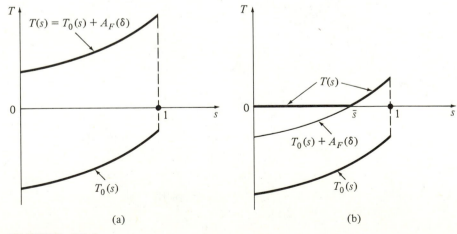

FIGURE 15.6

0, we have $[A_F + T_0(s)]' = T_0'(s) < 0$ for $0 \le s \le 1$, so the "interior solution" is not applicable. We must take $T^*(s) = \tau s + t_0 = t_0$ according to (15.27) as we have $\tau = 0$ for this section. The unknown constant t_0 is determined by (15.29) which simplifies for the present case to

$$\frac{\int_0^1 \dot{V}[t_0 - T_0(s)]\,ds}{\int_0^1 V[t_0 - T_0(s)]\,ds} = \delta \qquad (15.32)$$

The optimal policy is therefore to clear-cut the entire forest at t_0 which is the only root of (15.32) (see Figure 15.7a). If t_0 is negative, then we do the best we can and clear-cut now.

For cases with $T_0'(s) = 0$ over one or more segments of the logging path, as in Figure 15.7b, we still have $T^*(s) = t_0$ with t_0 determined by (15.32). This can be seen from the example of $T_0(s)$ in Figure 15.7b. For this case, we must have $T^*(s) = \bar{t}_0$ for at least the interval $s_l \le s \le s_r$ with $\bar{t}_0$ determined by a condition over (s_l, s_r) corresponding to (15.32) (as we have $T_0'(s) < 0$ there). Along $0 \le s \le s_l$, the optimal harvest schedule is $A_F + (-t_s)$ [see Figure (15.7b)]. However, this is not allowed because $A_F - t_s > \bar{t}_0$ violates the ordered site access constraint; so we compromise by harvesting a little sooner in $0 \le s \le s_l$ and a little later in $s_l \le s \le s_r$. Given $\bar{t}_0 < A_F + (-t_s)$, the optimal adjustment should be $T^*(s) = \hat{t}_0 > \bar{t}_0$ for $0 \le s \le s_r$. For $s_r \le s \le 1$, the optimal policy should be $A_F + (-t_b) < \bar{t}_0 < \hat{t}_0$ which violates the ordered site constraint; so we adjust again by harvesting a little later in $s_r \le s \le 1$

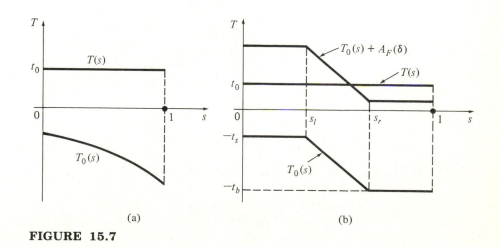

(a)   (b)

FIGURE 15.7

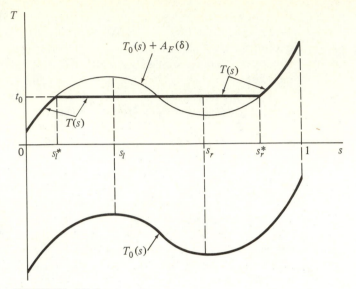

FIGURE 15.8

and a little sooner in $0 \leq s \leq s_r$. Since $\hat{t}_0 > A_F + (-t_b)$, it would not be optimal unless we make T^* continuous. This then gives $T^*(s) = t_0$ with t_0 determined by the necessary conditions for optimality summarized by (15.32).

3. *An Oscillating $T_0(s)$:* A simple example of this class of initial age distribution is given in Figure 15.8. For this example, $A_F + T_0(s)$ is less than $A_F + T_0(s_l)$ for $s_l < s \leq s_0$ and hence violates the ordered site access constraint. The optimal schedule is therefore a combination of interior and corner solutions in the form indicated by the heavy solid curve in Figure 15.8. It deviates from the interior solution over the stretch $s_l^* \leq s \leq s_r^* \leq s_0$ (with $s_l^* \leq s_l$ and $s_r^* \geq s_r$) where $T^*(s) = t_0$. The determination of t_0 as well as the locations s_l^* and s_r^* will be left as an exercise. Analyses of other more complicated $T_0(s)$ can also be found in the exercises and in Wan and Anderson (1983).

15.8   A Maximum Feasible Harvesting Rate

As we noted earlier, a logging company is usually faced with a maximum feasible harvesting rate corresponding to a positive lower bound τ on $T'(s)$. Whether the optimal harvesting policy is given by the interior solution now depends on whether the inequality constraint $T' \geq \tau > 0$ holds. Again, we limit ourselves in this section to the case $V = V(A)$, so that we have $\partial H/\partial u \equiv 0$ and $\lambda \equiv 0$. Thus, we have $T(s) = A_F + T_0(s)$ for the interior solution.

Consider first the case of a uniform initial age distribution so that $T'_0(s) \equiv 0$. For this case, the interior solution with $T'_0(s) = T'(s) \equiv 0$ violates the harvesting rate constraint; the optimal harvest schedule is therefore given by the corner solution (15.27), namely $T^*(s) = \tau s + t_0$ for $0 \leq s \leq 1$. For simplicity, we discuss only the case $t_0 \geq 0$, as the case $t_0 < 0$ can be analyzed in a way analogous to the same situation for $\tau = 0$ in Section 15.7. The constant t_0 is determined by (15.29) which (with $A = T - T_0 = \tau s + t_0 - T_0$) simplifies to read

$$\int_0^1 [\dot{V}(A) - \delta V(A)]e^{-\delta(\tau s + t_0)}\, ds = \int_0^1 \frac{d}{dA}[V(A)e^{-\delta A}]e^{-\delta T_0}\, ds$$

$$= \frac{1}{\tau e^{\delta T_0}}\left[V(A)e^{-\delta A} \right]_{A=t_0-T_0}^{\tau + t_0 - T_0} = 0$$

or

$$V(\tau + t_0 - T_0)e^{-\delta(\tau + t_0)} = V(t_0 - T_0)e^{-\delta t_0} \qquad (15.33)$$

Thus, for the more realistic case of limited harvesting capacity, the optimal harvest schedule is no longer Fisher's rule even for a uniform initial age distribution. Instead,

15G: The optimal schedule is to harvest at the maximum feasible rate $1/\tau$, starting at a time t_0 when the present value of net revenue of the first tree cut equals that of the last tree cut.

It is not difficult to see that we have $t_0 < A_F$ in order for trees at the far end to be logged nearer to their Fisher age.

When the initial age distribution of the forest is not uniform, the situation is more complicated but similar to the unconstrained-harvesting-rate case discussed in the last section. We again distinguish and treat separately the three cases: (1) $T'_0(s) \geq \tau$, (2) $T'_0(s) \leq \tau$, and (3) $T'_0(s) - \tau$ changes sign in $0 \leq s \leq 1$.

1. $T'_0(s) \geq \tau$: The optimal harvest schedule in this case is given by the interior solution $T^*(s) = T_0(s) + A_F$, with the same forest economic interpretation. The optimal time to harvest a particular tree is at its Fisher age.

2. $T'_0(s) \le \tau$: For this case, the inequality constraint on harvesting rate is binding throughout the solution domain and we have $T^*(s) = \tau s + t_0$ with t_0 chosen so that the integral condition (15.29) is satisfied. As such, the optimal harvest schedule for this case is qualitatively identical to a forest with a uniform initial age distribution. As in that case, the condition (15.29) may be written as

$$\int_0^1 \dot{V}[\tau s + t_0 - T_0(s)]e^{-\delta(\tau s + t_0)} \, ds$$

$$= \delta \int_0^1 V[\tau s + t_0 - T_0(s)]e^{-\delta(\tau s + t_0)} \, ds \qquad (15.34)$$

But unlike the case $T_0(s) \equiv$ constant, the integrals cannot be evaluated explicitly except for simple forms of $T_0(s)$.

3. $T'_0(s) - \tau$ *changes sign along the logging path:* It should be evident now that the solution for this class of $T_0(s)$ when there is a limited harvesting capacity is exactly the same as that for the $\tau = 0$ case. The corner solution here is the same as the results of Section 15.7 with $T_0(s)$ replaced by $T_0(s) - \tau s$.

Thus, we have effectively described the optimal harvest schedule for all possible continuous distributions of $T_0(s)$ for age-dependent revenue functions. A discussion of discontinuous initial age distributions can be found in Wan and Anderson (1983).

15.9 Harvesting Cost Varying with Harvesting Rate

Instead of c being a constant, we consider in this section unit harvesting cost functions which depend only on T' with the conventional U-shaped graph, that is, $c(T') > 0$ is convex in T' with a minimum at $T' = \tau_{min} > 0$ {so that $\min[c(T')] = c(\tau_{min})$}. This class of unit cost functions includes both the effect of a fixed cost component and an overload cost component. We limit ourselves here to cost functions which become unbounded as $u \equiv T'$ tends to infinity.

With $p = p(A) \equiv p(T - T_0)$ as before, we have from (15.23) and $V = p - c$ that $\partial H/\partial u = -e^{-\delta T}\dot{c}(u) + \lambda$, where $u = T'$. An interior solution of the optimal control problem requires

$$\lambda(s) = e^{-\delta T}\dot{c}(u) = e^{-\delta T(s)}\dot{c}[T'(s)] \qquad (15.35)$$

with the boundary conditions $\lambda(0) = \lambda(1) = 0$ satisfied by taking

$$T'(0) = T'(1) = \tau_{min} \qquad (15.36)$$

No other choice of harvesting rate at the two ends is possible, as c has a unique minimum point and $e^{-\delta T}$ never vanishes. Thus, for the type of harvesting cost functions considered in this section and for a price function depending only on tree age,

> **15H:**   The optimal harvest schedule should start and end with a harvesting rate corresponding to a minimum unit harvesting cost (provided of course $\tau_{\min} \geq \tau$).

For an interior solution, the optimal schedule itself is determined by the two-point BVP defined by (15.19), (15.22), (15.36) [or (15.24)], and (15.35) [or, upon solving for u,

$$u = T' = f(\lambda e^{\delta T}) \tag{15.37}$$

where f is the inverse of $\dot{c}$.] An exact solution for the special case of $c(u) = c_f(u) + c_0 + c_h u^{-1}$ is outlined in the exercises [see also Wan and Anderson (1983)].

When there is an upper limit to the harvesting capacity (a lower bound $\tau > 0$ on T'), the optimal harvest schedule depends on the sign of $T' - \tau$. It continues to be the interior solution if $T' \geq \tau$ for the entire forest. The situation is more complicated if $T' < \tau$ for some portion of the path. For example, if $\tau_{\min} < \tau$, then the inequality constraint $T' \geq \tau$ is binding for an initial segment of the logging path $0 \leq s \leq \bar{s}$, so that we have $T(s) = \tau s + t_0$ there and

$$\lambda(s) = -\int_0^s [\dot{p}(\tau s + t_0 - T_0) - \delta p(\tau s + t_0 - T_0) + \delta c(\tau)] e^{-\delta(\tau s + t_0)} \, ds \tag{15.38}$$

for $0 \leq s \leq \bar{s}$. However, for $\bar{s} \leq s \leq 1$, the condition (15.35) holds and the optimal harvest policy satisfies (15.19), (15.22), and (15.37) with $\lambda(1) = 0$. The two unknown parameters t_0 and $\bar{s}$ are determined by the continuity[1] of λ and T at the junction $s = \bar{s}$. A similar procedure for determining the optimal harvest schedule applies when $T' - \tau$ becomes negative in one or more segments of the logging path which may or may not include an end point. Of course, the firm has the option of increasing the harvesting capacity to the level of $1/\tau_{\min}$ to take advantage of the optimal strategy offered by the interior solution.

---

[1] With $c(u) \to \infty$ as $u \to \infty$, T must be continuous for all s in [0.1]; the continuity of λ follows.

15.10   The Ongoing Forest with Ordered Site Access

A harvested forest may be replanted for future lumber supply. Clearly, the longer the logging of the existing forest is delayed, the longer it takes to acquire revenue from future harvests. The significance of the opportunity cost associated with not logging sooner (than the Fisher age, say) was recognized by Faustmann, who first examined the optimal harvesting policy for an ongoing forest (with repeated harvesting and replanting, as in Section 15.3). However, more realistic models for problems in this area are needed. Analysis of harvesting policy for ongoing forests are meaningful only for long-term planning over a span of centuries, given that replanted trees have no net commercial value during the first few decades after germination. The fluctuation of price and cost with chronological time should be significant over such a planning period and should be included in a realistic mathematical model. On the other hand, the incorporation of such fluctuations in a model is certain to make the mathematical problems much less tractable, as we shall soon see. Therefore, it is important for us to seek as simple a mathematical formulation of the model problem as we possibly can. A formulation for an ongoing forest similar to that of Section 15.6 is even more attractive compared to other formulations proposed in recent years from this viewpoint [e.g., Heaps and Neher (1979); Davidson and Hellsten (1980)].

Within the framework of our ordered site access formulation, we let $T_k(s)$ be the time (measured from now) when the tree site at location s along the logging path is harvested during the kth harvest, $k = 1, 2, \ldots$. The initial age distribution of the trees in the forest is again denoted by $-T_0(s)$ where $T_0(s) \leq 0$ is the germination time distribution of the existing trees. The tree at location s will be $A_k(s) \equiv T_k(s) - T_{k-1}(s)$ years old when it is logged during the kth harvest. By construction, $T_k' \equiv dT_k/ds$ is nonnegative along the path with $T_k' = 0$ only if instantaneous harvesting is possible (given unlimited harvesting capacity), as T_k' is a measure of the time consumed in logging a particular tree site during the kth harvest (and $1/T_k'$ is therefore a measure of the harvesting rate h_k at location s for the kth harvest).

Similar to the case of "once-and-for-all forests," we let $p_k\,ds$ and $c_k\,ds$ be the commercial price and harvesting cost of the timber of the kth harvest over the incremental path strip $(s,\ s + ds)$. For reasons already explained, both p_k and c_k may vary with location s, logging time T_k, and tree age A_k, as well as current and previous harvest rates, $1/T_j'$, $j = 1, 2, \ldots, k$. The present value of the discounted future net revenue for the stumpage along an incremental path $(s,\ s + ds)$ from the kth harvest is $e^{-\delta_k T_k(s)}(p_k - c_k)\,ds$ where δ_k is the constant discount rate at the time of the kth harvest. The present value of the discounted future net revenue from the entire forest at the end of the Nth harvest is

$$P_N \equiv \sum_{k=1}^{N} \int_0^1 (p_k - c_k) e^{-\delta_k T_k} \, ds \qquad (15.39)$$

where N is ∞ if the forest is to be harvested repeatedly for the whole future. The management problem for the logging company is to choose a sequence of harvest schedules $\{ T_1, T_2, \ldots \}$ for the forest so that P_N is a maximum.

As in the once-and-for-all forest case, we introduce a new set of controls by the defining equations (of state)

$$T'_k \equiv u_k \qquad (k = 1, 2, \ldots) \qquad (15.40)$$

and write P_N as

$$P_N = \sum_{k=1}^{N} \int_0^1 e^{-\delta_k T_k} V_k(s, T_k, A_k, u_1, \ldots, u_k) \, ds \qquad (15.41)$$

where $V_k \equiv p_k - c_k$ is the net revenue per unit path length. The first-order necessary conditions for a maximum P_N may be expressed in terms of the Hamiltonian

$$H \equiv \sum_{k=1}^{N} e^{-\delta_k T_k} V_k + \lambda_k u_k \qquad (15.42)$$

The Euler differential equations for the problem are

$$\lambda'_k = -\frac{\partial H}{\partial T_k} = -\frac{\partial V_k}{\partial T_k} + \frac{\partial V_k}{\partial A_k} - \delta_k V_k e^{-\delta_k T_k} + \frac{\partial V_{k+1}}{\partial A_{k+1}} e^{-\delta_{k+1} T_{k+1}}$$

$$(15.43)$$

$$\frac{\partial H}{\partial u_k} = \lambda_k + \sum_{j=1}^{N} e^{-\delta_j T_j} \frac{\partial V_j}{\partial u_k} = 0 \qquad (15.44)$$

for $k = 1, 2, \ldots, N$ with $V_{N+1} \equiv 0$. The Euler boundary conditions are

$$\lambda_k(0) = \lambda_k(1) = 0 \qquad (k = 1, 2, \ldots) \qquad (15.45)$$

The optimization is subject to the inequality constraints on harvesting rates,

$$u_k \geq \tau_k \, (\geq 0) \qquad (k = 1, 2, \ldots) \qquad (15.46)$$

the "inaccessible past" constraint $T_k(0) \geq 0$ and possibly also $T_k(0) > T_{k-1}(1)$. When all the inequality constraints are satisfied by the solution of the BVP defined by (15.40), (15.43), (15.44), and (15.45), we have a (feasible) interior solution for our optimal control problem. In order to focus our attention on the main issues of interest here, we assume that the various convexity and concavity conditions are satisfied so that the necessary conditions for optimality are also sufficient.

At the other extreme, when all of the inequality constraints on u_k in (15.46) are not satisfied by the solution of the BVP, we must do the best we can by taking $u_k = \tau_k$ so that

$$T_k(s) = \tau_k s + t_k \tag{15.47}$$

where the constants of integration t_k, $k = 1, 2, \ldots$, are to be determined by (15.43) and (15.45).

Intermediate situations with some of the inequality constraints (on u_k) in (15.46) not met by the solution of the BVP are also possible and can be treated in a straightforward manner. Regardless of whether one or more of (15.46) are binding, we get from (15.43) and the boundary conditions $\lambda_k(0) = 0$

$$\lambda_k(s) = -\int_0^s \left[\left(\frac{\partial V_k}{\partial A_k} + \frac{\partial V_k}{\partial T_k} - \delta_k V_k \right) e^{-\delta_k T_k} - \frac{\partial V_{k+1}}{\partial A_{k+1}} e^{-\delta_{k+1} T_{k+1}} \right] ds$$

$$(k = 1, 2, \ldots) \tag{15.48}$$

The remaining boundary conditions $\lambda_k(1) = 0$ give

$$\int_0^1 \left[\left(\frac{\partial V_k}{\partial A_k} + \frac{\partial V_k}{\partial T_k} \right) e^{-\delta_k T_k} - \frac{\partial V_{k+1}}{\partial A_{k+1}} e^{-\delta_{k+1} T_{k+1}} \right] ds = \delta_k \int_0^1 V_k e^{-\delta_k T_k} ds$$

$$\tag{15.49}$$

The economic content of (15.49) may be stated as follows:

15I: At maximum profit, the discounted additional revenue from delaying each harvest by a unit time equals the *opportunity cost* consisting of the sum of discounted additional revenue of the replanted forest from the same delay and the "interest" earned on the discounted revenue of the harvested forest.

15.11 A Finite Sequence of Age-Dependent Net Revenue Functions

The rather complicated general results of the last section simplify considerably if $V_k \equiv p_k - c_k$, $k = 1, 2, 3, \ldots$, are monotone increasing concave functions of tree age $A_k \equiv T_k - T_{k-1}$ only. In that case (15.44) gives

$$\lambda_k \equiv 0 \qquad (k = 1, 2, 3, \dots, N) \tag{15.50}$$

which satisfy the boundary conditions (15.45), and the Euler differential equations in (15.43) become an algebraic system for $\{A_k\}$. For simplicity, let $\delta_k = \delta$ for all $k = 1, 2, \dots, N$ so that this system may be written as

$$\dot{V}_k(A_k) - \delta V_k(A_k) = \dot{V}_{k+1}(A_{k+1})e^{-\delta A_{k+1}} \qquad (k = 1, 2, \dots, N) \tag{15.51}$$

with $V_{N+1}(\cdot) \equiv 0$.

The system (15.51) may be solved by noting that the Nth equation,

$$\dot{V}_N(A_N) - \delta V_N(A_N) = 0 \tag{15.52}$$

involves only one unknown A_N and its unique solution is the well-known Fisher age $\alpha_N(\delta) \equiv A_F(\delta)$. Hence, we have

$$A_N(s) \equiv T_N(s) - T_{N-1}(s) = \alpha_N(\delta) \qquad \frac{\dot{V}_N[\alpha_N(\delta)]}{V_N[\alpha_N(\delta)]} = \delta \tag{15.53}$$

Having determined the optimal harvest age for the last harvest, $A_N(s)$, the $(N-1)$th equation

$$\dot{V}_{N-1}(A_{N-1}) - \delta V_{N-1}(A_{N-1}) = \dot{V}_N(\alpha_N)e^{-\delta \alpha_N} \tag{15.54}$$

involves only one unknown, and may be solved to get the unique solution $\alpha^*_{N-1}(\delta)$ for $A_{N-1}(s)$. The process is repeated to get $A_{N-2} = \alpha^*_{N-2}$, $A_{N-3} = \alpha^*_{N-3}, \dots$, with the first equation giving $A_1(s) = \alpha^*_1(\delta)$. These results for tree age distributions during the different harvests are then used to determine the optimal harvest schedules $\{T_k(s)\}$:

$$T_1(s) = T_0(s) + \alpha^*_1(\delta)$$

$$T_2(s) = T_1(s) + \alpha^*_2(\delta) = T_0(s) + \alpha^*_1(\delta) + \alpha^*_2(\delta)$$

$$T_k(s) = T_{k-1}(s) + \alpha^*_k(\delta) = T_0(s) + \sum_{j=1}^{k} \alpha^*_j(\delta) \qquad (k = 3, \dots, N-1)$$

$$T_N(s) = T_{N-1}(s) + \alpha_N(\delta) = T_0(s) + \sum_{j=1}^{N-1} \alpha^*_j(\delta) + \alpha_N(\delta) \tag{15.55}$$

For the case of a uniform initial age distribution and unlimited harvesting capacity so that $\tau_k = 0$, $k = 1, 2, 3, \dots, N$, the optimal harvesting policy is to clear-cut the entire forest instantaneously when the trees reach the age $\alpha^*_k(\delta)$ during the kth harvest ($k = 1, 2, \dots, N-1$) and when they reach the Fisher age $\alpha_N(\delta) \equiv A_F(\delta)$ for the last harvest (as shown in Figure 15.9a). [For simplicity, we will discuss only the case $T_0(s) + \alpha_1^*(\delta) \geq 0$.] This policy is optimal in view of the concavity of V_k. For a nonuniform $T_0(s)$ with $T_0'(s) \geq 0$, the harvesting policy is to cut trees when they reach the age α_k^* for the kth harvest and the Fisher age α_N for the last harvest (as shown in Figure 15.9b).

If $T_0'(s) \leq 0$, the harvest schedule $T_0(s) + \alpha_1^*(\delta)$ violates the inequality constraint $u_1 = T_1' \geq 0$ and we must replace (15.44) by $T_1(s) = t_1$, with t_1 determined by

$$\lambda_1' = -\{ \dot{V}_1[t_1 - T_0(s)] - \delta V_1[t_1 - T_0(s)] - \dot{V}_2(\alpha_2^*)e^{-\delta\dot{\alpha}_2} \}e^{-\delta t_1}$$

$$(15.56)$$

and $\lambda_1(0) = \lambda_1(1) = 0$. The second harvest has a uniform initial age distribution, and the results of (15.55) for $k \geq 2$ now apply from there on (see Figure 15.10) so that [after $T_1(s) = t_1$]

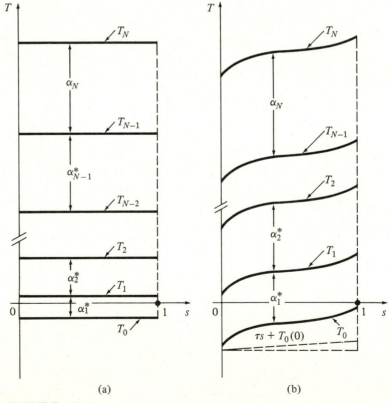

(a)   (b)

FIGURE 15.9

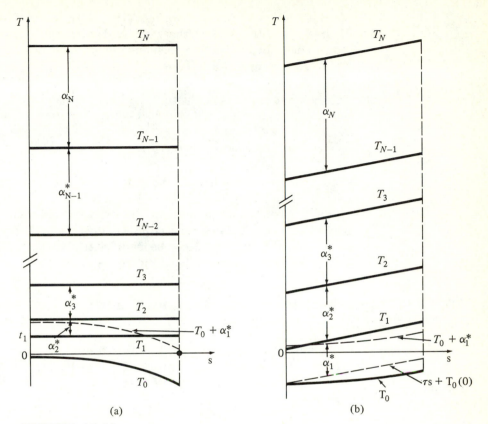

(a)   (b)

FIGURE 15.10

$$T_k(s) = T_1(s) + \sum_{j=2}^{k} \alpha_j^*(\delta) \qquad (k = 2, \ldots, N-1)$$

$$(15.57)$$

$$T_N(s) = T_1(s) + \sum_{j=2}^{N-1} \alpha_j^*(\delta) + \alpha_N(\delta)$$

If $T_0'(s)$ changes sign in the interval $(0,1)$, the schedule $T_0(s)$ $+ \alpha_1^*(\delta)$ again violates the inequality constraint $T_1' \geq 0$ in some segments of the interval. The optimal policy $T_1(s)$ for the first harvest is then determined by a procedure outlined in Section 15.7. This optimal first harvest schedule serves as the germination time for the second harvest. The results of (15.57) for $k \geq 2$ now apply as $T_1' \geq 0$ in $(0, 1)$.

The determination of the optimal harvest schedule for the limited-harvesting-capacity case does not differ substantially from the case of un-limited harvesting capacity discussed above. Assume $\tau_k = \tau$, $k = 1, 2, \ldots,$

N for simplicity. The results of (15.55) again apply if $T_0'(s) \geq \tau > 0$ (see Figure 15.9b with $\tau > 0$). Otherwise, we determine $T_1(s)$ according to procedures described in Section 15.7, and apply the results of (15.57) to T_k for $k \geq 2$ (see Figure 15.10b). Thus

> **15J:** An ongoing forest with a uniform age distribution and unlimited harvesting capacity should be clear-cut instantaneously at regular intervals, while a limited harvesting capacity should give for each harvest a sustained yield over a period of time.

Should this period last beyond the start of the next harvest in the schedules given by (15.57), say $T_k(1) = T_0(1) + \sum_{j=1}^k \alpha_j^*(\delta) > T_{k+1}(0)$, an adjustment for the two contiguous periods (and possibly beyond them) can be worked out to meet the stipulated constraints (if any). Note that both the solution process and the conclusions can be modified to allow for a different discount rate δ_k during each interval between harvests.

15.12 An Infinite Harvesting Sequence and Faustmann's Rotation

To harvest repeatedly through the entire future, we take $N = \infty$. Consider first the conventional problem with $V_k(A) \equiv V(A)$, and $\delta_k = \delta$, $k = 1, 2,$ $3, \ldots$. The system (15.51) simplifies to

$$\dot{V}(A_k) - \delta V(A_k) = \dot{V}(A_{k+1})e^{-\delta A_{k+1}} \qquad (k = 1, 2, \ldots) \qquad (15.58)$$

A solution of this system of an infinite number of coupled equations is known to be the Faustmann rotation period $R_F(\delta)$, that is,

$$A_k = R_F(\delta) \qquad (k = 1, 2, 3, \ldots) \qquad (15.59)$$

where $R_F(\delta)$ is a root of (15.10). This solution is unique if V is a monotone increasing concave function of its argument. Thus, the optimal policy for the case of no constraint on the harvesting rate is to cut a tree when it reaches its Faustmann age R_F (or immediately if it is already older).

It is important to note that, on the basis of the results for a finite sequence of N harvests, the infinitely many coupled equations (15.58) are to be solved simultaneously as a system and not as recursive relations starting with some A_1. In fact, unless A_1 is taken to be R_F, the quantities A_2, A_3, etc., obtained recursively from (15.58) either decrease monotonically without bound if $A_1 < R_F$ or increase monotonically without bound if $A_1 > R_F$, as found in Heaps and Neher (1979). Neither situation could be optimal. For the decreasing sequence of $\{A_k\}$, the trees in all later harvests are logged immediately after replanting. For the increasing sequence, the growth rate (measured in dollars) of trees of later harvests will eventually be less than the "interest rate" δ. We can do better in both cases. An immediate consequence of the above observation is the *instability* of the optimal schedule.

> **15K:**   If the trees cut during the kth harvest are not at their Faustmann age, the subsequent harvest schedules obtained from the schedule (15.58) will not be the optimal Faustmann rotation and will in fact diverge from it. (They actually lead to a negative net-profit policy.)

It is important to emphasize that the determination of the optimal harvest policy and the stability of this policy are two distinctly different issues. We saw from the last section that for the optimal policy in the case of a finite N, the quantities $\{A_1, A_2, \ldots\}$ are determined recursively backward starting from the last harvest schedule T_N. We will presently show that the solution for finite N tends to Faustmann rotation as $N \to \infty$.

For a finite harvesting sequence, we have $V_{N+1}(A) \equiv 0$; the last (Nth) harvest should therefore take place when the trees reach their Fisher age, so that $A_N = \alpha_N(\delta) \equiv A_F(\delta)$ with

$$\dot{V}[A_F(\delta)] = \delta V[A_F(\delta)] \qquad (15.60)$$

In other words, $A_F(\delta)$ is located by the intersection of the graph of the monotone increasing concave function $\delta V(A)$ and the graph of the positive,

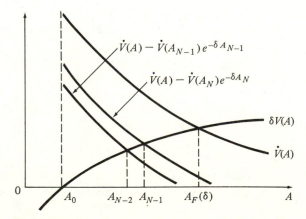

FIGURE 15.11

monotone decreasing convex function $\dot{V}(A)$ (see Figure 15.11). The optimal $(N-1)$th harvest schedule requires A_{N-1} to satisfy [see (15.58)]

$$\dot{V}(A_{N-1}) - \dot{V}(A_N)e^{-\delta A_N} = \delta V(A_{N-1}) \tag{15.61}$$

That is, A_{N-1} is located by the intersection of the graph of $\dot{V}(A)$, translated downward by a known distance $\dot{V}(A_N)e^{-\delta A_N}$, and the graph of $\delta V(A)$ (see Figure 15.11). Similarly A_k is located by the intersection of $\delta V(A)$ and a vertically translated $\dot{V}(A)$, with the amount of downward translation of $\dot{V}(A)$ depending on a previously determined A_{k+1}, namely, $\dot{V}(A_{k+1})e^{-\delta A_{k+1}}$. Keep in mind that $\dot{V}(A)e^{-\delta A}$ is a decreasing function of A for $A < A_F(\delta)$ and that $V(A) < 0$ for $A < A_0$. Clearly, we have $A_{N-1} < A_N \equiv A(\delta)$ and hence $\dot{V}(A_{N-1})e^{-\delta A_{N-1}} > \dot{V}(A_N)e^{-\delta A_N}$. It follows that $A_{N-2} < A_{N-1}$. Repeat the argument to get $A_k < A_{k+1}$ for $k = N-3, N-4, \ldots, 2, 1$ with $A_k > A_0$ since $\dot{V}(A_{k+1})e^{-\delta A_{k+1}} < \dot{V}(A_{k+1}) < \dot{V}(A_k)$ (see Figure 15.11). Thus, A_{N-1}, $A_{N-2}, \ldots, A_2, A_1$ form a monotone decreasing sequence inside the interval $(A_0, A_F(\delta))$, where A_0 is the unique zero crossing of $V(A)$. Thus, starting with $A_N = A_F(\delta)$, the relation (15.58) generates a monotone decreasing sequence $\{A_N, A_{N-1}, A_{N-2}, \ldots, A_{k+1}, A_k, A_{k-1}, \ldots, A_2, A_1\}$, which is bounded below by A_0 and bounded above by $A_F(\delta)$. Therefore, we have the following theorem and its important corollary:

15L: As $N \to \infty$, the harvest age sequence $\{A_N, \ldots, A_1\}$ generated by (15.58) with $A_N = A_F(\delta)$ tends to a limit $\hat{A}$, that is, $A_1 \to \hat{A}$ as $N \to \infty$.

To prove this theorem, let $a_k \equiv A_{N-k+1}$, $k = 1, 2, \ldots$. Then $\{a_k\}$ is a monotone decreasing sequence bounded from below by A_0 and therefore has a limit, denoted by $\hat{A}$. Given any $\epsilon > 0$ however small, there is always an $M < N$, sufficiently large depending on ϵ, for which $|a_k - \hat{A}| < \epsilon$ for $k > M$. It follows that for all $N > M$, $|A_j - \hat{A}| < \epsilon$ for all $j < N - M + 1$.

The following corollary is a consequence of the above theorem:

15M: If $\delta V(A) = \dot{V}(A)$ has a unique solution, the limit $\hat{A}$ is unique and identical to the Faustmann rotation $R_F(\delta)$.

Very briefly, we have, for a sufficiently large N, $A_{k+1} \simeq A_k \simeq \hat{A}$ for $k < N - M + 1$, so that (15.58) may be written as

$$\dot{V}(\hat{A})(1 - e^{-\delta \hat{A}}) = \delta V(\hat{A}) + \text{error} \tag{15.62}$$

where the error term can be made as small as we wish. Hence we have, in the limit as $N \to \infty$, $\hat{A} = R_F(\delta)$ with the Faustmann age R_F given by the solution of (15.10).

With the above corollary, we have reproduced the classical solution for the forest rotation for an ongoing forest. This policy applies to uniform or nonuniform initial distribution of tree ages as long as the constraint on the harvesting rate and the requirement of ordered access to tree sites are not violated. Otherwise, a solution process similar to that described in earlier sections for corner solutions applies.

15.13   Changing Net Revenue Functions

With decades separating two consecutive harvests, it may be unrealistic for the net revenue functions $V_k \equiv p_k - c_k$ to remain the same for all harvests. With technical progress, cheaper and/or more attractive substitutes will continue to reduce the demand for lumber and thereby its commercial value. On the other hand, new usages of timber may create a significant increase in demand in the future. In short, there is really no way for any forest manager to know for sure what V_k (in constant dollars) will be a century from now, not to mention the entire future. Under the circumstances, a reasonable approach by the logging company would be to plan only for the period with an accurate estimate of the net revenue for each of the N harvests and to assign a residual value to the forest land beyond the Nth harvest. In that case, the result of Section 15.10 (suitably modified if there is a nonzero residual value) applies.

In view of the instability of the optimal policy for time-invariant net revenue functions (see Sections 15.11 and 15.12), it would be prudent not to plan beyond the period with a reasonably definite net revenue structure. This often means planning only for a few harvests with the same net revenue function. Fortunately, the convergence of $\{\alpha_k^*(\delta)\}$ to the Faustmann rotation is extremely rapid in all cases tested. For example, we have for

$$V(A) = 950 - 1500e^{-A/60} \tag{15.63}$$

the optimal harvest schedules given in Table 15.1 for several values of the discount rate δ in the interval $(0.05, 0.1)$.

Should a time-varying net revenue structure be specified (known or imposed) for an infinite sequence of harvests, the optimal harvest policy is determined by the solution of the simultaneous equations (15.51) with $N = \infty$. For example, with

$$V_k(A) = (1 + \gamma)^k V(A) \qquad k = 1, 2, 3, \ldots \tag{15.64}$$

where γ is a known constant, an exact solution of the infinite system (15.51) is a modified Faustmann rotation $A_k = A_\gamma(\delta)$, $k = 1, 2, 3, \ldots$, where $A_\gamma(\delta)$ is the solution of (see exercises)

$$\frac{\dot{V}(A_\gamma)}{V(A_\gamma)} = \frac{\delta}{1 - (1 + \gamma)e^{-\delta A}} \tag{15.65}$$

TABLE 15.1
Optimal Tree Ages $\{\alpha_k^*\}$ for N Harvests and for $V(A) = 950 - 1500e^{-A/60}$

α_k^{δ}	0.05	0.06	0.07	0.08	0.09	0.10
$\alpha_N \equiv A_F(\delta)$	44.6664	42.1129	40.2200	38.7600	37.5994	36.6545
α_{N-1}^*	43.0800	41.0794	39.5330	38.2962	37.2823	36.4356
α_{N-2}^*	42.9056	40.9950	39.4911	38.2750	37.2715	36.4299
α_{N-3}^*	42.8853	40.9878	39.4884	38.2740	37.2711	36.4298
α_{N-4}^*	42.8829	40.9872	39.4883	38.2739		
α_{N-5}^*	42.8827	40.9871				
α_{N-6}^*	42.8826	$\vdots$	$\vdots$	$\vdots$	$\vdots$	$\vdots$
$\vdots$	$\vdots$					
$R_F(\delta)$	42.8826	40.9871	39.4883	38.2739	37.2711	36.4298

It follows that we have $A_\gamma(\delta) \le R_F(\delta)$ since future harvests now yield higher net return.

EXERCISES

1. $$V(A) \equiv p - c = V_0(1 - \sigma e^{-A/\bar{A}}) \qquad A \equiv \text{tree age}$$

   (a) Find the Fisher age A_F for the above V in terms of V_0, σ, and $\bar{A}$.
   (b) Find $T(s)$ for $T_0(s) \equiv 0$ and $u \ge \tau > 0$.

2. $$V(A) \equiv p - c = 950 - 1500e^{-A/60}$$

   (a) Find the Fisher age.
   (b) For a finite sequence of N harvests (of a forest) with the possibility of instantaneous clear-cutting, show that the optimal harvest schedule for the last four harvests (in an ordered site access formulation where the constraints on the control u and the state variable T are not binding) is $\{A_N = 36.6545\ldots, A_{N-1} = 36.4356\ldots, A_{N-2} = 36.4299\ldots, A_{N-3} = 36.4298\ldots\}$ where A_k is the tree age at the time of the kth harvest and $\delta = 0.1$.

3. (a) Suppose the lumber value function varies from harvest to harvest in the form $V_k(A) = (1 + \gamma)^k V(A)$, $k = 1, 2, \ldots$, where γ is a positive constant and $V(A)$ is a (unimodal) function like those

given in Exercises 1 and 2. Show that the necessary conditions for the optimal harvest schedule of an infinite harvesting sequence (when constraints on u and T are not binding) is satisfied by a modified Faustmann rotation $\hat{R}_F(\delta, \gamma)$ which is the root of

$$\frac{\dot{V}(A)}{V(A)} = \frac{\delta}{1 - (1 + \gamma)e^{-\delta A}}$$

(b) Find the modified Faustmann rotation $\hat{R}_F$ for $V(A) = 950 - 1500e^{-A/60}$ with $\delta = 0.1$ and $\gamma = 0.05$.

(c) For $\gamma > 0$, is $\hat{R}_F$ greater or less than R_F?

4. For a once-and-for-all forest with a nonuniform initial age distribution

$$T_0(s) = \begin{cases} -30(1 - s) & (0 \le s \le s_i) \\[2mm] -30(1 - s_i)\dfrac{s}{s_i} & (s_i \le s \le 1) \end{cases} \qquad (s_i \ge \tfrac{1}{2})$$

find the optimal harvest schedule if $V(A)$ is as given in Exercise 2 and $\delta = 0.1$. (You may take $s_i = \tfrac{1}{2}$ if you wish.)

5.
$$V = p(A) - c(u) \qquad c = c_f u + c_0 + \frac{c_l}{u}$$

(a) Show that the optimal harvest schedule $T(s)$ is the solution of

$$T'' + \delta(T')^2 = \frac{1}{2c_l}(T')^3\{\delta[p(T - T_0) - c_0] - \dot{p}(T - T_0)\}$$

$$T'(0) = T'(1) = u_{\min} \equiv \sqrt{\frac{c_l}{c_f}}$$

where $(\ )' \equiv d(\ )/ds$ and $(\dot{\ })$ indicates differentiation with respect to the argument of $(\ )$, assuming all constraints are not binding, that is, the interior solution is effective.

(b) For $T_0 = $ constant, show that a first integral of the above boundary-value problem (BVP) is

$$2c_l\delta s + 2c_l\left(\frac{1}{u_{\min}} - \frac{1}{T'}\right)$$

$$= p(T(0) - T_0) - p(T(s) - T_0) + \delta \int_{T(0)}^{T} [p(\zeta - T_0) - c_0]d\zeta$$

(c)   Show by the boundary condition $T'(1) = u_{\min}$ that

$$\frac{\left[p(T(s) - T)\right]_{T(0)}^{T(1)}}{\int_{T(0)}^{T(1)} [p(\zeta - T_0) - c_0]d\zeta - 2c_l} = \delta$$

which is a condition relating $T(0)$ and $T(1)$.

(d)   Obtain the solution of the BVP for $T(s)$ defined in part (a) in the form of $s(T)$:

$$(2\delta c_l)s = \left[\frac{2c_l}{u_{\min}} - p(T(0) - T_0) + c_0\right][e^{\delta(T-T(0))} - 1]$$

$$+ \delta \int_{T(0)}^{T} [p(\zeta - T_0) - c_0]d\zeta$$

(e)   Show that the constant $T(0)$ is determined by the condition

$$2c_l\delta = \left[\frac{2c_l}{u_{\min}} - p(T(0) - T_0) + c_0\right][e^{\delta(T(1)-T(0))} - 1]$$

$$+ \delta[P(T(1) - T_0) - P(T(0) - T_0)]$$

with

$$P_0(x) \equiv \int_0^x [P(t) - c_0]dt$$

and with $T(1)$ given in terms of $T(0)$ by the relation in part (c).

Appendix:

The Environmental Cost

Except for a brief digression in Section 15.2, this chapter has been concerned exclusively with the net profit from the sale of timber for the forest owner. The treatment in no way reflects a lack of substantive and interesting mathematical problems related to social issues of forest harvesting. One important social issue is the environmental cost of logging the forest, particularly in the case of no replanting. We may include this in our model by adding an opportunity cost term $\psi(T(0))$ to (15.18) to get

$$P = -\psi(T(0)) + \int_0^1 e^{-\delta T}(p - c)\,ds \qquad (15A.1)$$

where for a genuine opportunity cost of alternative land use, we may take $\psi(t)$ as given in (15.5a). Instead, we take $\psi(t)$ in the form given by (15.6a) as we are interested in the environmental cost of clear-cutting. In that case, all the first-order necessary conditions except $\lambda(0) = 0$ for the case $\psi \equiv 0$ continues to hold and the Euler boundary condition at $s = 0$ is replaced by

$$\lambda(0) = \dot{\psi}(T(0)) = -2e^{-\delta T(0)}I(T(0)) \qquad (15A.2)$$

As long as V depends on u, we again have a two-point boundary problem as in Section 15.6. In particular, we have from (15.22) and (15A.2)

$$\lambda(s) = \dot{\psi}(T(0)) - \int_0^s (V_T + V_A - \delta V)e^{-\delta \tau}\,ds \qquad (15A.3)$$

with the second Euler boundary condition $\lambda(1) = 0$ giving

$$\int_0^1 (V_T + V_A)e^{-\delta T}\,ds = \dot{\psi}(T(0)) + \delta \int_0^1 V e^{-\delta T}\,ds \qquad (15A.4)$$

or

$$\int_0^1 (V_T + V_A)e^{-\delta T}\,ds = -2e^{-\delta T(0)}I(T(0)) + \delta \int_0^1 V e^{-\delta T}\,ds \qquad (15A.5)$$

Even without an explicit solution of the optimal harvest schedule, we see from (15A.5) that

15N: At maximum profit, the incremental gain in discounted net revenue of all sites in the forest from delaying harvesting by one time unit equals the interest earned on the present value of the net income from the total harvest according to the optimal schedule minus twice the benefit of a better environment with the uncut forest per unit time (measured in dollars) at the time of the harvest.

The interior or corner solution of the two-point boundary-value problem proceeds as in the case of $\psi \equiv 0$.

For the case $V_u \equiv 0$, the problem with a nonvanishing opportunity cost term is more complex. We have from (15.23) $\lambda(s) \equiv 0$ which does not satisfy the inhomogeneous transversality condition (15A.2) unless $I(T(0)) = 0$

which generally will not be the case. We need then what is called a *singular* solution for our optimal control problem.

As $\lambda(0) = -2I(T(0))e^{-\delta T(0)} < 0$ and $\lambda(s)$ is continuous, we have $\lambda(s) < 0, 0 \leq s \leq \bar{s}$ for some $\bar{s} > 0$. In that range, we must have $u \equiv 0$ in order that H be a maximum. It follows that

$$T(s) = T(0) \quad \text{and} \quad \lambda' = -e^{-\delta T}(V_A + V_T - \delta V) \equiv \phi(T(0)) \quad (0 \leq s \leq \bar{s})$$

$$(15A.6)$$

and therewith $\lambda = \lambda(0) + \phi(T(0))s = -2I(T(0))e^{-\delta T(0)} + \phi(T(0))s$ for $0 \leq s \leq \bar{s}$. For the continuity of $\lambda(s)$, we take

$$\bar{s} = \frac{2I(T(0))e^{-\delta T(0)}}{\phi(T(0))} \tag{15A.7}$$

so that $\lambda(\bar{s}) = 0$. We can now take $\lambda(s) \equiv 0$ for $\bar{s} \leq s \leq 1$. From (15.22), we get

$$(V_T + V_A - \delta V)e^{-\delta T} = 0 \qquad (\bar{s} \leq s \leq 1) \tag{15A.8}$$

or

$$\frac{V_T + V_A}{V} = \delta$$

which gives $T = T^*(s; T_0, \delta)$. If V does not depend on s explicitly, this relation determines the constant $T(s) \equiv T(0)$.

Bibliography

Abramowitz, M., and I. A. Stegun, *Handbook of Mathematical Functions,* AMS 55, National Bureau of Standards, U.S. Dept. of Commerce, 1965.

Arnott, R. J., and J. G. MacKinnon, "Market and Shadow Land Rents with Congestion," *Am. Econ. Rev., 68,* 1978, 588–600.

Aronson, D. G., and H. F. Weinberger, "Nonlinear Diffusion in Population in Mathematics," vol. 446, *Partial Differential Equations and Related Topics,* Springer, Berlin, 1975, pp. 5–49.

Ascher, U., J. Christiansen, and R. D. Russell, "A Collocation Solver for Mixed Order Systems of Boundary Value Problems," *Math. Comp., 33,* 1979, 659–679.

Ascher, U., and F. Y. M. Wan, "Numerical Solution for Maximum Sustainable Consumption Growth with a Multi-Grade Exhaustible Resource," *SIAM J. Sci. Statis. Computing, 1,* 1980, 160–172.

Ashton, W. D., *The Theory of Road Traffic Flow,* Methuen, London, 1966.

Barenblatt, G. I., *Similarity, Self-Similarity and Intermediate Asymptotics,* Consultant Bureau (Div. of Plenum), New York, 1979.

Bellman, R., *Perturbation Techniques in Mathematics, Physics and Engineering,* Holt, Rinehart & Winston, New York, 1966.

Bender, E. A., *An Introduction to Mathematical Modeling,* John Wiley & Sons (Wiley-Interscience), New York, 1978.

Bluman, G. W., and J. D. Cole, *Similarity Methods for Differential Equations,* Springer-Verlag, Berlin, 1974.

Boyce, W. E., and R. C. DiPrima, *Elementary Differential Equations and Boundary Value Problems,* 3rd Ed., John Wiley & Sons, New York, 1976.

Bradley, R., R. D. Gibson, and M. Cross, *Case Studies in Mathematical Modeling,* John Wiley & Sons, New York, 1981.

Bryson, A., and Y. C. Ho, *Applied Optimal Control,* Ginn, Lexington, MA, 1969.

Burghes, D. N., and A. Graham, *Introduction to Control Theory Including Optimal Control,* John Wiley & Sons, New York, 1980.

Burghes, D. N., and M. S. Borrie, *Modelling with Differential Equations,* Ellis Horwood Ltd. Publisher, Chichester, distributed by Halsted Press (Division of John Wiley & Sons), 1981.

Burmeister, E., and A. R. Dobell, *Mathematical Theories of Economic Growth,* Macmillan, New York, 1970.

Chandler, R. E., R. Herman, and E. W. Montroll, "Traffic Dynamics: Studies in Car Following," *Oper. Res., 6,* 1958, 165–184.

Clark, C. W., *Mathematical Bio-Economics,* John Wiley & Sons, New York, 1976.

Colley, S. J., "The Tumbling Box," *Amer. Math. Monthly 94,* 1987, 63–68.

Cross, M., and A. D. Moscardini, *Learning the Arts of Mathematical Modeling,* Ellis Horwood Ltd./Halsted Press (Division of John Wiley & Sons), New York, 1985.

Dasgupta, P. S., and G. M. Heal, *Economic Theory and Exhaustible Resources,* Nisbet-Cambridge Univ. Press, Cambridge, U.K., 1979.

Davidson, R., and M. Hellsten, "Optimal Forest Rotation with Costly Planting and Harvesting," presented at Fifth Canadian Conference on Economic Theory in Vancouver, B.C., May 1980.

Dym, C. L., and E. S. Ivey, *Principles of Mathematical Modeling,* Academic Press, San Diego, CA, 1980.

Edelstein-Keshet, *Mathematical Models in Biology,* Random House, New York, 1988.

Faustmann, M., "Berechnung des Werthes, welchen Weldboden sowie nach nicht haubare Holzbestande für die Weldwirtschaft besitzen," *Allgemeine Forst und Jagd Zeitung, 25,* 1849, 441.

————, "On the Determination of the Value Which Forest Land and Immature Stands Pose for Forestry," *Martin Faustmann and the Evolution of Discounted Cash Flow,* M. Gane (ed.), Oxford Inst. Paper No. 42, Oxford, 1968.

Fisher, I., *The Theory of Interest,* Macmillan, New York, 1930.

Fung, Y. C., *Foundations of Solid Mechanics,* Prentice-Hall, New York, 1965.

Giordano, F. R., and M. D. Weir, *A First Course in Mathematical Modeling,* Brooks/Cole Publishing Co., Monterey, CA, 1985.

Greenspan, H. P., "Applied Mathematics at MIT," *Amer. Math. Monthly, 80,* 1973, 67–72.

Greenspan, H. P., and D. J. Benney, *Calculus: An Introduction to Applied Mathematics,* McGraw-Hill, New York, 1973.

Haberman, R., *Mathematical Models: Mechanical Vibrations, Population Dynamics and Traffic Flow (An Introduction to Applied Mathematics),* Prentice-Hall, New York, 1977.

————, *Elementary Applied Partial Differential Equations,* Prentice-Hall, New York, 1983.

Hartwick, J., "Intergenerational Equity and the Investing of Rents from Exhaustible Resources," *Amer. Econ. Rev., 67,* 1977, 972–974.

Heaps, T., and P. A. Neher, "The Economics of Forestry When the Rate of Harvest Is Constrained," *J. Environ. Econ. Manage., 6,* 1979, 297–319.

Herman, R., E. W. Montroll, R. B. Potts, and R. W. Rothery, "Traffic Dynamics: Analysis of Stability in Car Following," *Oper. Research., 7,* 1959, 86–106.

Hildebrand, F. B., *Advanced Calculus for Applications,* 3rd Ed., Prentice-Hall, New York, 1976.

Hillier, F. S., and G. J. Lieberman, *Operations Research,* 2nd Ed., Holden-Day, Oakland, CA, 1974.

Kanemoto, Y., "Cost-Benefit Analysis and the Second Best Land Use for Transportation," *J. Urban Econ., 4,* 1977, 483–503.

———, *Theories of Urban Externalities,* North Holland, NY, 1980.

Kevorkian, J., *Partial Differential Equations: Analytical Solution Techniques,* Wadsworth, Belmont, CA, 1989.

Kevorkian, J., and J. D. Cole, *Perturbation Methods in Applied Mathematics,* Springer-Verlag, Berlin, 1981.

Langhaar, H. L., *Dimensional Analysis and the Theory of Models,* John Wiley & Sons, New York, 1951.

Larkin, P. A., "Scientific Technology Needs for Canadian Shelf-Seas Fisheries," Interim Report, Fisheries Research Board of Canada, Ottawa, February 1975.

Lighthill, M. J., and G. B. Whitham, "On Kinematic Waves, II: A Theory of Traffic Flow on Long Crowded Roads," *Proc. Roy. Soc., A 229,* 1955, 317–345.

Lin, C. C., and L. Segel, *Mathematics Applied to Deterministic Problems in the Natural Sciences,* Macmillan, New York, 1974.

Love, A. E. H., *A Treatise on the Mathematical Theory of Elasticity,* 4th Ed., Dover, New York, 1944.

Luce, R. D., and H. Raiffa, *Games and Decisions,* John Wiley & Sons, New York, 1957.

Ludwig, D., "Some Mathematical Problems in the Management of Biological Resources," *Appl. Math. Notes, 2,* 1976, 39–56.

Massey, B. S., *Units, Dimensional Analysis and Physical Similarity,* Van Nostrand Reinhold, New York, 1971.

Minorsky, N., *Nonlinear Oscillation,* Van Nostrand Reinhold, New York, 1962.

Murray, J. D., *Nonlinear Differential Equation Models in Biology,* Clarendon Press, Oxford, 1977.

O'Malley, R. E., *Introduction to Singular Perturbations,* Academic Press, San Diego, CA, 1974.

Pavlides, T., *Biological Oscillators: Their Mathematical Analysis,* Academic Press, San Diego, CA, 1973.

Pearse, P., "The Optimum Forest Rotation," *Forestry Chron., 2,* 1967, 178–195.

Rawls, J., *A Theory of Justice,* The Belknap Press of Harvard University Press, Cambridge, MA, 1971.

Rubinow, S. I., *Mathematical Problems in the Biological Sciences,* SIAM-CBMS Publications, Philadelphia, 1973.

Samuelson, P. A., "Economics of Forestry in an Evolving Society," *Econ. Inqu., IXV,* 1976, 466–492.

Schmidt, R. M., "A Centrifugal Cratering Experiment: Development of a Gravity-Scaled Parameter," *Impact and Explosion Crater,* D. J. Roody et al. (eds.), Pergamon Press, Elmsford, NY, 1977, pp. 1261–1278.

Schmidt, R. M., and K. A. Holsapple, "Crater Ejecta Scaling Laws 1: Fundamental Forms Based on Dimensional Analysis," *J. Geophys. Res., 88,* 1983.

Segel, L. A., *Modeling Dynamic Phenomena in Molecular and Cellular Biology,* Cambridge University Press, Cambridge, U.K., 1984.

Smith, D. R., *Singular-Perturbation Theory: An Introduction with Applications,* Cambridge University Press, Cambridge, U.K., 1985.

Smith, J. M., *Mathematical Ideas in Biology,* Cambridge University Press, Cambridge, U.K., 1968.

Sokolnikoff, I. S., *Tensor Analysis,* John Wiley & Sons, New York, 1964.

Solow, R. M., "Congestion, Density and the Use of Land in Transportation," *Swedish J. Econ., 74,* 1972, 161–173.

———, "Intergenerational Equity and Exhaustible Resources," *Rev. Econ. Studies* (Symp. issue), 1974, 29–46.

Solow, R. M., and F. Y. M. Wan, "Extraction Costs in the Theory of Exhaustible Resources," *Bell J. Econom., 7,* 1976, 359–370.

Stoker, J. J., *Water Waves,* Wiley-Interscience, New York, 1957.

Taylor, G. I., "The Formation of a Blast Wave by a Very Intense Explosion, II: The Atomic Explosion of 1945," *Proc. Roy. Soc. A, 201,* 1950, 175.

Tou, J. T., *Modern Control Theory,* McGraw-Hill, New York, 1964.

Walters, A. A., "Production and Cost Functions," *Econometrica,* January 1963.

Wan, F. Y. M., "Perturbation and Asymptotic Solutions for Problems in the Theory of Urban Land Rent," *Studies Appl. Math., 56,* 1977, 219–239.

———, "Bifurcation Theory and the Two Hundred Mile Fishing Limit," *Appl. Math. Notes, 4,* 1978, 74–87.

———, "Accurate Solutions for the Second Best Land Use Problem I: Absentee Ownership," *I.A.M.S. Tech. Report* 79-30, University of British Columbia, Vancouver, July 1979.

———, "Eulerian Wobble," *Appl. Math. Notes 4,* 1979, 33–41.

———, "Constant Sustainable Consumption Rate in Optimal Growth with a Multi-Grade Exhaustible Resource," *Studies in Appl. Math., 63,* 1980, 47–66.

———, "Mathematical Models and Their Formulation," in *Handbook of Applied Mathematics,* 2nd Ed., C. E. Pearson (ed.), Van Nostrand Reinhold, New York, 1983, pp. 1044–1139.

———, "Ordered Site Access and Optimal Forest Rotation," *Studies in Appl. Math., 73,* 1985, 155–175.

Wan, F. Y. M., and K. Anderson, "Optimal Forest Harvesting with Ordered Site Access", *Stud. Appl. Math., 68,* 1983, 189–226.

Wang, C.-T., *Applied Elasticity,* McGraw-Hill, New York, 1953.

Whitham, G. B., *Linear and Nonlinear Waves,* Wiley-Interscience, New York, 1974.

Whittaker, E. T., and G. N. Watson, *A Course of Modern Analysis,* 4th Ed., Cambridge University Press, Cambridge, U.K., 1952.

Index